THEORETICAL PROBLEMS IN THE SPECTROSCOPY AND GAS DYNAMICS OF LASERS

TEORETICHESKIE PROBLEMY SPEKTROSKOPII I GAZODINAMICHESKIKH LAZEROV

ТЕОРЕТИЧЕСКИЕ ПРОБЛЕМЫ СПЕКТРОСКОПИИ И ГАЗОДИНАМИЧЕСКИХ ЛАЗЕРОВ

The Lebedev Physics Institute Series

Editors: Academicians D. V. Skobel'tsyn and N. G. Basov

P. N. Lebedev Physics Institute, Academy of Sciences of the USSR

Recent Volumes in this Series

Volume 35	Electronic and Vibrational Spectra of Molecules
Volume 36	Photodisintegration of Nuclei in the Giant Resonance Region
Volume 37	Electrical and Optical Properties of Semiconductors
Volume 38	Wideband Cruciform Radio Telescope Research
Volume 39	Optical Studies in Liquids and Solids
Volume 40	Experimental Physics: Methods and Apparatus
Volume 41	The Nucleon Compton Effect at Low and Medium Energies
Volume 42	Electronics in Experimental Physics
Volume 43	Nonlinear Optics
Volume 44	Nuclear Physics and Interaction of Particles with Matter
Volume 45	Programming and Computer Techniques in Experimental Physics
Volume 46	Cosmic Rays and Nuclear Interactions at High Energies
Volume 47	Radio Astronomy: Instruments and Observations
Volume 48	Surface Properties of Semiconductors and Dynamics of Ionic Crystals
Volume 49	Quantum Electronics and Paramagnetic Resonance
Volume 50	Electroluminescence
Volume 51	Physics of Atomic Collisions
Volume 52	Quantum Electronics in Lasers and Masers, Part 2
Volume 53	Studies in Nuclear Physics
Volume 54	Photomesic and Photonuclear Reactions and Investigation Method with Synchrotrons
Volume 55	Optical Properties of Metals and Intermolecular Interactions
Volume 56	Physical Processes in Lasers
Volume 57	Theory of Interaction of Elementary Particles at High Energies
Volume 58	Investigations in Nonlinear Optics and Hyperacoustics
Volume 59	Luminescence and Nonlinear Optics
Volume 60	Spectroscopy of Laser Crystals with Ionic Structure
Volume 61	Theory of Plasmas
Volume 62	Methods in Stellar Atmosphere and Interplanetary Plasma Research
Volume 63	Nuclear Reactions and Interaction of Neutrons and Matter
Volume 64	Primary Cosmic Radiation
Volume 65	Stellarators
Volume 66	Theory of Collective Particle Acceleration and Relativistic Electron Beam Emission
Volume 67	Physical Investigations in Strong Magnetic Fields
Volume 68	Radiative Recombination in Semiconducting Crystals
Volume 69	Nuclear Reactions and Charged-Particle Accelerators
Volume 70	Group-Theoretical Methods in Physics
Volume 71	Photonuclear and Photomesic Processes
Volume 72	Physical Acoustics and Optics: Molecular Scattering of Light; Propagation of Hypersound; Metal Optics
Volume 73	Microwave–Plasma Interactions
Volume 74	Neutral Current Sheets in Plasmas
Volume 75	Optical Properties of Semiconductors
Volume 76	Lasers and Their Applications
Volume 78	Research in Molecular Laser Plasmas
Volume 77	Radio, Submillimeter, and X-Ray Telescopes
Volume 79	Luminescence Centers in Crystals
Volume 80	Synchrotron Radiation
Volume 81	Pulse Gas-Discharge Atomic and Molecular Lasers
Volume 82	Electronic Characteristics and Electron–Phonon Interaction in Superconducting Metals and Alloys
Volume 83	Theoretical Problems in the Spectroscopy and Gas Dynamics of Lasers
Volume 84	Temporal Characteristics of Laser Pulses and Interaction of Laser Radiation with Matter
Volume 86	Superconductivity
Volume 89	Electrical and Optical Properties of III–IV Semiconductors

In preparation

Volume 85	High-Power Lasers and Laser Plasmas
Volume 87	Coherent Cooperative Phenomena
Volume 88	Cosmic Rays in the Stratosphere and in Near Space

Proceedings (Trudy) of the P. N. Lebedev Physics Institute

Volume 83

Theoretical Problems in the Spectroscopy and Gas Dynamics of Lasers

Edited by

N. G. Basov

P. N. Lebedev Physics Institute
Academy of Sciences of the USSR
Moscow, USSR

Translated from Russian by
Donald H. McNeill

SPRINGER SCIENCE+BUSINESS MEDIA, LLC

Library of Congress Cataloging in Publication Data

Main entry under title:

Theoretical problems in the spectroscopy and gas dynamics of lasers.

(Proceedings (Trudy) of the P. N. Lebedev Physics Institute; v. 83)
Translation of Teoreticheskie problemy spektroskopii i gazodinamicheskikh lazerov.
Includes bibliographical references.
1. Gas lasers–Addresses, essays, lectures. 2. Plasma lasers–Addresses, essays, lectures. I. Title: Spectroscopy and gas dynamics of lasers. II. Basov, Nikolaĭ Gennadievich, 1922- III. Series: Akademii͡a nauk SSSR. Fizicheskiĭ institut. Proceedings; v. 83.
QCl.A4114 vol. 83 [QC689] 530'.08s
[535.5'8] 77-17616

ISBN 978-1-4757-6821-3 ISBN 978-1-4757-6819-0 (eBook)
DOI 10.1007/978-1-4757-6819-0

The original Russian text was published by Nauka Press in Moscow in 1975 for the Academy of Sciences of the USSR as Volume 83 of the Proceedings of the P. N. Lebedev Physics Institute. This translation is published under an agreement with the Copyright Agency of the USSR (VAAP).

PREFACE

This anthology is devoted to the theory of gas and plasma lasers. A considerable portion is allotted to gas-dynamic lasers. The dependence of their operating characteristics on the composition of the mixture, the pressure, and the nozzle shape are investigated in detail. Chemical gas-dynamic and electrical gas-dynamic lasers, as well as high-temperature electrical gas-dynamic lasers with free electrons formed as a result of thermal ionization, are discussed. Far-infrared lasers and the conditions for increasing their efficiency are studied. The feasibility of plasma lasers which operate due to rapid cooling of free plasma electrons is demonstrated. The basic types of plasma lasers are examined: electrical (including those using electron beams), moving plasma (plasma-dynamic and pinch discharge lasers), and plasma chemical (including the use of molecules with dissociating lower terms). The prospects for using a multiply charged ion plasma as an intense light source are discussed. Considerable attention is given to the physical kinetics of the processes involved.

This publication is intended for a large circle of researchers and engineers specializing in quantum electronics.

CONTENTS

SOME TRENDS IN LASER DEVELOPMENT

L. A. Shelepin

An analysis of the current state of gas and plasma lasers is given. The basic trends and factors limiting the efficiency of these lasers are described. The quasiequilibrium distributions over the vibrational, rotational, and electronic levels which play an important role in research on the physical kinetics of lasers are discussed.

Introduction

This anthology is devoted to a number of the most promising approaches to developing gas and plasma lasers. These lasers operate over a wide range of the spectrum, from submillimeter wavelengths to the vacuum ultraviolet, and have become widespread. In recent years there has been a rapid growth in their output power. This is primarily true of molecular lasers using vibrational–rotational transitions operating in the near infrared. The highest powers achieved for continuous operation are about 10^5 W. It is expected that in the near future powers of up to 10^6 W will be reached. High powers are combined with high efficiencies (up to 20-30%) in this type of laser. Depending on how these lasers are pumped, they are referred to as electrical, chemical, or gas-dynamic lasers.

Lasers using electronic transitions of atoms and molecules have substantially lower energies than those using vibrational–rotational transitions. However, this situation is now beginning to change. There has been a sharp increase in the efficiency of plasma lasers using electronic transitions, which is very important from the standpoint of the wavelength range involved (plasma lasers operate in the visible and ultraviolet).

The purpose of the present article, which serves as an introduction to this collection, is twofold: to trace the principal trends of recent years and point out the main difficulties and factors limiting the power of lasers, and to systematize the material in order that the principal articles of this anthology [1-4], which cover a wide range of current problems, may be regarded as a unit.

Lasers using vibrational transitions have been examined in detail in a review [5] that summarizes data obtained up to the beginning of 1972. This anthology mainly contains results not included in that review, such as treatments of the theory of lasers operating in the far infrared and plasma lasers using electronic transitions. In addition, some aspects of the interaction of laser radiation with matter are discussed.

The references given in this paper are not complete and are given only to illustrate the basic ideas.

1. The Physical Kinetics of Lasers

The basis of the theory of lasers and the interaction of laser radiation with matter is the physical kinetics of the processes in the corresponding medium. Knowledge of the relaxation properties of the system and of the level population kinetics is required to analyze the operation of lasers, optimize them, and increase their power. Among this group of problems some fundamentally new aspects have recently appeared which reveal the basic trends in the kinetics of gas and plasma lasers.

The original approach to the problem of creating an inverted population consisted of analyzing two- and three-level schemes with external pumping to the upper working level and depopulation of the lower level. This approach was limited. The distribution of populations (including the population inversion) together with the external pumping are primarily determined by the internal relaxation properties of the system. A multilevel approach was necessary, such as that proposed by Bates et al. [6] for atomic recombination or by Gudzenko and Shelepin [7, 8] for the theory of lasers. It meant including the populations of many levels and solving (mainly by computer) the corresponding population balance equations. In such calculations it is important to choose relaxation models in accordance with the physical requirements. A number of models and approximations in atomic kinetics have been examined by Biberman et al. [9]. The multilevel approach based on solution of the balance equations has yielded a series of important results (see [3]), but its awkwardness is a disadvantage.

At the present time general techniques for using quasiequilibrium population distributions to describe relaxation processes are included in atomic and molecular kinetics. These distributions arise due to differences among some of the characteristic relaxation times in a complicated system with a large number of levels. A quasiequilibrium distribution is established over certain degrees of freedom (or over groups of levels) and depends on a small number of time-dependent parameters. These distributions have an obvious physical significance. They greatly simplify solution of the relaxation problem and analysis of the kinetics of a laser medium. An appreciation of the generality of the method of quasiequilibrium distributions is appropriate for an understanding of the current state of the theory. (The questions referred to here are discussed in detail in [4].)

We now discuss briefly some specific distributions. One of the first examples of a quasiequilibrium distribution was the introduction of vibrational and rotational temperatures in the analysis of relaxation in diatomic molecules. For multiatomic molecules the so-called thermodynamic model, based on the introduction of partial vibrational temperatures T_i for each type (i) of vibration [10], was proposed. It is founded on the fact that the characteristic vibrational (energy) exchange time for this type of vibration (V−V process) is much less than the vibrational−translational relaxation time (V−T process) and the energy exchange time between various types of vibrations (V−V' process), i.e.,

$$\tau_{VV} \ll \tau_{VT}, \ \tau_{VV'}. \tag{1}$$

The partial vibrational temperatures T_i characterize the vibrational energy stored in each type of vibration and are functions of time. They differ from one another and from the gas temperature T. This model has been used to study the kinetics of CO_2 [11] and other multiatomic molecular lasers. However, the equations for T_i in most studies are only valid for low pumping levels. Equations for T_i that are valid for the general case and include an arbitrary number of types of vibrations and degeneracy of vibrational levels are given in Chapter I of Biryukov's paper [1]. They make it possible to correctly describe the kinetics of gas-dynamic lasers, where the initial temperatures are very high.

An important advance in our understanding of the structure of quasiequilibrium distributions was the Treanor distribution for an anharmonic oscillator (discussed in detail in [12]). It gives the population x_n of a level of the anharmonic oscillator as

$$x_n = x_0 \exp(\gamma n) \exp(-E_n/T) \tag{2}$$

which is determined by two parameters, the gas temperature T and the parameter γ, which is independent of the level number. The parameter γ may be expressed in terms of the vibrational temperature of the first level $T_1 = E_1/\ln (x_0/x_1)$ as

$$\gamma=\frac{E_1}{T_1}-\frac{E_1}{T}. \tag{3}$$

The vibrational temperature $\theta_n = E_n/\ln (x_0/x_n)$, as opposed to the Boltzmann distribution case, depends on the level number. The physical significance of the Treanor distribution involves conservation of the number of vibrational quanta. It is valid for single-quantum transitions even when

$$\tau_{VV}\ll\tau_{VT}. \tag{4}$$

For $T_1/T > 1$ the upper levels are overpopulated compared to the Boltzmann distribution. However, at the highest levels of the anharmonic oscillator V–T processes dominate and a Boltzmann distribution at the gas temperature is valid. The population distribution function for the intermediate levels including V–V and V–T processes along with radiative transitions is examined in [5]. Nonresonance vibrational energy exchange, which produces a Treanor distribution in an anharmonic oscillator, plays an important role in gaseous mixtures when

$$\tau_{VV},\ \tau_{VV'}\ll\tau_{VT}. \tag{5}$$

If two molecules, A and B, are modeled by harmonic oscillators, then when the stored vibrational energy deviates from an equilibrium state corresponding to a gas temperature of T, the vibrational temperatures T_A and T_B will be different. Taking single-quantum exchange into account, the relation between these temperatures is given by

$$kT_B=\frac{E_B}{E_A/kT_A-(E_A-E_B)/kT}. \tag{6}$$

Under nonequilibrium conditions it is possible to redistribute the vibrational energy in gaseous mixtures by changing the gas temperature. The effects associated with such a redistribution due to cooling as a gas expands are discussed in Chapter III of Biryukov's paper [1]. Gordiets et al. [13] found the vibrational energy distribution in two- and three-component mixtures. It seems that when $\tau_{VV'} \sim \tau_{VT}$ the distribution functions depend strongly on the concentrations of the molecular components; that is, a change in the concentration of an impurity may have a significant effect on the vibrational energy distribution and, thus, on the operation of a laser.

In general, nonresonance vibrational exchange phenomena are very important for lasers. Inclusion of effects due to the redistribution of the vibrational energy has made it possible to find the conditions for intense dissociation at low gas temperatures. For a substantial disequilibrium (when $T_1 > T$) the upper vibrational levels (modeled here by an anharmonic oscillator) become overpopulated due to nonresonance vibrational energy exchange. In the case of diatomic molecules this usually has relatively little effect on the dissociation rate since the populations of levels lying near the dissociation limit are determined by the gas temperature (that is, by V–T processes).

Multiatomic molecules may decay by means of predissociation if the vibrational mode being pumped has a dissociation limit energy larger than the minimum. The dissociation rate is determined by the population of the level nearest the predissociation limit. According to Eq. (2) this population may be increased by reducing the gas temperature. Thus, dissociation may be controlled by changing either the amount of stored vibrational energy or the gas temperature. Nonequilibrium dissociation may be very intense and ensure a large quantity of free atoms (or radicals). This is, however, of fundamental importance for chemical lasers which require free atoms to initiate chain reactions.

All these results, which have changed the field of vibrational kinetics in recent years, are by no means isolated. Rotational kinetics is discussed in Chapter I of Reshetnyak's paper [4]. By analogy with V−T, V−V, and V−V' processes, it is possible to consider R−T, R−R, and R−R' processes. In the adiabatic region of the rotational spectrum, quasistationary distributions have much in common with vibrational kinetics. Quasistationary distributions over rotational level populations yield important information for optimization of lasers using rotational transitions.

Quasistationary distributions of electronic level populations in atoms are discussed in Chapter II of Reshetnyak's paper [4]. They are represented by a series of time derivatives of the ground state population. The first approximation involves the constant sink method [6]. These distributions describe, in particular, radiative-collisional recombination and ionization in an atomic plasma. For a hydrogen atom the temperature θ_n between neighboring levels in the first approximation is given by

$$\theta_n = \frac{T_e}{1 + \frac{T_e}{E_{n+1,\,n}} \ln \frac{\beta_n \varkappa + 1}{\beta_{n+1} \varkappa + 1}}, \tag{7}$$

where T_e is the electron temperature, the constants β_n and β_{n+1} are expressed in terms of the probabilities of collisional transitions between neighboring levels, and $\varkappa$ is expressed in terms of the temperature θ_1 between the lower levels. An important advantage of this distribution is that the populations (as in the case of the Treanor distributions) are described by two parameters T_e and θ_1, thus eliminating the necessity of solving the balance equations for a large number of levels. This makes it possible to solve the complicated self-consistent problems which arise in an analysis of the physical kinetics of plasma lasers.

Reshetnyak [4] also analyzes the quasistationary population distributions of atomic and ionic levels in a plasma of complex chemical composition containing a mixture of elements. If in the first approximation the distribution function for each ion (atom) is determined by collisions with electrons, the overall distribution function is the result of collisions among atoms. Charge exchange plays an important role.

In Chapter IV of his article [4] Reshetnyak derives quasistationary distributions over the charge states of multiply charged ions formed by the interaction of laser radiation with matter. Thus, the method of quasistationary distributions is now acquiring a very general character. It includes distributions over vibrational levels (of di- and multiatomic molecules, and of gaseous mixtures), over rotational levels, over electronic levels (hydrogen atom, complex atoms, and mixtures of elements), and over the charge states of ions. Since atomic and molecular kinetics are the theoretical basis for studies of the operating mechanisms of lasers, for optimizing existing systems, and for developing new lasers, the further development of the method of quasistationary distributions is of paramount importance.

2. Electrical Lasers Using Vibrational − Rotational Molecular Transitions

Of the three principal types of lasers using vibrational−rotational transitions (electrical or electrical discharge, gas-dynamic, and chemical) electrical lasers are the most used in practice at this time. We now discuss briefly their potentials and limitations.

(a) Continuous Electrical Lasers. The first CO_2 molecular laser was built in 1964 [14] and had a cw power of about 1 mV. In 1969 the output power of this type of laser was already 10 kW. The physical kinetics of CO_2 lasers was examined by Gordiets et al. [11] using a model of partial vibrational temperatures in a multiatomic molecule. The basic factor limiting the power of a cw electrical laser is heat release during relaxation. Most of the energy pumped by electrons into a CO_2 molecule goes into heat. As the temperature is increased, relaxation of the upper working level speeds up and the population of the lower level is increased.

The situation is similar for other lasers using multiatomic molecules, but their power is lower than that of the CO_2 laser. The kinetics of electrical lasers using CO and the hydrogen halides are analyzed using the Treanor distribution (see Gordiets et al. [5]), and it is found that heat release reduces the difference between the gas and vibrational temperatures and increases the rate of V – T processes. Thus, increases in the density of the active medium and in the power of lasers using vibrational transitions are limited by heat release (and by a corresponding increase in the rate of relaxation of the upper working level). Two approaches have been considered for removing this constraint: going to pulsed operation and effectively increasing the rate of pumping the working gas.

(b) Pulsed Lasers. Pulsed electrical lasers at pressures of less than 100 torr are discussed in Chapter IV of Biryukov's paper [1]. When comparatively short pulses (shorter than the relaxation time for the upper laser level) are used, the maximum population inversion is independent of the relaxation time and is determined by the total vibrational energy obtained from the electrons. By keeping the product of the pulse duration and the magnitude of the current constant, it is possible to increase the working gas pressure indefinitely. However, at pressures above 100 torr difficulties arise in connection with an increased breakdown voltage and filamentation of the discharge, both of which limit the growth in the average power [15].

(c) Electroionization Lasers. By using an external source (electron beam, photoionization) to preionize the working gas, it is possible to bypass the above-mentioned difficulties and use a pressure greater than atmospheric [16, 17]. The required concentration of charged particles is maintained by the electron beam, and the electron temperature can be controlled independently. Currently an energy of 0.1 J/cm^3 can be obtained (with a power of 1 MW) at pressures of greater than 10 atm using a CO_2-N_2-He mixture [18]. This type of laser is very promising from the standpoint of obtaining high-power pulses. The average powers are still small because of low repetition rates.

(d) TEA Lasers. A transverse discharge configuration has been used to reduce the working voltage at high gas pressures. Then the cavity length is much greater than the interelectrode separation. Filamentation of the discharge is prevented by using pulses that are much shorter than the time required for an arc to develop (or by reducing the current density) [19]. Technical difficulties have appeared in TEA lasers operating on this principle because of the need to decrease the risetime when shortening the overall duration of the pulse and because of the need to increase the voltage. Preionization substantially improves the characteristics of TEA lasers [20] as it permits the use of large gas volumes. At atmospheric gas pressures, pulses shorter than 10^{-9} sec with powers of about 1 MW have been obtained [21]. It is expected that the average power will also be increased [22].

3. Chemical Lasers

The pumping source in chemical lasers [5, 23] is chemical reactions. During the early stages of development of these lasers it was typical to search for reactions in which the rate of formation of products in the upper excited level was greater than in the lower. However, as follows from vibrational kinetics, there is a rapid redistribution of vibrational energy in a system of molecules due to relaxation processes. During this redistribution an inverted population may be produced. The requirements on the amount of and characteristic time for chemical pumping dictate that relaxation processes be taken into account. Exothermic reactions are required for an efficient chemical laser so that a substantial amount of energy will be pumped into vibrational degrees of freedom. Examples of such reactions are the formation of hydrogen halides and combustion reactions. It is also necessary that the reaction take place sufficiently rapidly, over a time comparable to the relaxation time. This requirement is satisfied by chain reactions (especially branch chain reactions).

The basic difficulty restraining the development of high-power chemical lasers was the problem of initiating chain reactions. There was a sharp jump in the power of chemical lasers in 1969 when large quantities of free atoms were obtained as a result of nonequilibrium dissociation of multiatomic molecules (discussed in Section 1). Once the problem of initiation was solved, there was a rapid growth in the power of chemical lasers. At present lasers using CO (initial mixture $CS_2 + O_2$), CO_2 (initial mixture $DF^* + CO_2$), and HF (initial mixture $H_2 + F$) are widely used. An output power of 4500 W has been obtained from an HF laser (free flourine is formed by nonequilibrium dissociation of SF_6) [24]. A further increase in the power of chemical lasers has been due to solving the technical problems of rapidly mixing the reagents and removing the depleted gas. It is expected that the power of HF chemical lasers will be increased to about 10^5 W [25].

Pumping of lasers by three-body recombination reactions (see Chapter III of Biryukov's paper [1]) is of interest from the standpoint of substantially increasing the density of the reacting materials.

4. Gas-Dynamic Lasers

In 1966 it was predicted that a population inversion in the vibrational levels of the CO_2 molecule would be obtained upon expansion of the gas [26]. An analogous effect on atomic electronic transitions is discussed elsewhere [27, 28]. The first detailed calculations were done for expansion from a nozzle [29] and from a slit [30]. Lasing was observed experimentally in 1970 [40-42] and since then there has been a rapid increase in the powers obtained. Continuous-wave powers of about 10^5 W [34] have been achieved and are the highest (cw) powers for lasers of any type. A number of problems associated with gas-dynamic lasers are discussed in detail in Biryukov's article [1].

An inversion is produced in gas-dynamic lasers because of the difference in the relaxation times of the upper and lower working levels as the gas expands. At realistically attainable gas cooling rates the densities of CO_2 molecules in the laser cavity cannot exceed 10^{18} cm^{-3}, since as the density is increased the relaxation time of the upper laser level during adiabatic expansion of a preheated gas becomes less than the characteristic time for changes in the density and temperature of the gas. To remove this limitation combination methods have been proposed for producing the inversion.

(a) Chemical Gas-Dynamic Lasers. These lasers use chemical reactions to pump the upper laser level. With them it is possible to greatly increase the effective relaxation time of the upper laser level.

(b) Electro-Gas-Dynamic Lasers. A supplementary electrical discharge in the gas-dynamic flow is utilized in this type of laser. Lasers with transverse electrical excitation in expanding supersonic flows seem promising. In these devices the maximum inversion may be raised by more than an order of magnitude above that in ordinary gas-dynamic lasers.

(c) High-Temperature Lasers. An important question for electro-gas-dynamic lasers is initiation of the electrical discharge. Here either electron or laser beams may be used. However, the most promising approach at present is the high-temperature electro-gas-dynamic laser. At temperatures above 3500°C there is a substantial amount of free electrons due to thermal ionization. Because of these thermal electrons such lasers do not require supplementary initiation and have a number of advantages. They are analyzed in the article by Gudzenko et al. [3].

5. Far-Infrared Lasers

Lasers in the far infrared and submillimeter wavelength ranges have their peculiarities compared with those in the near infrared. Although lasing is obtained on a large number of vibrational–rotational transitions, the cw powers are small [35]. (The pulsed power from H_2O vapor lasers is 15 kW [36].) The primary problem with increasing the power is the high relaxation rate. In order to obtain significant probabilities for radiative vibrational–rotational transitions, dipole molecules must be used; however, the relaxation rates for dipole molecules are large, as in the case of rotational transitions. In addition, the undeveloped state of the theory of rotational kinetics and the kinetics of dipole molecules has played a major role here.

The problems of obtaining efficient lasing in the far infrared from both diatomic and multiatomic molecules are examined in detail in Chapter I of Reshetnyak's article [4] using the quasistationary distributions derived there. From the standpoint of increasing the power, the following schemes offer possibilities: an electrical water vapor laser with a transverse discharge and a high pumpthrough rate, gas-dynamic lasers (in particular, the combustion reaction $H_2 + O_2 \rightarrow H_2O$ with subsequent supersonic expansion from a nozzle), and chemical lasers (with formation of rotationally excited molecules).

6. Plasma Lasers

Lasers using plasmas as the active medium are discussed in detail by Gudzenko et al. [3] and Reshetnyak [4]. They operate in the infrared, visible, and ultraviolet. The cw powers achieved are 150 W [37]. This is much lower than for lasers using vibrational transitions of molecules. Here the basic difficulties are the rapid relaxation of electronic levels and the high pumping intensities required, as well as the existence of a radiative mechanism for emptying the lower working level. The radiative mechanism forbids use of large plasma volumes because of increased population of the lower level due to reabsorption. In 1968 it was shown that under recombining plasma conditions a collisional mechanism may be realized, thus opening the prospect of using large volumes.

Depending on the direction of the flow of electrons moving over the atomic levels, plasma lasers are referred to as ionization (or gas) or recombination (or plasma) lasers. In recent years there has only been a slight increase in the power of lasers with an ionization flow (these include most lasers using electronic transitions). The operation of plasma lasers, first proposed by Gudzenko et al. [7], is based on rapid cooling of the free electrons in the plasma. Despite the fact that there are still comparatively few operating lasers at present (see, for example, [38]), the efficiency of plasma lasers is being rapidly increased [3]. We now list the basic types of plasma laser.

(a) Electrical Plasma Lasers. These include pulsed recombination lasers operating in the afterglow of an electrical pulse. Plasma lasers can operate in a steady state using electron beams. Here there are prospects of obtaining very efficient lasers [39].

(b) Moving Plasma Lasers. These include the plasma-dynamic lasers proposed by Gudzenko et al. [28] which operate due to cooling of the free electrons as the plasma expands through a slit or nozzle [40, 41], as well as pinch discharge lasers, which operate as the plasma is compressed. (Most of the existing devices work in the ionization regime with radiative depopulation of the lower working level [42].)

(c) Plasma Chemical Lasers. Plasma chemical processes proceed rapidly, and their use seems promising. Right now there is special interest in the development of lasers using electronic transitions of molecules which exist only in excited states with a lower repulsive term [43]. It is possible that plasma lasers may acquire a dominant position in view of the possibility of using large volumes (because of the collisional mechanism) with maximal energy yield per unit volume and efficient pumping to the upper level and in view of the existence of great progress in the short wavelength (even x-ray) region of the spectrum [3].

7. The Interaction of Laser Radiation with Matter. The Limits of Applicability of Population Kinetics

Modern physical kinetics plays a significant role in the analysis of the interaction of laser radiation with matter as well as in the analysis of the mechanisms of lasers. Here there are four distinct groups of questions: technological problems (welding, cutting, drilling holes, etc.), stimulation of chemical reactions, atmospheric investigations, and heating of matter to produce thermonuclear reactions. The use of the molecular and atomic kinetics discussed in Section 1 may be very fruitful in all these problems. Thus, research into the possibility of stimulating chemical reactions is inseparably related to vibrational kinetics.

The same may be said about the effects of a laser beam passing through the atmosphere. In Chapter VI of Biryukov's paper [1] molecular kinetics is used to examine the radiation in the wake of a fast body flying through the atmosphere. The kinetics of the processes within a laser spark are especially complicated. A series of questions related to the kinetics of a multiply charged ionic plasma and the use of such a plasma as a powerful source of radiation are discussed by Reshetnyak [4], and its use as an active medium for an x-ray laser is discussed by Gudzenko et al. in their article [3].

The population kinetics discussed above and the analyses of laser mechanisms and laser interactions with matter made using it are completely correct only for characteristic times that are not too small. In the general case of nonstationary rapid processes, particles of matter cannot be regarded as completely independent. Because of the perturbation of the medium there are correlations among the particles which are retained over a time T_2, the characteristic time for destruction of coherence. Usually this time is much less than T_1, the characteristic relaxation time for the populations. Thus, the above results apply to times greater than T_2. For shorter times it is no longer possible to neglect the nondiagonal elements of the density matrix. It is necessary to consider the kinetics of particles together with radiation and to include the resulting coherent effects [44-46]. From a brief listing of the basic trends we can see, therefore, that despite the difficulties and limitations there are many possibilities for further increasing the efficiency of lasers. Many of these possibilities are discussed in this book.

Literature Cited

1. A. S. Biryukov, Tr. FIAN, 83:13 (1975) (this volume).
2. A. S. Biryukov, V. M. Marchenko, and L. A. Shelepin, Tr. FIAN, 83:87 (1975) (this volume).
3. L. I. Gudzenko, L. A. Shelepin, and S. I. Yakovlenko, Tr. FIAN, 83:100 (1975) (this volume).
4. S. A. Reshetnyak, Tr. FIAN, 83:146 (1975) (this volume).
5. B. F. Gordiets, A. I. Osipov, E. V. Stupochenko, and L. A. Shelepin, Usp. Fiz. Nauk 108:655 (1972).

6. D. R. Bates, A. E. Kingston, and R. W. P. McWhirter, Proc. Roy. Soc., A267:297 (1963).
7. L. I. Gudzenko and L. A. Shelepin, Zh. Éksp. Teor. Fiz., 45:1445 (1963).
8. L. I. Gudzenko and L. A. Shelepin, Dokl. Akad. Nauk SSSR, 160:1296 (1963).
9. L. M. Biberman, V. S. Vorobev, and I. T. Yakubov, Usp. Fiz. Nauk, 107:353 (1972).
10. B. F. Gordiets, N. N. Sobolev, V. V. Sokovikov, and L. A. Shelepin, Phys. Lett., A125:173 (1967).
11. B. F. Gordiets, N. N. Sobolev, and L. A. Shelepin, Zh. Éksp. Teor. Fiz., 53:1822 (1967).
12. C. E. Treanor, J. W. Rich, and R. G. Rehm, J. Chem. Phys., 48:1798 (1968).
13. B. F. Gordiets, Sh. S. O. Mamedov, A. I. Osipov, and L. A. Shelepin, Teor. Eksp. Khim., 9:460 (1973).
14. C. V. N. Patel, Phys. Rev. Lett., 12:588 (1964).
15. A. E. Hill, Appl. Phys. Lett., 12:324 (1968).
16. N. G. Basov, E. M. Belenov, V. A. Danilychev, O. M. Kerimov, and I. B. Kovsh, Zh. Éksp. Teor. Fiz., 63:2010 (1972).
17. C. A. Fenstermacher, M. J. Nutter, W. T. Leland, and K. Boyer, Appl. Phys. Lett., 20:56 (1972).
18. N. G. Basov, E. M. Belenov, and V. A. Danilychev, Zh. Éksp. Teor. Fiz., 64:108 (1973).
19. J. A. Beaulieu, Appl. Phys. Lett., 16:504 (1970).
20. P. J. Berger and D. C. Smith, Appl. Phys. Lett., 21:167 (1972).
21. J. Gilbert and A. L. Lachambre, Appl. Phys. Lett., 18:187 (1971).
22. J. A. Beaulieu, IEEE J. Quantum Electron., QE-7:495 (1971).
23. M. S. Dzhidzhoev, V. T. Platonenko, and R. V. Khokhlov, Usp. Fiz. Nauk, 100:641 (1970).
24. L. D. Hess, J. Chem. Phys., 55:466 (1971).
25. A. N. Chester, Laser Focus, 7:25 (1971).
26. V. K. Konyukhov and A. M. Prokhorov, Pis'ma Zh. Éksp. Teor. Fiz., 3:436 (1966).
27. I. R. Hurle and A. Herzberg, Phys. Fluids, 8:1601 (1965).
28. L. I. Gudzenko, S. S. Filippov, and L. A. Shelepin Zh. Éksp. Teor. Fiz., 51:1115 (1966).
29. N. G. Basov, V. G. Mikhailov, A. N. Oraevskii, and V. A. Shcheglov, Zh. Tekh. Fiz., 38:2031 (1968).
30. A. S. Biryukov, B. F. Gordiets, and L. A. Shelepin, Zh. Éksp. Teor. Fiz., 57:585 (1969).
31. E. T. Gerry, Laser Focus, 6:27 (1970).
32. A. P. Dronov, A. S. D'yakov, E. M. Kudryavtsev, and N. N. Sobolev, Pis'ma Zh. Éksp. Teor. Fiz., 11:516 (1970).
33. V. K. Konyukhov, I. V. Matrosov, A. M. Prokhorov, D. T. Shalunov, and N. N. Shirokov, Pis'ma Zh. Éksp. Teor. Fiz., 12:461 (1970).
34. E. T. Gerry, SPIE J., 9:61 (1971).
35. A. F. Krupnov, Izv. Vuzov: Radiofizika, 13:961 (1970).
36. R. A. McFarlane and L. M. Feetz, Appl. Phys. Lett., 14:385 (1969).
37. W. H. Seeling and K. V. Banse, Laser Focus, 6:33 (1970).
38. E. L. Latush and M. F. Sem, Zh. Éksp. Teor. Fiz., 64:2017 (1973).
39. L. I. Gudzenko, M. V. Nezlin, and S. I. Yakovlenko, Zh. Tekh. Fiz., 43:1931 (1973).
40. V. M. Goldfarb, E. V. Ilina, G. A. Lukyanov, and V. V. Sachin, Proc. 11th Intern. Conf. on Phenomena in Ionized Gases, Prague, Contributed Papers (1973), p. 16.
41. S. W. Bowen and C. Park, AIAA J., 10:522 (1972).
42. E. V. M. Likhachev, M. S. Rabinovich, and V. M. Sutovskii, Pis'ma Zh. Éksp. Teor. Fiz., 5:55 (1967).
43. L. I. Gudzenko and S. I. Yakovlenko, Dokl. Akad. Nauk SSSR, 207:1085 (1972).
44. G. A. Askaryan, Zh. Éksp. Teor. Fiz., 46:403 (1964); 48:666 (1965).
45. T. M. Makhviladze and L. A. Shelepin, Izv. Akad. Nauk SSSR, Ser. Fiz., 37:2190 (1973).
46. J. Stone and E. Thiel, J. Chem. Phys., 59:2909 (1973).

THE PHYSICAL KINETICS OF GAS-DYNAMIC LASERS†

A. S. Biryukov

A wide range of problems associated with gas-dynamic lasers is examined. Equations describing the kinetics of vibrational relaxation in mixtures of multiatomic molecular gases are derived. These equations are used to explain the conditions for inversions in gas-dynamic lasers with supersonic nozzles and with adiabatic expansion of the gas from a slit. The optimum initial pressures and temperatures, nozzle shapes and sizes, and mixture compositions for these lasers are found. Gas-dynamic lasers with supplementary disequilibrium sources, including nonresonance vibrational exchange, chemical recombination reactions, and electronic excitation, are discussed in detail.

INTRODUCTION

In recent years the attention of many researchers has been drawn to molecular lasers, with greatest interest in gas-dynamic lasers. This is explained by the high output powers (about 10^2 kW) obtained from molecular gases with rapid cooling and by the relative simplicity of constructing such systems. In these systems an inversion is produced due to a rapid drop in the gas temperature upon expansion. In this respect CO_2 lasers have the best characteristics. In CO_2 lasers an inversion is produced between the 00^01-10^00 (or 00^01-02^00) vibrational states of two vibrational modes owing to a difference in the relaxation times of the laser levels. Thus far, a large number of specific systems have been built and a substantial amount of experimental and theoretical development has taken place.

The rapid growth of research on gas-dynamic lasers raises a number of important problems. First there is the problem of optimizing the lasers: design of apparatus and analysis of the effect of the nozzle components on the population inversion, efficient choice of the composition of the active medium and investigation of the roles of each component of the mixutre, and choice of the best means for producing the initial conditions (heating, combustion, explosion, etc.). Second, there is the development and analysis of new gas-dynamic laser systems that use composite excitation techniques, in particular, the combination of gas-dynamic expansion and chemical reactions (chemical gas-dynamic lasers) and the combination of electrical excitation and gas-dynamic expansion (electrical gas-dynamic lasers). And, third, there is the study of the physical kinetics in various types of gas-dynamic lasers. This third area includes developing a method of computing the vibrational relaxation and finding and correctly utilizing the collisional transition probabilities.

The task of the present article is both to study the physical kinetics of gas-dynamic lasers and find the optimum operating parameters and regimes, and to investigate the possibilities of increasing the efficiency of lasers by using combination methods of producing inversion (chemical gas-dynamic lasers and electrical gas-dynamic lasers).

† This article includes material from a dissertation for the degree of candidate of Physical and Mathematical Sciences defended on October 29, 1973. The research adviser was L. A. Shelepin.

In the first chapter, after a review of published data on gas-dynamic lasers, we present some original results from a theoretical study of vibrational relaxation in gaseous mixtures of multiatomic molecules and an analysis of collisional transition probabilities. These results are used in formulating a technique for designing gas-dynamic lasers.

In the second chapter the kinetics of processes in low-temperature (initial temperatures, $T_0 \leqslant 2500°K$) gas-dynamic lasers are studied, and a detailed investigation is made of the effect of the shape of a supersonic nozzle on the laser parameters.

The third chapter deals with the combined effect of chemical reactions and nonresonance vibrational energy exchange on the gas dynamics for equilibrium and nonequilibrium initial conditions, and the chemical gas-dynamic laser is analyzed on this basis. A recombination molecular laser using vibrational–rotational transitions (with three-body recombination) is proposed.

In the fourth chapter a combination of the gas-dynamic method of producing a nonequilibrium state and a transverse electrical discharge (i.e., the electrical gas-dynamic laser) is analyzed. The basic features of such a system are first illustrated in terms of a model of a pulsed gas-discharge laser. Both subsonic and supersonic gas flow rates are considered.

The fifth chapter treats the problem of high-temperature gas-dynamic lasers ($T_0 \geq 3200°K$). The role of nonequilibrium chemical reactions is examined, and the possibility of constructing an electrical gas-dynamic laser using thermal ionization is pointed out.

In the sixth chapter the method of computing vibrational relaxation developed in the early chapters for analyzing laser systems is used to study the mechanism for production of infrared radiation in the wakes of objects moving at high supersonic velocities in the upper atmosphere.

CHAPTER I

THE PROBLEMS OF MOLECULAR KINETICS. A REVIEW OF THE LITERATURE

1. A Brief Review of Published Data on Gas-Dynamic Lasers

The operating principle of the gas-dynamic laser is based on the fact that when the temperature of the translational degrees of freedom of the gas particles is changed rapidly a population inversion may be produced due to a difference in the relaxation times of the upper and lower working levels. The use of thermal pumping for creating an inversion was pointed out in [1, 2]. The possibility of lasing on the 00^01-10^00 levels of carbon dioxide gas in the mixture CO_2-N_2 after it is cooled by expansion through a nozzle was first discussed in [3]. The use of expansion to produce an inversion in the electronic transitions of atoms in a plasma was considered in [4, 5].

The practical realization of a gas-dynamic laser was preceded by theoretical studies [6-8] in which the basic operating features of such lasers were first derived semiquantitatively. Thus, [6] was devoted to a study of vibrational relaxation and the possibility of producing a population inversion in CO_2 in a mixture with nitrogen by expanding preheated (to temperature T_0) gas through an axially symmetric Laval nozzle. A calculation indicated the existence of an optimum initial temperature of about 1500°K. The analogous problem was considered in [7] taking into account the possible dissociation of CO_2, and the vibrational temperature distribution along the axis of the nozzle was calculated for a five-component mixture (CO_2, CO, O_2, O,

$^{13}CO_2$). In a previous article [8] we analyzed the relaxation of a CO_2-N_2-He mixture as it expanded adiabatically into a vacuum. Here the model made it possible to examine both flow in a plane-parallel channel and free expansion through a slit. The results indicated that there are optimum values of the system parameters: initial temperatures and pressures, mixture composition, etc.

Later, a number of calculations [9-11] were made analogous to those of [6] for CO_2 mixtures containing water vapor.

The existence of an inverted population due to thermal pumping was first observed experimentally in [12], when radiation from a CO_2 laser was amplified in carbon dioxide gas behind a shock reflected from the end of a shock tube. Lasing was first observed later [13-15], also using shock tubes with a nozzle [13, 14] or a slit [15] at the end to couple the tubes with a low-pressure receiver. A diaphragm covering the entrance to the slit or nozzle was easily ruptured when the shock arrived at the end of the tube and the high-pressure, shock-heated, working mixture flowed into the receiver. The optical cavity was positioned perpendicular to the axis of the flow. Further studies have mainly been aimed at optimizing gas-dynamic lasers and producing high output powers [16-37]. Here we note [17] in which it was proposed that chemical reaction products be used as the active medium of a gas-dynamic laser. Such systems using combustion and explosion of carbon compounds (and other reactions) have been constructed experimentally many times [10, 14, 22-29].

While low-temperature ($T_0 \leqslant 2500$°K) lasers have been the subject of a large number of investigations, high-temperature gas-dynamic lasers have been studied little. The physical kinetics in them may differ strongly from the case of low temperatures ($T_0 < 2500$°K); in particular, CO_2 may be more than half dissociated and thermal ionization begins to play a role (for $T_0 > 3000$°K). The deviation of the chemical composition from equilibrium during expansion was noted in [7], and the effect of chemical reactions on the vibrational kinetics in the low-temperature region for $T_0 \leqslant 3000$°K was discussed in [33, 38]. The role of nonequilibrium recombination reactions was studied in [39]. In [10] the inversion distribution along the axis of a nozzle was computed for temperatures all the way up to $T_0 \approx 4000$°K; however, the effect of chemical reactions was neglected. The output power for high initial temperatures (up to ~4000°K) was measured experimentally in [31], but the measurements were not explained.

The problem that has been least studied in the area of low-temperature lasers is still the effect of the nozzle shape on the kinetics. In most work on this topic one or another type of nozzle has been examined, but, even when they are brought together, these articles [6, 11, 31, 32] do not enable us to draw a final conclusion on the role of the nozzle shape in the production of inversion.

Additional sources of disequilibrium may be employed to improve the operation of gas-dynamic lasers. Electronic excitation of molecular vibrational levels must be included among such sources along with nonequilibrium chemical reactions. Historically, low-pressure electrical discharge lasers were the first to appear. The basic physical processes in lasers using molecular vibrational levels, in particular those of CO_2, were analyzed theoretically for the steady-state case in [40-43]. However, later on much more attention began to be devoted to the pulsed regime since it was possible to obtain substantial output powers there. Population inversions and lasing in CO_2 gas in the pulsed regime were studied experimentally in [44-50].

The increased working pressure in early laser designs was accompanied by a need for very high fields to break down the gas. However, this difficulty was soon bypassed by using a transverse discharge [51, 52] and pumpthrough of the gas [53-56]. Pumpthrough at high pressures made it possible to increase the pulse repetition rate and thereby ensured large average powers.

These advantages of pulsed over steady-state excitation at high gas feed rates makes it possible in principle to combine the gas-dynamic laser with continuous excitation in a discharge. This arrangement was first considered in [57]. In a coordinate system attached to the moving gas the situation is completely analogous to a pulsed electrical discharge laser with all its advantages, and in a coordinate system attached to the fixed discharge gap, it is analogous to a transverse discharge laser with a supersonic feed rate. Here a continuous regime is realized at powers higher than in ordinary gas-dynamic lasers and comparable to the power from transverse discharge lasers.

From this brief review it is clear that, as gas-dynamic lasers were developed and the physical processes taking place in the expanding gas were studied, the need arose to work out a series of fundamental questions associated both with finding the optimum operating conditions for these lasers (for example, examining the effect of nozzle shapes) and with analyzing the prospects for increasing the power by using combination methods of producing a population inversion (such as chemical and electrical gas-dynamic lasers).

The basis of a theoretical analysis of the physical processes in gas-dynamic lasers is vibrational kinetics. (Vibrational relaxation in gases is reviewed in [37].). However, until recently all the work [6-11, 58, 59] involving the relaxation equations included some errors. Hence it is important to obtain the correct kinetic equations and to correctly use the available experimental and theoretical data on the probabilities of elementary processes in them. In Sections 2 and 3 of this chapter we shall examine a group of problems worked out previously by us [60-61] which make it possible to analyze the operation of gas-dynamic lasers.

2. The Kinetic Equations for Vibrational Relaxation

At the basis of any analysis of the processes for creating a population inversion in gas-dynamic lasers lie the equations of vibrational kinetics. In general they are a system of balance equations for a large number of molecular levels. At present it is impossible to solve such a system of equations; hence, simplifying model assumptions are used. According to the currently generally accepted model [41] in most practically important cases it is possible to introduce a vibrational temperature for each normal vibrational mode in multiatomic molecules. This is because the exchange of quanta within a specific vibrational mode (V – V process) takes place over a time much less than that for transfer of energy to translational degrees of freedom (V – T process) or for energy exchange between different vibrational modes (V – V' process).

Because it greatly simplifies population kinetic calculations (a large set of equations for the individual levels is replaced by a single equation for the vibrational temperature), this model has been fruitful for analyzing the population-inversion mechanisms in many actual laser systems [6-11, 39-43]. Thus, a number of authors [6, 58, 59] have posed the problem of deriving the relaxation kinetic equations for multiatomic molecules. In some of the first work in this field [62, 63] an equation was introduced describing the equilibration between the symmetric and deformation vibrations of CO_2 including two quantum transitions. Relaxation equations are found in [6] for all the types of vibrations in gaseous CO_2, and in [58] a derivation is given of the kinetic equations for vibrational relaxation of multiatomic molecules due to single quantum transitions in different modes. Analogous equations are derived in [59] including multiple quantum exchange to describe relaxation with the participation of three vibrational modes. Nevertheless, in many practically important cases the use of these equations leads to errors since in their derivation the often-encountered case of degenerate vibrational modes was either neglected or incorrectly treated. In addition, several peculiarities of relaxation in a gas of identical molecules were not taken into account (or were, but incorrectly) in several papers (e.g., [58, 59]).

Another interesting aspect of the theory of vibrational relaxation of multicomponent molecular gaseous mixtures is the distribution of the energy among the different vibrational modes during quasiequilibrium when the rate of energy exchange between these modes is much greater than the rate of vibrational–translational relaxation. The relation between the average amount of stored vibrational quanta and the vibrational temperatures during rapid single quantum energy exchange in a binary mixture of diatomic gases was first obtained in [64, 65]. Later it was generalized to the special cases of multicomponent mixture [66] and of a binary mixture of diatomic gases with two channels for fast exchange [67, 68] and with a single channel for multiquantum exchange [69]. Thus, in a binary mixture of harmonic oscillators A and B, with quantum energies E_A and E_B, for rapid exchange of n quanta from oscillator A with m quanta from oscillator B the vibrational temperatures T_A and T_B and the gas temperature T are related by

$$\frac{nE_A}{T_A}-\frac{mE_B}{T_B}=\frac{nE_A-mE_B}{T}. \tag{1.1}$$

We note, however, that expressions such as Eq. (1.1) and others found in [64-69] only couple the temperatures of two vibrational modes and do not reflect the situation which may occur during rapid vibrational energy exchange in a mixture of multiatomic molecular gases.

The purpose of this section is to obtain kinetic equations for the most general case that will describe energy relaxation of a vibrational mode of a multiatomic molecule modeled by a collection of harmonic oscillators. It is assumed that a change in the energy may be caused by V–T processes and multiquantum V–V' and V–V processes involving an arbitrary number of degenerate modes. The resulting equations are used to describe vibrational relaxation in the particular gas mixture CO_2-N_2. The relationship among the vibrational temperatures during rapid vibrational exchange between modes is also found as a generalization of Eq. (1.1).

We shall examine vibrational relaxation in a system of multiatomic molecules using a binary mixture of gases A and B as an example. Here we shall be interested only in transitions due to collisions between A and B molecules (this corresponds to the type of gas of interest to us being a small additive to the other). Relaxation in a single-component gas will be examined later. Let the molecules of A and B have a total of L vibrational modes modeled by harmonic oscillators. In this case an arbitrary vibrational state of the system A + B can be described by a set of L vibrational quantum numbers $(v_1, v_2, \ldots, v_L)$. The overall process in one of the possible channels for relaxation of the vibrational level populations due to a collision between A and B molecules can be written in the form

$$\begin{gathered}(v_1, \ldots, v_p;\ v_{p+1}, \ldots, v_L)\equiv\\ \equiv\{v_i;\ v_j\}\to(v_1\pm l_1, \ldots, v_p\pm l_p;\ v_{p+1}\mp l_{p+1}, \ldots, v_L\mp l_L)\equiv\{v_i\pm l_i;\ v_j\mp l_j\},\end{gathered} \tag{1.2}$$

where $i = 1, \ldots, p$; $j = (p+1), \ldots, L$. This notation means that due to a collision the system A + B went from some initial state $\{v_i;\ v_j\}$ to a new state in a way such that in the modes labeled i there was a jump upward (downward) by l_i quanta and in the remaining states, labeled j, there was a jump downward (upward) by l_j quanta. Among both groups of modes, i and j, there may be modes that belong to different molecules.

Let us choose an arbitrary mode ξ of the A molecule belonging, for example, to the i-th group of modes. We shall be interested in the changes in the average amount, ε_ξ, of vibrational quanta stored in this mode. We first write the balance equation describing the change in the populations $N_A(v_\xi)$ of the levels v_ξ due to process (1.2) during a collision between A and B molecules:

$$\frac{dN_A(v_\xi)}{dt}=\frac{Z_{AB}}{N_B}\sum_{\{v_i;v_j\}}\Bigg[P\begin{Bmatrix}v_i-l_i\to v_i\\ v_j+l_j\to v_j\end{Bmatrix}N_A\{v_{i_A}-l_{i_A};v_{j_A}+l_{j_A}\}N_B\{v_{i_B}-l_{i_B};v_{j_B}+l_{j_B}\}-$$
$$-P\begin{Bmatrix}v_i\to v_i-l_i\\ v_j\to v_j+l_j\end{Bmatrix}N_A\{v_{i_A};\ v_{j_A}\}N_B\{v_{i_B};\ v_{j_B}\}+$$
$$+P\begin{Bmatrix}v_i+l_i\to v_i\\ v_j-l_j\to v_j\end{Bmatrix}N_A\{v_{i_A}+l_{i_A};\ v_{j_A}-l_{j_A}\}N_B\{v_{i_B}+l_{i_B};\ v_{j_B}-l_{j_B}\}-$$
$$-P\begin{Bmatrix}v_i\to v_i+l_i\\ v_j\to v_j-l_j\end{Bmatrix}N_A\{v_{i_A};\ v_{j_A}\}N_B\{v_{i_B};\ v_{j_B}\}\Bigg]. \tag{1.3}$$

Here $N_A\{v_{iA};\ v_{jA}\}$ and $N_B\{v_{iB};\ v_{jB}\}$ are the populations of the vibrational states $\{v_{iA};\ v_{jA}\}$ and $\{v_{iB};\ v_{jB}\}$ of A and B molecules, respectively, $N_A(v_\xi)=\sum_{\{v_{i_A};\ v_{j_A}\}\neq v_\xi}N_A\{v_{i_A};\ v_{j_A}\}$, N_A and N_B are the total number of A and B molecules per unit volume, Z_{AB} is the number of collisions of an A molecule with B molecules per unit time, and P is the probability of a composite vibrational transition in the system A + B due to a single collision. The summation in Eq. (1.3) is done over all levels of all modes in both types of molecule except for the chosen mode ξ in A.

We assume that each vibrational mode i(or j) can be r_i- (or r_j-) fold degenerate. Then according to the principle of detailed balance, the probabilities of the direct and reverse processes are related by equations of this type:

$$P\begin{Bmatrix}v_i-l_i\to v_i\\ v_j+l_j\to v_j\end{Bmatrix}=\frac{g\{v_i;\ v_j\}}{g\{v_i-l_i;\ v_j+l_j\}}P\begin{Bmatrix}v_i\to v_i-l_i\\ v_j\to v_j+l_j\end{Bmatrix}\prod_{i=1}^{p}x_{0i}^{l_i}\prod_{j=p+1}^{L}x_{0j}^{-l_j}. \tag{1.4}$$

Here

$$x_{0s}=\exp\left(-\frac{h\nu_s}{kT}\right),\qquad s=i=1,\ldots,p;\qquad s=j=p+1,\ldots,L; \tag{1.5}$$

ν_s is the frequency of the normal vibrations of the s-th mode, T is the gas temperature, and $g\{v_i;\ v_j\}$ is the statistical weight of the vibrational state $\{v_i;\ v_j\}$ of the system A + B and is given by

$$g\{v_i;\ v_j\}=g(v_1)\cdots g(v_p)\,g(v_{p+1})\cdots g(v_L)=$$
$$=\prod_{i=1}^{p}\frac{(v_i+r_i-1)!}{v_i!\,(r_i-1)!}\prod_{j=p+1}^{L}\frac{(v_j+r_j-1)!}{v_j!\,(r_j-1)!}\equiv\prod_{i=1}^{p}\binom{v_i+r_i-1}{v_i}\prod_{j=p+1}^{L}\binom{v_j+r_j-1}{v_j}. \tag{1.6}$$

In order to obtain the desired equation for the vibrational energy in the ξ-th mode from Eq. (1.3), we shall make some transformations on its right-hand side. First of all, we shall express the probabilities P of the composite transitions between arbitrary upper states in terms of the corresponding probabilities of transitions between the lowest levels taking part in the transitions. In this case

$$P\begin{Bmatrix}v_i+l_i\to v_i\\ v_j-l_j\to v_j\end{Bmatrix}=\frac{g\{v_j\}}{g\{v_j-l_j\}\,g\{l_j\}}P\begin{Bmatrix}l_i\to 0\\ 0\to l_j\end{Bmatrix}\prod_{i=1}^{p}\binom{v_i+l_i}{v_i}\prod_{j=p+1}^{L}\binom{v_j}{v_j-l_j} \tag{1.7}$$

(for a derivation of this equation see Section 3).

Further transformation of Eq. (1.3) is greatly simplified if, as shown above, for each type of normal vibration (mode) we introduce a vibrational temperature T_s which characterizes the vibrational energy stored in the given mode. The temperatures T_s will generally differ from one another and from the gas temperature T. Now the populations in Eq. (1.3) may be written in the form

$$N_A\{v_{i_A};\ v_{j_A}\} = N_A g\{v_{i_A};\ v_{j_A}\} \prod_{i_A} x_{i_A}^{v_{i_A}} \left(1 - x_{i_A}\right)^{r_{i_A}} \prod_{j_A} x_{j_A}^{v_{j_A}} \left(1 - x_{j_A}\right)^{r_{j_A}}, \tag{1.8}$$

or more compactly, leaving out the subscripts A and B, as

$$N_A\{v_{i_A};\ v_{j_A}\} N_B\{v_{i_B};\ v_{j_B}\} = N_A N_B g\{v_i;\ v_j\} \prod_{i=1}^{p} x_i^{v_i} (1 - x_i)^{r_i} \prod_{j=p+1}^{L} x_j^{v_j} (1 - x_j)^{r_j}. \tag{1.9}$$

Here

$$x_s = \exp\left(-\frac{h\nu_s}{kT_s}\right), \quad s = i = 1, \ldots, p; \quad s = j = p+1, \ldots, L. \tag{1.10}$$

If some mode s belongs, for example, to an A molecule, then the average amount ε_s of vibrational quanta stored in that mode per molecule may be expressed in terms of x_s. To do this, it is necessary to multiply Eq. (1.8) by v_s and to sum over all $\{v_{iA};\ v_{jA}\}$, including v_s, from 0 to ∞:

$$\varepsilon_{s_A} = \prod_{i_A} \left(1 - x_{i_A}\right)^{r_{i_A}} \prod_{j_A} \left(1 - x_{j_A}\right)^{r_{j_A}} \sum_{\{v_{i_A};\ v_{j_A}\}} \left[v_s g\{v_{i_A};\ v_{j_A}\} \prod_{i_A} x_{i_A}^{v_{i_A}} \prod_{j_A} x_{j_A}^{v_{j_A}} \right]. \tag{1.11}$$

Summing, using Eqs. (1.6) and (1.10), and leaving out the subscript A, we obtain

$$\varepsilon_s = r_s \frac{x_s}{1 - x_s}. \tag{1.12}$$

We note that by summing over all $\{v_{iA};\ v_{jA}\}$ the contribution of combination levels to the total amount of vibrational quanta ε_s stored in mode s has been taken into account.

We now return directly to our derivation of the kinetic equation for the variation in the vibrational energy in the chosen mode ξ. Multiplying both parts of Eq. (1.3) by v_ξ, then summing over all $\{v_i;\ v_j\}$, including v_ξ, from 0 to ∞ with the aid of Eq. (1.4)-(1.12), we find after a series of transformations that

$$\begin{aligned} \frac{d\varepsilon_\xi}{dt} = Z_{AB} P \begin{Bmatrix} l_i \to 0 \\ 0 \to l_j \end{Bmatrix} l_\xi \prod_{i=1}^{p} (1 - x_i)^{r_i} \prod_{j=p+1}^{L} (1 - x_j)^{r_j} \binom{l_j + r_j - 1}{l_j}^{-1} \times \\ \times \left[\prod_{i=1}^{p} x_{0i}^{l_i} \prod_{j=p+1}^{L} \left(\frac{x_j}{x_{0j}}\right)^{l_j} - \prod_{i=1}^{p} x_i^{l_i} \right] \prod_{i=1}^{p} \binom{l_i + r_i - 1}{l_i} \times \\ \times F(l_i + r_i,\ 1;\ 1;\ x_i) \prod_{j=p+1}^{L} \binom{l_j + r_j - 1}{l_j} F(l_j + r_j,\ 1;\ 1;\ x_j). \end{aligned} \tag{1.13}$$

From this equation and the explicit form of the hypergeometric function we find the final form of the equation:

$$\frac{d\varepsilon_\xi}{dt} = Z_{AB} P \begin{Bmatrix} l_i \to 0 \\ 0 \to l_j \end{Bmatrix} l_\xi \prod_{i=1}^{p} \binom{l_i + r_i - 1}{l_i} (1 - x_i)^{-l_i} \prod_{j=p+1}^{L} (1 - x_j)^{-l_j} \left[\prod_{i=1}^{p} x_{0i}^{l_i} \prod_{j=p+1}^{L} \left(\frac{x_j}{x_{0j}}\right)^{l_j} - \prod_{i=1}^{p} x_i^{l_i} \right]. \tag{1.14a}$$

Using Eq. (1.12) and the equilibrium value of ε_{0s} for gas temperature T,

$$\varepsilon_{0s} = r_s \frac{x_{0s}}{1 - x_{0s}},$$

we rewrite Eq. (1.14a) in the following form:

$$\frac{d\varepsilon_\xi}{dt}=Z_{AB}P_{AB}\begin{Bmatrix}l_i\to 0\\ 0\to l_j\end{Bmatrix} l_\xi \prod_{i=1}^{p}\binom{l_i+r_i-1}{l_i}[r_i(r_i+\varepsilon_{0i})]^{-l_i}\prod_{j=p+1}^{L}(r_j\varepsilon_{0j})^{-l_j}\times$$

$$\times\left\{\prod_{i=1}^{p}[\varepsilon_{0i}(r_i+\varepsilon_i)]^{l_i}\prod_{j=p+1}^{L}[\varepsilon_j(r_j+\varepsilon_{0j})]^{l_j}-\prod_{i=1}^{p}[\varepsilon_i(r_i+\varepsilon_{0i})]^{l_i}\prod_{j=p+1}^{L}[\varepsilon_{0j}(r_j+\varepsilon_j)]^{l_j}\right\}. \qquad (1.14b)$$

The probability $P_{AB}\begin{Bmatrix}l_i\to 0\\ 0\to l_j\end{Bmatrix}$ in Eqs. (1.14a) and (1.14b) depends on the degrees of degeneracy of the modes i and j which take part in the transition. As is shown in the following section,

$$P_{AB}\begin{Bmatrix}l_i\to 0\\ 0\to l_j\end{Bmatrix}=P_{AB}^{SSH}\begin{Bmatrix}l_i\to 0\\ 0\to l_j\end{Bmatrix}\prod_{i=1}^{p}r_i^{l_i}\binom{l_i+r_i-1}{l_i}^{-1}\prod_{j=p+1}^{L}r_j^{l_j}, \qquad (1.15)$$

where P_{AB}^{SSH} is the probability of the transition calculated without including the degeneracy using the usual method of Schwartz, Slawsky, and Herzfeld (SSH) [62, 63]. Substituting Eq. (1.15) in Eq. (1.14b) we obtain yet another form of the kinetic equation we have been seeking:

$$\frac{d\varepsilon_\xi}{dt}=Z_{AB}P_{AB}^{SSH}\begin{Bmatrix}l_i\to 0\\ 0\to l_j\end{Bmatrix} l_\xi\prod_{i=1}^{p}(r_i+\varepsilon_{0i})^{-l_i}\prod_{j=p+1}^{L}\varepsilon_{0j}^{-l_j}\left\{\prod_{i=1}^{p}[\varepsilon_{0i}(r_i+\varepsilon_i)]^{l_i}\times\right.$$

$$\left.\times\prod_{j=p+1}^{L}[\varepsilon_j(r_j+\varepsilon_{0j})]^{l_j}-\prod_{i=1}^{p}[\varepsilon_i(r_i+\varepsilon_{0i})]^{l_i}\prod_{j=p+1}^{L}[\varepsilon_{0j}(r_j+\varepsilon_j)]^{l_j}\right\}. \qquad (1.16)$$

Equations (1.14a), (1.14b), and (1.16) describe the relaxation of vibrational quanta (or energy) in some arbitrary mode ξ in a single channel which is determined by specifying the numbers $\{l_i; l_j\}$. When there are several channels, the relaxation rates for each of them calculated using Eqs. (1.14a), (1.14b), and (1.16) must be added together. Of course, the equation for ε_ξ found in this way must generally be solved together with the equations describing the energy relaxation of the remaining modes.

Equations (1.14a), (1.14b), and (1.16) are the most general form of the equations for the vibrational energy (or stored quanta) in a mixture of harmonic oscillators and describe relaxation due to collisions between different types of molecules. For $l_i = l_j = 0$ (for $i \neq \xi$) and $l_\xi = 1$ Eqs. (1.14a), (1.14b), and (1.16) transform into the usual Landau–Teller equation for vibrational–translational relaxation. For $L = 2$, $l_\xi = 1$, $l_j = 1$, $r_\xi = 1$, and $r_j = 1$ they correspond to single quantum exchange in a binary mixture of diatomic molecules. In the special case $\{l_i; l_i\} \leq 1$ these equations become the formula obtained in [58] for $\{r_i; r_i\} = 1$ describing single-quantum exchange with the participation of many modes. However, in [58], and also in [6, 59], the case of degenerate oscillators has been incorrectly treated and the peculiarities of collisional relaxation of identical particles have not been properly taken into account.

These peculiarities are due to the fact that, as opposed to the case of collisions between different molecules A + B, an arbitrary energy state in the system A + A is not determined uniquely by the set of quantum numbers $\{v_i; v_j\}$ of two molecules alone, but is the same if, for example, we consider the level $v_i + k$ instead of v_i for some mode and in the other molecule we consider the level $w_i - k$ in place of w_i for the same mode. In fact, this means that in A + A collision part of the energy (or quanta) lost by mode ξ in one of the A molecules may be distributed differently in both molecules. Formally, this case corresponds to a doubling of the degree of degeneracy of the mode. Thus, the relaxation equation for the vibrational energy stored in mode ξ in a single-component multiatomic gas may be obtained using the results for a binary mixture. In this case, instead of Eq. (1.14b), we have

$$\frac{d\varepsilon_\xi}{dt}=\frac{1}{2}Z_{AA}P_{AA}\begin{Bmatrix} l_i\to 0\\ 0\to l_j\end{Bmatrix} l_\xi \prod_{i=1}^{p}\binom{l_i+2r_i-1}{l_i}[r_i(r_i+\varepsilon_{0i})]^{-l_i}\prod_{j=p+1}^{L}(r_j\varepsilon_{0j})^{-l_j}\times$$
$$\times\left\{\prod_{i=1}^{p}[\varepsilon_{0i}(r_i+\varepsilon_i)]^{l_i}\prod_{j=p+1}^{L}[\varepsilon_j(r_j+\varepsilon_{0j})]^{l_j}-\prod_{i=1}^{p}[\varepsilon_i(r_i+\varepsilon_{0i})]^{l_i}\prod_{j=p+1}^{L}[\varepsilon_{0j}(r_j+\varepsilon_j)]^{l_j}\right\}. \quad (1.17)$$

Here the notation is the same as in Eq. (1.14b), but the probabilities are determined by (see Section 3)

$$P_{AA}\begin{Bmatrix} l_i\to 0\\ 0\to l_j\end{Bmatrix}=P_{AA}^{SSH}\begin{Bmatrix} l_i\to 0\\ 0\to l_j\end{Bmatrix}\prod_{i=1}^{p}(2r_i)^{l_i}\binom{l_i+2r_i-1}{l_i}^{-1}\prod_{j=p+1}^{L}(2r_j)^{l_j}. \quad (1.18)$$

Substituting Eq. (1.18) into (1.17) we obtain an equation analogous to Eq. (1.16):

$$\frac{d\varepsilon_\xi}{dt}=\frac{1}{2}\prod_{i=1}^{p}2^{l_i}\prod_{j=p+1}^{L}2^{l_j}\cdot Z_{AA}P_{AA}^{SSH}\begin{Bmatrix} l_i\to 0\\ 0\to l_j\end{Bmatrix} l_\xi\prod_{i=1}^{p}(r_i+\varepsilon_{0i})^{-l_i}\prod_{j=p+1}^{L}\varepsilon_{0j}^{-l_j}\times$$
$$\times\left\{\prod_{i=1}^{p}[\varepsilon_{0i}(r_i+\varepsilon_i)]^{l_i}\prod_{j=p+1}^{L}[\varepsilon_j(r_j+\varepsilon_{0j})]^{l_j}-\prod_{i=1}^{p}[\varepsilon_i(r_i+\varepsilon_{0i})]^{l_i}\prod_{j=p+1}^{L}[\varepsilon_{0j}(r_j+\varepsilon_j)]^{l_j}\right\}. \quad (1.19)$$

It is clear from a comparison of Eqs. (1.16) and (1.19) that the relaxation equation for ε_ξ in a binary mixture of gases A + B has the same form as for a single-component gas. In the latter case, however, all other conditions being the same, an additional factor $\frac{1}{2}\prod_{i=1}^{p}2^{l_i}\prod_{j=p+1}^{L}2^{l_j}$ appears which increases the relaxation rate. Its presence is due to the appearance of additional channels for relaxation of the energy of mode ξ during A + A collisions because of different possible mechanisms for redistribution of the vibrational energy before and after collisions between identical molecules. Thus, Eq. (1.19) may be obtained directly by summing Eqs. (1.16) for each possible relaxation channel involving two identical modes from groups i and j belonging to different molecules. These channels differ in that for transitions in identical modes the jumps may be different, changing within the limits of 0 to l_i (l_j); however, the sum of the jumps in these modes must always be l_i (l_j) for the given energy transition.

Now we shall use the general equations derived here to describe vibrational relaxation in the mixture CO_2-N_2 as an example.

We shall denote the symmetric, deformed, and asymmetric modes of CO_2 and vibrations of N_2 by the numbers 1 to 4 (Fig. 1): $h\nu_1/k = 2000°K$, $h\nu_2/k = 960°K$, $h\nu_3/k = 3380°K$, and $h\nu_4/k = 3353°K$. Here $r_1 = r_3 = r_4 = 1$ and $r_2 = 2$ (the deformed oscillation modes are doubly degenerate).

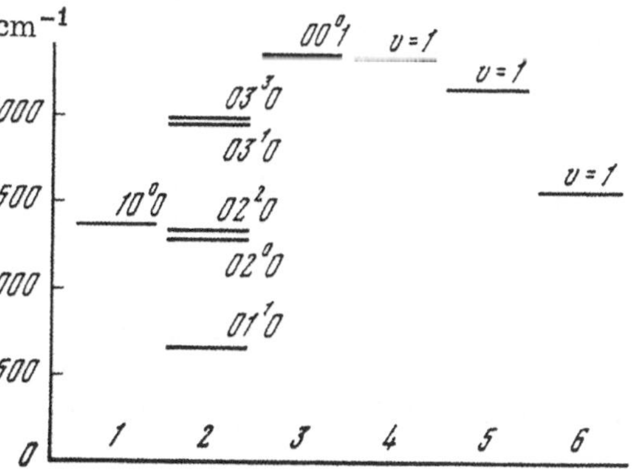

Fig. 1. Diagram of the lower vibrational levels of the CO_2 (curves 1-3), N_2 (curve 4), CO (curve 5), and O_2 (curve 6) molecules.

The basic channels for energy relaxation in modes 1-4 are due to the following processes (the notation for the corresponding probabilities is given on the right):

$$\begin{array}{lll}
N_2(v=1)+CO_2(00^00) \rightleftarrows CO_2(00^01)+N_2(v=0), & & P(4\to 3),\\
N_2(v=1)+CO_2(00^00) \rightleftarrows \begin{cases} N_2(v=0)+CO_2(11^10), \\ N_2(v=0)+CO_2(03^10), \end{cases} & & \begin{array}{l} P(4\to 1,\ 2),\\ P(4\to 2),\end{array}\\
CO_2(00^01)+M \rightleftarrows \begin{cases} CO_2(11^10)+M, \\ CO_2(03^10)+M, \end{cases} & & \begin{array}{l} P^M(3\to 1,2),\\ P^M(3\to 2),\end{array}\\
CO_2(10^00)+M \rightleftarrows CO_2(02^00)+M, & & P^M(1\to 2),\\
CO_2(01^10)+M \rightleftarrows CO_2(00^00)+M, & & P^M(2\to 0).
\end{array} \tag{1.20}$$

Here M = N_2 or CO_2. We note that for clarity in Eq. (1.20) only transitions among the lowest levels have been written. Including the processes of Eq. (1.20) and using Eqs. (1.14b) and (1.17), the relaxation equations for ε_1 to ε_4 may be written in the form

$$\left.\begin{aligned}
\frac{d\varepsilon_4}{dt} &= Z_{N_2,CO_2}\Big\{P(4\to 3)\Big[e^{\frac{27}{T}}\varepsilon_3(1+\varepsilon_4)-\varepsilon_4(1+\varepsilon_3)\Big]+\\
&+\frac{1}{2}P(4\to 1,\ 2)\Big[e^{-\frac{393}{T}}\varepsilon_1\varepsilon_2(1+\varepsilon_4)-\varepsilon_4(1+\varepsilon_1)(2+\varepsilon_2)\Big]+\\
&+\frac{1}{8}P(4\to 2)\Big[e^{-\frac{473}{T}}\varepsilon_2^3(1+\varepsilon_4)-\varepsilon_4(2+\varepsilon_2)^3\Big]\Big\},\\
\frac{d\varepsilon_3}{dt} &= -Z_{CO_2,N_2}P(4\to 3)\Big[e^{\frac{27}{T}}\varepsilon_3(1+\varepsilon_4)-\varepsilon_4(1+\varepsilon_3)\Big]+\\
&+\frac{1}{8}[Z_{CO_2,CO_2}P^{CO_2}(3\to 2)+Z_{CO_2,N_2}P^{N_2}(3\to 2)]\Big[e^{-\frac{500}{T}}\varepsilon_2^3(1+\varepsilon_3)-\\
&-\varepsilon_3(2+\varepsilon_2)^3\Big]+\frac{1}{2}[Z_{CO_2,CO_2}P^{CO_2}(3\to 1,\ 2)+Z_{CO_2,N_2}P^{N_2}(3\to 1,\ 2)]\times\\
&\times\Big[e^{-\frac{420}{T}}\varepsilon_1\varepsilon_2(1+\varepsilon_3)-\varepsilon_3(1+\varepsilon_1)(2+\varepsilon_2)\Big],\\
\frac{d\varepsilon_2}{dt} &= -\frac{3}{8}[Z_{CO_2,CO_2}P^{CO_2}(3\to 2)+Z_{CO_2,N_2}P^{N_2}(3\to 2)]\Big[e^{-\frac{500}{T}}\varepsilon_2^3(1+\varepsilon_3)-\\
&-\varepsilon_3(2+\varepsilon_2)^3\Big]-\frac{1}{2}[Z_{CO_2,CO_2}P^{CO_2}(3\to 1,2)+Z_{CO_2,N_2}P^{N_2}(3\to 1,2)]\times\\
&\times\Big[e^{-\frac{420}{T}}\varepsilon_1\varepsilon_2(1+\varepsilon_3)-\varepsilon_3(1+\varepsilon_1)(2+\varepsilon_2)\Big]-\frac{3}{8}Z_{CO_2,N_2}P(4\to 2)\times\\
&\times\Big[e^{-\frac{473}{T}}\varepsilon_2^3(1+\varepsilon_4)-\varepsilon_4(2+\varepsilon_2)^3\Big]-\frac{1}{2}Z_{CO_2,N_2}P(4\to 1,\ 2)\times\\
&\times\Big[e^{-\frac{393}{T}}\varepsilon_1\varepsilon_2(1+\varepsilon_4)-\varepsilon_4(1+\varepsilon_1)(2+\varepsilon_2)\Big]-\frac{1}{2}[Z_{CO_2,CO_2}P^{CO_2}(1\to 2)+\\
&+Z_{CO_2,N_2}P^{N_2}(1\to 2)]\Big[e^{-\frac{80}{T}}\varepsilon_2^2(1+\varepsilon_1)-\varepsilon_1(2+\varepsilon_2)^2\Big]+\\
&+[Z_{CO_2,CO_2}P^{CO_2}(2\to 0)+Z_{CO_2,N_2}P^{N_2}(2\to 0)](1-x_{02})(\varepsilon_{02}-\varepsilon_2),\\
\frac{d\varepsilon_1}{dt} &= -\frac{1}{2}[Z_{CO_2,CO_2}P^{CO_2}(3\to 1,\ 2)+Z_{CO_2,N_2}P^{N_2}(3\to 1,\ 2)]\Big[e^{-\frac{420}{T}}\varepsilon_1\varepsilon_2\times\\
&\times(1+\varepsilon_3)-\varepsilon_3(1+\varepsilon_1)(2+\varepsilon_2)\Big]-\frac{1}{2}Z_{CO_2,N_2}P(4\to 1,\ 2)\times\\
&\times\Big[e^{-\frac{393}{T}}\varepsilon_1\varepsilon_2(1+\varepsilon_4)-\varepsilon_4(1+\varepsilon_1)(2+\varepsilon_2)\Big]+\frac{1}{4}[Z_{CO_2,CO_2}P^{CO_2}(1\to 2)+\\
&+Z_{CO_2,N_2}P^{N_2}(1\to 2)]\Big[e^{-\frac{80}{T}}\varepsilon_2^2(1+\varepsilon_1)-\varepsilon_1(2+\varepsilon_2)^2\Big].
\end{aligned}\right\} \tag{1.21}$$

In practice the system of Eqs. (1.21) is usually solved with simplifying assumptions, specifically: $h\nu_3 = h\nu_4$, $h\nu_1 = 2h\nu_2$, $(1+\varepsilon_{02})/(1+\varepsilon_2) \approx 1$, and $T_1 = T_2$. In most cases the latter is justified in view of the rapid exchange of vibrational quanta between the symmetric and deformed modes since $P^M(1 \to 2)$ is much greater than the other probabilities. Then we may rewrite Eq. (1.21) in the form

$$\left.\begin{aligned}
\frac{d\varepsilon_4}{dt} &= Z_{N_2,CO_2}\Big\{P(4\to 3)(\varepsilon_3-\varepsilon_4)+P(4\to\Sigma)\times\\
&\times\Big[e^{-\frac{500}{T}}\Big(\frac{\varepsilon_2}{2}\Big)^3(1+\varepsilon_4)-\varepsilon_4\Big(1+\frac{\varepsilon_2}{2}\Big)^3\Big]\Big\},\\
\frac{d\varepsilon_3}{dt} &= -Z_{CO_2,N_2}P(4\to 3)(\varepsilon_3-\varepsilon_4)+[Z_{CO_2,CO_2}P^{CO_2}(3\to\Sigma)+\\
&+Z_{CO_2,N_2}P^{N_2}(3\to\Sigma)]\Big[e^{-\frac{500}{T}}\Big(\frac{\varepsilon_2}{2}\Big)^3(1+\varepsilon_3)-\varepsilon_3\Big(1+\frac{\varepsilon_2}{2}\Big)^3\Big],\\
\frac{d(\varepsilon_2+2\varepsilon_1)}{dt} &= -3[Z_{CO_2,CO_2}P^{CO_2}(3\to\Sigma)+Z_{CO_2,N_2}P^{N_2}(3\to\Sigma)]\times\\
&\times\Big[e^{-\frac{500}{T}}\Big(\frac{\varepsilon_2}{2}\Big)^3(1+\varepsilon_3)-\varepsilon_3\Big(1+\frac{\varepsilon_2}{2}\Big)^3\Big]-3Z_{CO_2,N_2}P(4\to\Sigma)\times\\
&\times\Big[e^{-\frac{500}{T}}\Big(\frac{\varepsilon_2}{2}\Big)^3(1+\varepsilon_4)-\varepsilon_4\Big(1+\frac{\varepsilon_2}{2}\Big)^3\Big]+[Z_{CO_2,CO_2}P^{CO_2}(2\to 0)+\\
&+Z_{CO_2,N_2}P^{N_2}(2\to 0)](1-x_{02})(\varepsilon_{02}-\varepsilon_2).
\end{aligned}\right\} \quad (1.21a)$$

Here we use the following simplifying notation:

$$P(4\to\Sigma)=(1+\varepsilon_{02})^{-1}P(4\to 1,\ 2)+P(4\to 2),$$
$$P^M(3\to\Sigma)=(1+\varepsilon_{02})^{-1}P^M(3\to 1,\ 2)+P^M(3\to 2).$$

In the following the probabilities of transitions in a single collision, P_{ij}, are often replaced by the rate coefficients for the corresponding process, W_{ij} (in cm^3/sec). The relation between P and W is given by expressions of the type $P_{ij} = W_{ij}N_B/Z_{AB}$.

We now examine the quasistationary energy distribution in a mixture of multiatomic gases. Such a distribution may occur when the rate of exchange of vibrational energy between different modes is much greater than the vibrational–translational rate. In this case, with an overall nonequilibrium vibrational energy in the system, the vibrational temperatures T_s of the separate modes are "rigidly" connected to one another, and they are determined only by this overall nonequilibrium energy and the gas temperature T. According to [64-69] the relationships among the vibrational temperatures for a quasistationary distribution may be found by considering the stationary ($d\varepsilon_\xi/dt = 0$) relaxation equations for the average amount of stored vibrational quanta in the different modes and eliminating terms from them that include the vibrational–translational relaxation channel. Beginning with this and limiting ourselves to consideration of vibrational exchange taking place in one channel only for each mode, we find from Eqs. (1.16) and (1.19) that for a quasistationary distribution

$$\prod_{i=1}^{p}[\varepsilon_{0i}(r_i+\varepsilon_i)]^{l_i}\prod_{j=p+1}^{L}[\varepsilon_j(r_j+\varepsilon_{0j})]^{l_j}=\prod_{i=1}^{p}[\varepsilon_i(r_i+\varepsilon_{0i})]^{l_i}\prod_{j=p+1}^{L}[\varepsilon_{0j}(r_j+\varepsilon_j)]^{l_j}. \quad (1.22)$$

From this we find the relationship among the vibrational temperatures to be

$$\sum_{i=1}^{p}\frac{l_ih\nu_i}{T_i}-\sum_{j=p+1}^{L}\frac{l_jh\nu_j}{T_j}=\frac{1}{T}\left(\sum_{i=1}^{p}l_ih\nu_i-\sum_{j=p+1}^{L}l_jh\nu_j\right). \quad (1.23)$$

Equation (1.23) for fast multiquantum vibrational exchange in a single channel is a generalization of the results in [64-69] to the case of an arbitrary number of modes participating in the exchange. We note that Eq. (1.23) is obtained from the condition $d\varepsilon_\xi/dt = 0$ for a single arbitrarily chosen mode, ξ. Analogous expressions (with other l_i and l_j) also hold for the remaining modes of this system. The L (in general) equations of the type of Eq. (1.23) obtained in this way may be regarded as a system of L linear equations in L unknowns, $1/T_i$ and $1/T_j$. When the determinant of this system is nonzero, it has the unique trivial solution $T_i = T_j = T$. This means that a nonequilibrium quasistationary distribution cannot be established. If, on the other hand, the number of independent equations L_1 is less than L, then a solution with $T_i, T_j \neq T$ always exists. This case corresponds to a quasistationary distribution where the value of L_1 determines the "rigidity" of the connection between the vibrational temperatures. Hence, for $L_1 = L - 1$ the temperatures of all the modes may be expressed in terms of the temperature of a single mode and the value of this temperature will be determined by the total nonequilibrium vibrational energy stored in the system. For $L_2 = L - 2$ the "rigidity" of the system is decreased and the temperatures of two modes will be independent. Therefore, for $L_2 \leq L - 2$ it is possible to speak only of a partial quasiequilibrium.

3. Calculation of Cross Sections for Vibrational Transitions Involving Degenerate Oscillators

During studies of vibrational transitions one often meets difficulties due to the absence of experimentally measured probabilities of a number of elementary processes. Thus, the missing cross sections are usually calculated. The technique for calculating the collisional probabilities of vibrational transitions in nondegenerate oscillators is discussed in detail in [62, 63, 43, 70]. For a large number of colliding molecular pairs it yields results that agree as to order of magnitude with the experimental data. Including long-range attractive forces [71] and the contribution of rotational motion to the vibrational transition probability [72, 73] has made it possible to greatly increase the accuracy of this theory. To use the probabilities found by the SSH method [62, 63] correctly in the systems of Eqs. (1.14b) and (1.17) derived in the preceding section, it is necessary to extend it to the case of transitions due to collisions of degenerate oscillators.

It is shown in [62, 63, 43] that the probability of a vibrational transition may be represented as the product of an orientational factor which takes into account the change in the translational energy during exchange with vibrational degrees of freedom and factors which are the squares of the matrix elements of the transitions of all the participating (i.e., which change their energy state) modes. For notational simiplicity we shall combine all factors which do not depend on the quantum numbers of the vibrational levels into one factor, $F(\mu, T, \Delta E)$, depending only on the reduced mass of colliding particles μ, the gas temperature T, the parameters of the intermolecular interaction potential, and the energy ΔE, converted into translational degrees of freedom during a collision (ΔE is independent of the level number for a harmonic oscillator).

Thus, the probability $P(n \to m)$ of a transition in one mode from state n to state m during a collision may be written in the form

$$P(n \to m) = F(\mu, T, \Delta E) V_{nm}^2. \quad (1.24)$$

Here V_{nm}^2 is the square of the matrix element of the vibrational transition,

$$V_{nm} = \int_{-\infty}^{\infty} \psi_n(\eta) V(\eta) \psi_m(\eta) d\eta; \quad (1.25)$$

$\psi_n(\eta)$ is the wave function of the harmonic oscillator, η is the normal coordinate of the vibrations, and $V(\eta)$ is the intermolecular interaction potential. According to [62, 63, 70] this potential may be represented in the form of a product of exponents depending on the translational and normal coordinates. Then a dependence $V(\eta) = \exp(-\eta d)$ is assumed (the preexponential factor may be included in F). An expression for d is given in [70].

We shall first examine the case of a nondegenerate mode in brief. The wave functions of a nondegenerate harmonic oscillator are well known to be

$$\psi_n(\eta) = \left(\frac{\alpha}{\sqrt{\pi}\,2^n n!}\right)^{1/2} H_n(\alpha\eta)\, e^{-\frac{\alpha^2\eta^2}{2}}, \tag{1.26}$$

where $\alpha = 2\pi(M\nu/h)^{1/2}$, M is the reduced mass of the oscillator, ν is its frequency, and H_n is a Hermite polynomial. Substituting $V(\eta)$ in Eq. (1.26) in Eq. (1.25) we obtain

$$V_{nm} = \sqrt{\frac{2^{n+m}}{n!\,m!}}\, e^{\frac{d^2}{4\alpha^2}} \left(-\frac{d}{2\alpha}\right)^{n+m} \sum_{t=0}^{\min(n,\,m)} t! \binom{m}{t}\binom{n}{t}\left(\frac{d^2}{2\alpha^2}\right)^{-t}.$$

Simple estimates show that $\Delta = d^2/2\alpha^2 < 1$ (more often $\Delta \ll 1$); hence, we have, to good accuracy,

$$V_{nm} = \sqrt{\frac{2^{n+m}}{n!\,m!}} \left(-\frac{d}{2\alpha}\right)^{n+m} [\min(n,m)]! \binom{m}{\min(n,\,m)} \binom{n}{\min(n,\,m)} \Delta^{-\min(n,\,m)}. \tag{1.27}$$

Substituting Eq. (1.27) in Eq. (1.24) we find

$$P^{SSH}(n \to m) = \frac{\Delta^{n-m}}{(n-m)!} \binom{n}{m} F_{n-m} \tag{1.28}$$

for the transition probability. Setting n = n − m, m = 0 in Eq. (1.28) we will have

$$P^{SSH}(n - m \to 0) = \frac{\Delta^{n-m}}{(n-m)!} F_{n-m}.$$

Now it is easy to obtain a recurrence relation which expresses the probabilities of transitions among upper states in terms of the transition probabilities among the lowest states:

$$P^{SSH}(n \to m) = P^{SSH}(n - m \to 0) \binom{n}{m}. \tag{1.29}$$

When several oscillators participate at once in a vibrational transition such that some of them involve a transition downward by l_i quanta and others a transition upward by l_j quanta [see Eq. (1.2)], Eq. (1.29) for the probability of the combined transition is replaced by

$$P^{SSH}\begin{Bmatrix} v_i + l_i \to v_i \\ v_j - l_j \to v_j \end{Bmatrix} = P^{SSH}\begin{Bmatrix} l_i \to 0 \\ 0 \to l_j \end{Bmatrix} \prod_{i=1}^{p} \binom{v_i + l_i}{v_i} \prod_{j=p+1}^{L} \binom{v_j}{v_j - l_j}. \tag{1.30}$$

We now turn to an analysis of the case of degenerate modes. We first consider collisions of different molecules A and B. Then the wave function of a degenerate harmonic oscillator is written in the form

$$\psi_n^{AB} = \left(\frac{\alpha^r}{\sqrt{\pi^r}\,2^n \prod_{i=1}^{r} k_i!}\right)^{1/2} e^{-\frac{\alpha^2}{2}\sum_{i=1}^{r}\eta_i^2} \prod_{i=1}^{r} H_{k_i}(\alpha\eta_i), \qquad \sum_{i=1}^{r} k_i = n. \tag{1.31}$$

Here r is the multiplicity of the degeneracy [for r = 1, Eq. (1.31) transforms to Eq. (1.26)]. Carrying out operations similar to those used to obtain Eq. (1.28), we find

$$P_{AB}(n \to m) = \frac{\Delta_{AB}^{n-m}}{g_n (n-m)!} \sum_{\{k_i\}=0}^{n} \sum_{\{q_i\}=0}^{m} \prod_{i=1}^{r} \binom{k_i}{q_i} \frac{(n-m)!}{(k_i - q_i)!} F_{n-m}^{AB}. \tag{1.32}$$

Here $g_n = \binom{n+r-1}{n}$ is the statistical weight of the initial state of the oscillator and $\sum_{i=1}^{r} q_i = m$. Summing in Eq. (1.32) yields

$$P_{AB}(n \to m) = \frac{\Delta_{AB}^{n-m}}{(n-m)!}\binom{n}{m}\frac{r^{n-m}}{g_{n-m}}F_{n-m}^{AB}. \tag{1.33}$$

Similarly, for $P_{AB}(m \to n)$, we find

$$P_{AB}(m \to n) = \frac{g_n}{g_m}\frac{\Delta_{AB}^{n-m}}{(n-m)!}\binom{n}{m}\frac{r^{n-m}}{g_{n-m}}F_{n-m}^{AB}\exp\left(-\frac{\Delta E}{kT}\right).$$

The corresponding recurrence relations take the form

$$P_{AB}(n \to m) = P_{AB}(n-m \to 0)\binom{n}{m},$$
$$P_{AB}(m \to n) = P_{AB}(0 \to n-m)\binom{n}{m}\frac{g_n}{g_m g_{n-m}}. \tag{1.34}$$

It is clear from a comparison of Eqs. (1.28) and (1.33) that inclusion of the degeneracy of the mode in the transition probability leads to an additional factor

$$P_{AB}(n \to m) = P_{AB}^{SSH}(n \to m)\frac{r^{n-m}}{g_{n-m}}.$$

Using Eq. (1.34) it is easy to write the recurrence relation for the probability of the composite vibrational transition:

$$P_{AB}\begin{Bmatrix} v_i \to v_i - l_i \\ v_j \to v_j + l_j \end{Bmatrix} = P_{AB}\begin{Bmatrix} l_i \to 0 \\ 0 \to l_j \end{Bmatrix}\frac{g\{v_j + l_j\}}{g\{v_j\}\,g\{l_j\}}\prod_{i=1}^{p}\binom{v_i}{v_i - l_i}\prod_{j=p+1}^{L}\binom{v_j + l_j}{v_j}, \tag{1.35}$$

where

$$P_{AB}\begin{Bmatrix} l_i \to 0 \\ 0 \to l_j \end{Bmatrix} = P_{AB}^{SSH}\begin{Bmatrix} l_i \to 0 \\ 0 \to l_j \end{Bmatrix}\prod_{i=1}^{p} r_i^{l_i}\binom{l_i + r_i - 1}{l_i}^{-1}\prod_{j=p+1}^{L} r_j^{l_j}. \tag{1.36}$$

For a collision of two identical particles A with degenerate vibrational modes the method for calculating the probabilities is the same as for a collision of a single A molecule with a structureless B particle. However, here the wave functions have the form

$$\psi_n^{AA} = \left(\frac{\alpha^{2r}}{\sqrt{\pi^{2r}}\,2^n\prod_{i=1}^{2r} k_i!}\right)^{1/2} e^{-\frac{\alpha^2}{2}\sum_{i=1}^{2r}\eta_i^2}\prod_{i=1}^{2r} \mathrm{H}_{k_i}(\alpha\eta_i), \qquad \sum_{i=1}^{2r} k_i = n, \tag{1.37}$$

as opposed to Eq. (1.31); that is, the degree of degeneracy of each oscillator is doubled. This is due to the fact that in this case both the initial and the final energy states of the system of identical particles may be made up of vibrational states of both of the colliding particles. Thus, even in summing the vibrational energy balance equations the statistical weight must be taken with a doubled degree of degeneracy.

Using Eq. (1.37) we obtain equations analogous to Eqs. (1.33) and (1.34) for the probability $P_{AA}(n \to m)$:

$$P_{AA}(n \to m) = \frac{\Delta^{n-m}}{(n-m)!}\binom{n}{m}\frac{(2r)^{n-m}}{G_{n-m}}F_{n-m}^{AA},$$
$$P_{AA}(n \to m) = P_{AA}(n-m \to 0)\binom{n}{m}, \tag{1.38}$$
$$P_{AA}(m \to n) = P_{AA}(0 \to n-m)\binom{n}{m}\frac{G_n}{G_m G_{n-m}},$$

where

$$P_{AA}(n\to m)=P_{AA}^{SSH}(n\to m)\frac{(2r)^{n-m}}{G_{n-m}};\qquad G_n=\binom{n+2r-1}{n}.$$

For a composite vibrational transition we have

$$P_{AA}\begin{Bmatrix}v_i\to v_i-l_i\\ v_j\to v_j+l_j\end{Bmatrix}=P_{AA}\begin{Bmatrix}l_i\to 0\\ 0\to l_j\end{Bmatrix}\frac{G\{v_j+l_j\}}{G\{v_j\}\,G\{l_j\}}\prod_{i=1}^{p}\binom{v_i}{v_i-l_i}\prod_{j=p+1}^{L}\binom{v_j+l_j}{v_j},\tag{1.39}$$

while

$$P_{AA}\begin{Bmatrix}l_i\to 0\\ 0\to l_j\end{Bmatrix}=P_{AA}^{SSH}\begin{Bmatrix}l_i\to 0\\ 0\to l_j\end{Bmatrix}\prod_{i=1}^{p}(2r_i)^{l_i}\binom{l_i+2r_i-1}{l_i}^{-1}\prod_{j=p+1}^{L}(2r_j)^{l_j}.$$

Thus, Eqs. (1.35), (1.36), and (1.39) make it possible to apply the probabilities calculated using the SSH method to transitions due to collisions of degenerate oscillators as well.

4. Utilization of Experimental Data on Vibrational Relaxation Times

During their analyses of the kinetics of vibrational relaxation many authors use experimentally measured transition probabilities in addition to the computed values. Thus, the question arises of correctly interpreting the experimental data and comparing computed and experimental values of the probabilities of composite transitions (involving a change in the energy state of several oscillators) both in a binary mixture and in a gas of identical molecules. The purpose of this section is, within the framework of the harmonic oscillator model, to find the relationship between the experimentally measured vibrational relaxation time and the probabilities of elementary processes in Eqs. (1.14b) and (1.17).

Most existing techniques for determining the vibrational relaxation time [47, 74-82] are based on observation of the time variation of the integrated emission from the system of vibrational–rotational bands being studied. In this way the energy relaxation time of a specific vibrational degree of freedom through all possible channels is measured. The simplest equation for the vibrational relaxation of a harmonic oscillator (pure diatomic gas) was derived in [83] and has the form

$$\frac{d\varepsilon}{dt}=-\frac{\varepsilon-\varepsilon_0}{\tau},\tag{1.40}$$

where τ is the relaxation time. The relation between the experimentally measured quantity τ and the probability P_{10} of a single quantum transition is given by

$$\tau=\left[ZP_{10}\left(1-e^{-\frac{h\nu}{kT}}\right)\right]^{-1}.\tag{1.41}$$

Equations (1.40) and (1.41) are easily obtained from Eq. (1.17) if we set $\{l_i\}_{i\neq\xi}=0$, $\{l_j\}=0$, $l_\xi=1$, and $P\begin{Bmatrix}l_i\to 0\\ 0\to l_j\end{Bmatrix}=P_{10}$ in it.

In the case of relaxation by means of V–V' exchange it is not possible to obtain such a simple expression for τ and the relaxation equation is only approximated by Eq. (1.40). Using the CO_2 molecule as an example we shall show that information about elementary processes is included in the measured relaxation time of, for example, an asymmetric mode.

In view of the large (~3400°K) vibrational quantum (energy) (see Fig. 1), one can, with reasonable accuracy up to temperatures of about 2000°K, consider the controlling mechanism in the time variation of ε_3 to be collisional relaxation of the 00^01 level.

As already noted above, each relaxation channel is characterized by its own set of numbers $\{l_i; l_j\}$. For notational simplicity we shall limit ourselves to transitions from the 00^01 level to the levels closest to it in resonance. The following transitions may be included among such channels:

l_1	1	0	1	0	0
l_2	1	3	0	2	4
l_3	1	1	1	1	1

On the other hand, theoretical calculations [62, 63, 84] indicate that the first two of these five transitions have equal probabilities to within an order of magnitude. Then the most probable of the remaining three transitions ($00^01 \rightarrow 04^00$) has roughly an order of magnitude less probability [at relatively low ($\lesssim 1000$°K) temperatures]. This permits reduction of the number of processes to be examined to two. For exact resonance ($h\nu_1 = 2h\nu_2$) and equilibrium between the symmetric and deformed modes we find $T_1 = T_2$ and $\varepsilon_1 = (\varepsilon_2/2)^2(1+\varepsilon_2)^{-1}$. Under these assumptions the system of Eqs. (1.17) takes the form [see also Eq. (1.20)]

$$\frac{d\varepsilon_3}{dt} = -Z\left[P(3\rightarrow 2) + \frac{P(3\rightarrow 1, 2)}{1+\varepsilon_2}\right]\left(1+\frac{\varepsilon_2}{2}\right)^3\left[\varepsilon_3 - e^{-\frac{500}{T}}\left(\frac{\varepsilon_2}{2+\varepsilon_2}\right)^3(1+\varepsilon_3)\right],$$
$$\frac{d(\varepsilon_2+2\varepsilon_1)}{dt} = -3\frac{d\varepsilon_3}{dt} - ZP(2\rightarrow 0)(1-x_{02})(\varepsilon_2-\varepsilon_{02}). \qquad (1.42)$$

It is clear from Eq. (1.42) that the equation for the asymmetric mode cannot in general be reduced to the form of Eq. (1.40) and the relaxation time (if in this case it is possible to introduce such a concept) is determined not only by the probabilities of the elementary processes $P(3 \rightarrow 2)$, $P(3 \rightarrow 1, 2)$ but also by the state of the deformed and symmetric vibrations [i.e., it also depends on $P(2 \rightarrow 0)$]. Thus, before using experimentally obtained values of the probabilities in calculations it is necessary to examine critically the conditions under which the measurements were made. There is only one case in which Eq. (1.42) transforms into Eq. (1.40). This takes place when the lower levels are almost in equilibrium with the translational degrees of freedom; i.e., $T_1 = T_2 \approx T$. Then

$$\frac{d\varepsilon_3}{dt} = -\frac{\varepsilon_3 - \delta\varepsilon_{03}}{\tau}, \qquad (1.43)$$

where

$$\delta = \frac{1-x_{03}}{1-x_3}; \qquad \tau = (1-x_{02})^3[ZP(3\rightarrow\Sigma)]^{-1};$$
$$P(3\rightarrow\Sigma) = P(3\rightarrow 2) + (1+\varepsilon_{02})^{-1}P(3\rightarrow 1, 2),$$

and it can be assumed that $\delta = 1$ over a wide range of temperatures with good accuracy. Thus, because there are two transitions with similar probabilities it is only possible to obtain information about their total effect from the experimentally measured value of τ for the CO_2 molecule. Physically, the factor $(1-x_{02})^3$ in the expression for τ is due to the fact that (as opposed to the case of V−T relaxation) escape of a quantum from an asymmetric mode is accompanied by the excitation of three deformed (or one deformed and one symmetric) mode quanta which in turn relax in a V−T channel. Hence, the higher the temperature the more easily these vibrations are excited, which leads† to a reduction in τ. Neglect of this factor when using experimental probabilities in Eqs. (1.14b) and (1.17) causes errors. In particular, this concerns the case of high temperatures, as is obvious from Eq. (1.43).

Fortunately, in most measurements of τ the situation is just such that the condition $T_1 = T_2 \approx T$ is well satisfied. In particular, the calculations in [61] confirmed this. There the vibrational relaxation of the 00^01 level of the CO_2 molecule was investigated using the phase method of [76, 77]. At present this method is regarded as most accurate for determining the vibrational relaxation time. The essence of the method is that the CO_2 (or a mixture of it with N_2, He or H_2O) to be examined is placed in a cuvette onto which is directed a CO_2 laser beam that is modulated at a frequency ω. The CO_2 molecules in the cuvette in states 10^00 or 02^00 are shifted to the 00^01 level by the laser, and radiation from the 00^0v band with a wavelength of about 4.3 μ due to a slight deviation in the population of the 00^0v levels from equilibrium is aimed at a detector. This radiation will be modulated at the same frequency ω, but due to a delay in the form of relaxation processes the output will be shifted in phase relative to the laser radiation. When Eq. (1.43) is applicable this phase shift is very simply related to the relaxation time τ and is reliably measured experimentally. A criterion for the applicability of Eq. (1.43) to the phase method was derived in [61]. The criterion is basically that the number of quanta absorbed by the molecules in the volume per unit time must not exceed the equilibrium flux of excited particles from the deformed mode in the ground state.

Thus, it has been shown that within the framework of the harmonic oscillator model the experimental data obtained when Eq. (1.43) is applicable have a simple physical interpretation. If, however, the deviations from equilibrium during the experiment are large, then the picture becomes considerably more complicated and the probabilities of deactivation of the 00^01 level cannot be found in a single experiment without making a parallel measurement of the vibrational temperature of the lowest levels ($v_1v_2^l0$).

CHAPTER II

THE PHYSICAL KINETICS OF GAS-DYNAMIC LASERS

1. Population Inversion in the Vibrational Levels of CO_2 during Adiabatic Expansion into a Vacuum

The operating principle of the gas-dynamic laser is based on the fact that during the rapid drop in the gas temperature due to expansion of a gas a population inversion may take place because of a difference in the relaxation times of the laser levels. The characteristic times for changes in the vibrational and gas temperatures are comparable, so it is necessary to examine the gas-dynamics and relaxation processes in molecular mixtures together. In our analysis of the operating features of low-temperature gas-dynamic lasers we shall first discuss the case of adiabatic expansion into a vacuum. In our discussion we shall follow [8].‡

† In other words, the reduction in τ with increasing temperature is due, besides the increased probability of the direct process $P(3 \rightarrow \Sigma)$, to the increasing population of the lower deformed levels, which leads to convergence of the direct and reverse transitions and to a drop in τ.

‡ An article by Biryukov et al. [8], published in 1969, was one of the first to discuss the mechanism of population inversion in a CO_2-N_2-He mixture. Although at that time there were no reliable experimental data on the probabilities of vibrational relaxation and their tempera-

The main channels for relaxation of vibrational energy in a CO_2-N_2-He mixture [8] may be assumed [see also Eq. (1.20) and Fig. 1] to be resonance exchange through the lowest vibrational level of the N_2 molecule and the 00^01 level of the CO_2 molecule (with probability $W_{4,3}$), transfer of the energy of asymmetric CO_2 vibrations through the 00^01 level to deformed and symmetric vibrations (with probability $W_{3,\Sigma}$), and joint relaxation of the energies $\mathcal{E}_1+\mathcal{E}_2$ to translational degrees of freedom through the $0v_2^l0$ levels (with probability $W_{2,0}$). Direct collisional relaxation of the energies $\mathcal{E}_1$, $\mathcal{E}_3$, and $\mathcal{E}_4$ to translational degrees of freedom and radiative transitions are not considered due to their small probabilities. Resonance exchange of vibrational quanta between N_2 and CO_2 and relaxation of the energy of asymmetric vibrations through the upper levels are insignificant (due to the reduced populations). However, use of experimentally measured probabilities makes it possible to effectively take even these transitions into account.

We note that N_2 molecules produce a unique energy reservoir from which the main energy outlet is through CO_2. This permits slowing down of the relaxation of the 00^01 level in the presence of nitrogen. Helium effectively depopulates the lower states of the deformed mode and promotes more rapid cooling of the gas (due to an increase in the adiabatic index when He is added).

The equations describing the time variation of the vibrational energies $\mathcal{E}_i$ of a single molecule under the above assumptions may be written in the form [8]:

$$\left.\begin{aligned}
&\frac{dx_4}{dt}=S(1-x_4)^3(x_3-x_4)W_{4,3}N_{CO_2},\\
&\frac{dx_3}{dt}=S(1-x_4)(1-x_3)^2(x_4-x_3)W_{4,3}N_{N_2}-S(1-x_3)^2\Big(x_3-\\
&\quad -x_2^3e^{-\frac{500}{T}}\Big)(W_{3,\Sigma}^{CO_2}N_{CO_2}+W_{3,\Sigma}^{N_2}N_{N_2}+W_{3,\Sigma}^{He}N_{He}),\\
&\frac{dx_2}{dt}=\frac{(1+x_2)^2(1-x_2)^2}{1+4x_2+x_2^2}\Big[\frac{3}{2}S\Big(x_3-x_2^3e^{-\frac{500}{T}}\Big)(W_{3,\Sigma}^{CO_2}N_{CO_2}+\\
&\quad +W_{3,\Sigma}^{N_2}N_{N_2}+W_{3,\Sigma}^{He}N_{He})-\frac{x_2-x_{02}}{1-x_2}(W_{2,0}^{CO_2}N_{CO_2}+W_{2,0}^{N_2}N_{N_2}+W_{2,0}^{He}N_{He})\Big],\\
&S=(1-x_3)(1-x_2)^2(1-x_1)=(1-x_3)(1-x_2)^3(1+x_2).
\end{aligned}\right\}\qquad(2.1)$$

In writing these equations it was assumed that $h\nu_3 = h\nu_4$ and expressions of the form

$$\mathcal{E}_i=\frac{h\nu_i x_i}{1-x_i},\quad i=1,\ 3,\ 4;\quad \mathcal{E}_2=\frac{2h\nu_2x_2}{1-x_2}$$

were used for the vibrational energies, as well as the obvious relation $\frac{d\mathcal{E}_i}{dt}=\frac{d\mathcal{E}_i}{dx_i}\frac{dx_i}{dt}$ and the known relation between the probabilities of the forward and reverse processes. To simplify later calculations we shall assume the gas temperature T and the particle densities of the gases N_{CO_2}, N_{N_2}, and N_{He} in Eq. (2.1) to be independent of the spatial coordinates; i.e., we shall operate with gas parameters averaged over the entire volume.

We now consider the cooling of the gas during expansion. Let the gas begin to expand freely into a vacuum at time $t = 0$ (temperature T_0, density N_0, and characteristic dimension R_0). The leading layers of gas expand [85] with a constant escape velocity $v_0 = f(\gamma, c_0)$, where

---

ture dependences or strict relaxation equations such as Eqs. (1.4b) and (1.17), the results are in good qualitative agreement with later experiments [15, 21]. Refinements due to the use of current values of the probabilities and equations for this low-temperature case do not change the semiquantitative picture of these phenomena.

γ is the adiabatic index and c_0 is the speed of sound in the unperturbed gas. The law of motion of the boundary of the gas is $r = R_0 + v_0 t$, and the average density N_{avg} varies with time according to

$$N_{avg} \sim N_0 (R_0 + v_0 t)^{-n}, \tag{2.2}$$

where n is the dimensionality of the problem. The value n = 1 corresponds to expansion of an extended plane layer of thickness $2R_0$ or to escape from a long narrow slit of half-width R_0 bounded by a plane parallel channel with the same dimensions (this may serve as a model of a plane nozzle with a small aperture angle); n = 2 corresponds to expansion of a long cylindrical burst of gas of radius R_0 or to approximately free flow out of a long narrow slit of half-width R_0, and n = 3 corresponds to expansion of a gas sphere of initial radius R_0.

The maximum expansion velocity in the nonstationary regime is defined as $v_0 = 2c_0/(\gamma - 1)$ [85], and for stationary flow from a slit with n = 1 it is $v_0 = c_0[2/(\gamma - 1)]^{1/2}$. Thus, averaging the gas parameters over the entire volume and using a constant value of v_0 in Eq. (2.2) is a good approximation in the case of a steady flow since the flow velocity changes by about two times overall (for $\gamma \approx 1.4$) from the beginning of the expansion to $t \to \infty$. This approximation is somewhat worse for a nonstationary flow.

We shall assume that each rotational degree of freedom of the CO_2 and N_2 molecules has energy $^1/_2 kT$, and we assume equilibrium between the rotational and translational degrees of freedom. Then, denoting the amount of vibrational energy per unit time going into translational and rotational degrees of freedom by Q (referred to a single particle of the gas) and using the energy-conservation equation, the ideal gas equation of state, and Eq. (2.2), we find

$$\frac{dT}{dt} = \left(\frac{Q}{k} - \frac{n v_0 T}{R_0 + v_0 t}\right)\left(\frac{3}{2} + \frac{1 + k_{N_2}}{1 + k_{N_2} + k_{He}}\right)^{-1} \tag{2.3}$$

for the average translational temperature of the gas, where $k_{N_2} = N_{N_2}/N_{CO_2}$ and $k_{He} = N_{He}/N_{CO_2}$ are the relative concentrations of N_2 and He, and k is the Boltzmann constant. Beginning with Eq. (2.1) we have

$$Q = -\sum_{i=1}^{4} \frac{d\mathcal{E}_i}{dt} = \Big[S\,(h\nu_3 - 3h\nu_2)\left(x_3 - x_2^3 e^{-\frac{500}{T}}\right)(W_{3,\Sigma}^{CO_2} + W_{3,\Sigma}^{N_2} k_{N_2} + W_{3,\Sigma}^{He} k_{He}) +$$

$$+ 2h\nu_2 \frac{x_2 - x_{02}}{1 - x_2}(W_{2,0}^{CO_2} + W_{2,0}^{N_2} k_{N_2} + W_{2,0}^{He} k_{He})\Big] \frac{N_{CO_2}}{1 + k_{N_2} + k_{He}}. \tag{2.4}$$

In evaluating the escape velocity $v_0 = f(\gamma, c_0)$, the average sound speed c_0 was calculated for the unperturbed gas mixture, while in finding the effective adiabatic index γ,† it was assumed that the vibrational degrees of freedom of CO_2 and N_2 were "frozen."

The vibrational transition probabilities in Eq. (2.1) depend strongly on the translation temperature. To find their values and temperature dependences we used the data of [41, 74, 75] and made additional calculations of the transition probabilities for high temperatures in the range 1000-2500°K.

To describe vibrational relaxation in a molecular gas (or mixture of gases) initially heated to temperature T_0 and cooled by free expansion into a vacuum, it is necessary to solve a closed system of relaxation and gas-dynamic equations. We have solved Eqs. (2.1) together

† We note that the change in the effective adiabatic index during cooling of the gas was taken into account by the first term on the right-hand side of Eq. (2.3).

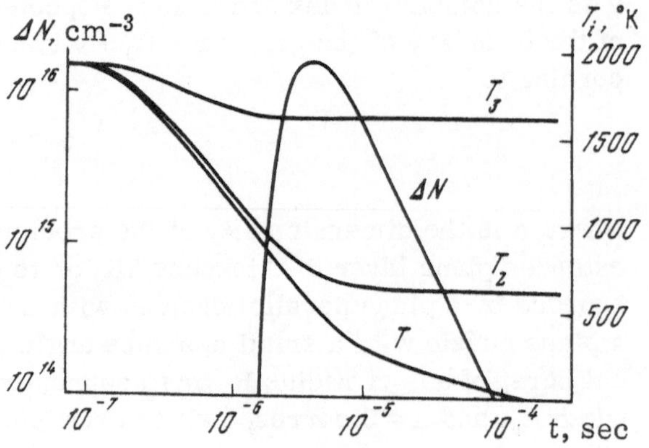

Fig. 2. Typical time variation of the gas temperature T, the vibrational temperatures T_i, and the inversion ΔN for adiabatic expansion of a gas mixture into a vacuum: CO_2:N_2 = 1:5; p_0 (CO_2) = 1 atm at 300°K; T_0 = 1900°K; R_0 = 0.3 cm; n = 2.

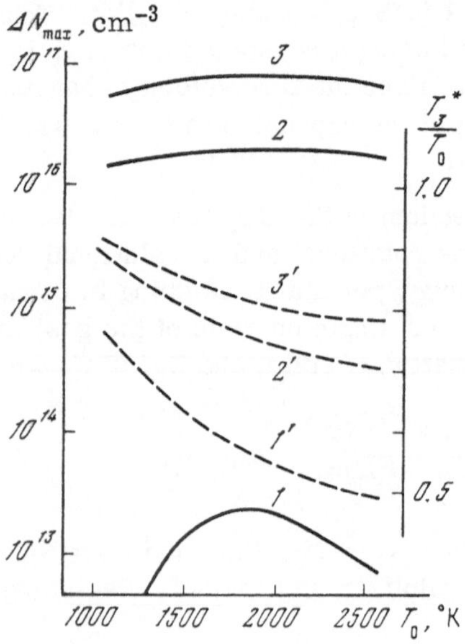

Fig. 3. Dependence of the maximum inversion (1-3) and the degree of "freezing" T_3^*/T_0 (1'-3') on the initial gas temperature T_0 for pure CO_2 (1, 1'), CO_2:N_2 = 1:5 (2, 2'), and CO_2:N_2:He = 1:5:10 (3, 3'). Everywhere $p_0(CO_2)$ = 1 atm (at 300°K); R_0 = 0.1 cm; n = 1.

with Eq. (2.3) including Eqs. (2.2) and (2.4) on an electronic computer for various mixtures of the gases CO_2, N_2, and He with different initial temperatures T_0, pressures, and characteristic dimensions R_0 for n = 1, 2, 3 as well. The results of this calculation are shown in Figs. 2-6.

Figure 2 shows typical time dependences for the vibrational temperatures T_i in N_2 and CO_2, for the temperature of the translational degrees of freedom T, and for the inverted population $\Delta N = N_{00^01} - N_{10^00}$ in an adiabatically expanding gas. It is clear that at the initial stage of expansion, when the gas temperature T and density are still large, the vibrational temperatures differ little from the gas temperature in view of the high relaxation rate. However, as the gas expands the vibrational relaxation rates decrease due to a rapid reduction in the gas temperature and density, and a gap develops between the various vibrational temperatures and the gas temperature.

At some time when the density and temperature of the gas have fallen significantly, the process of vibrational relaxation practically ceases and the remaining accumulation of vibrational energy does not change. There is a "freezing" of the vibrational energies.† Since the total relaxation rate of the 00^01 level in the mixture is less than the decay rate of the $0v_2^l0$

† This effect lasts until radiative decay of the vibrational levels sets in. For the CO_2 molecule this time is $\gtrsim 10^{-2}$ sec.

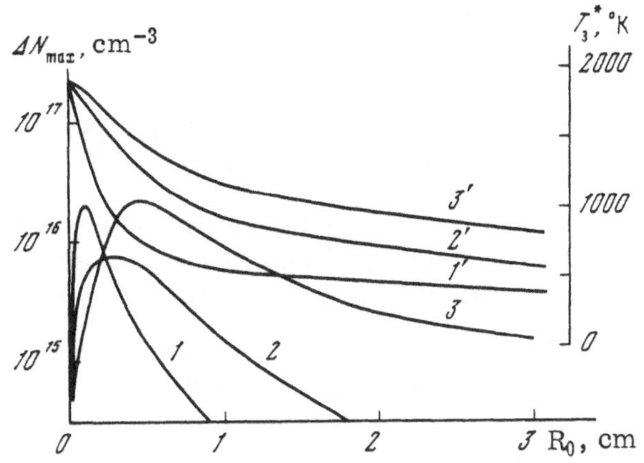

Fig. 4. Dependence of the maximum inversion (1-3) and "frozen" temperature T_3^* (1'-3') on the initial characteristic dimension of the gas R_0 for the mixture $CO_2:N_2 = 1:5$. $T_0 = 1900$°K; n = 1 (1, 1', 2, 2'), n = 2 (3, 3'); $p_0(CO_2) = 1$ atm (1, 1', 3, 3') and 0.3 atm (2, 2'). The pressure is reduced to that at 300°K.

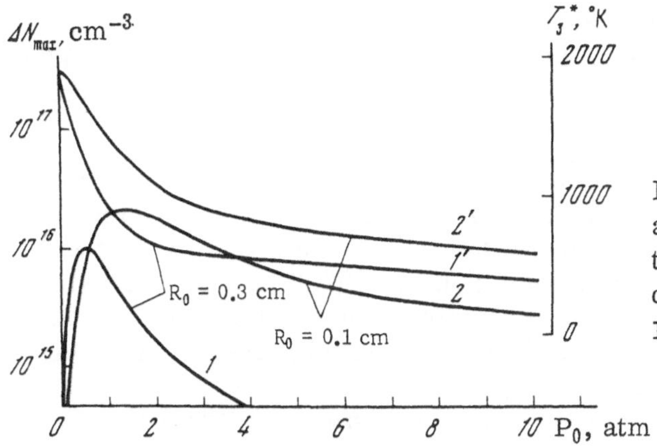

Fig. 5. The maximum inversion ΔN_{max} (1, 2) and "frozen" temperature T_3^* (1', 2') as functions of the initial partial pressure of CO_2 (reduced to that at 300°K) for various R_0: $CO_2:N_2 = 1:5$; $T_0 = 1900$°K; n = 1.

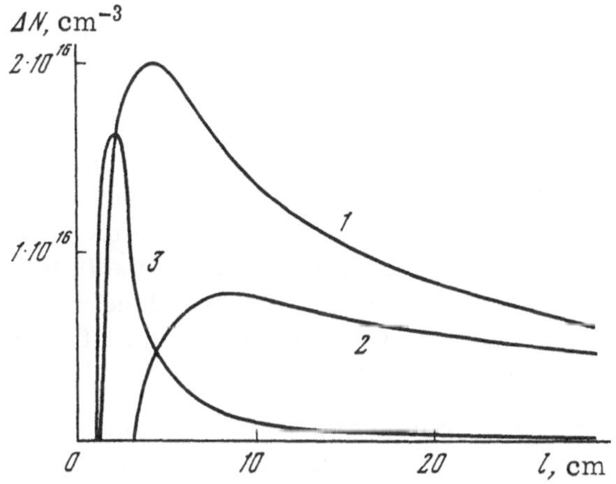

Fig. 6. Variation in the inversion along the axis of the expansion for several R_0, p (CO_2), and n. 1) $R_0 = 0.1$ cm, $p_0(CO_2) = 1$ atm, and n = 1; 2) $R_0 = 0.3$ cm, $p_0(CO_2) = 0.3$ atm, and n = 1; 3) $R_0 = 0.3$ cm, $p_0(CO_2) = 1$ atm, and n = 2. Mixture $CO_2:N_2 = 1:5$, $T_0 = 1900$°K (pressures are shown for 300°K).

level, the "frozen" temperature T_3^* may be much larger than T_2^* and the gap between T_3 and T_2 may become large enough to produce an inverted population of the $00^01 - 10^00$ levels of the CO_2 molecule. This inversion rapidly increases (due to a growth in the difference $x_3 - x_2^2$), reaches its maximum value, and then begins to decrease. Since up to this time $T_3 = T_3^* = \text{const}$ the fall in the inversion is due to a reduction in the density of CO_2 molecules because of the expansion of the gas. The rate of gas cooling, the magnitudes of the "frozen" vibrational temperatures T_i^*, the time at which the inversion reaches its maximum, and the amount of this inversion are determined by various parameters of the gaseous mixture before the expansion process begins.

Analysis has shown that adding nitrogen to CO_2 in the proportion $CO_2 : N_2 = 1:3$ to $1:5$ greatly increases the inversion [8].

When $W_{4,3} \gg W_{3,\Sigma}$ the results are independent of $W_{4,3}$. Thus, a calculation with the probability $W_{4,3}$ an order of magnitude smaller than that used in deriving the bulk of the results (the inequality $W_{4,3} > W_{3,\Sigma}$ is still satisfied) yielded a reduction in ΔN_{max} of only 25%.

In Fig. 3 ΔN_{max} and T_3^*/T_0 (the degree of "freezing") are plotted as a function of the gas temperature T_0 for various mixtures. From this figure it is clear that for the mixtures CO_2-N_2 and CO_2-N_2-He the inversion is insensitive to the initial temperature of the gases in the range 1100-2000°K. The optimal values, $T_0 = 1800$-2000°K, are explained, on the one hand, by an increase in the number of vibrational excited molecules and in the initial rate of cooling of the gas as T_0 is increased, and, on the other hand, by an increase in the relaxation rate of the energy of the asymmetric vibrations of the CO_2 molecule. For this same reason, as T_0 increases, the degree of "freezing" T_3^*/T_0 also falls. It is also apparent from Fig. 3 that adding helium to the mixture CO_2-N_2 increases the inversion threefold (by increasing the gas cooling rate and speeding up the relaxation of the deformed modes).

The effect of the characteristic dimension R_0 on the inversion and the "frozen" temperature T_3^* is shown in Fig. 4. The optimum values R_0^{opt} are 0.1 and 0.3 cm for the one-dimensional case and 0.5 cm for the two-dimensional case. The existence of these optimal values is explained as follows. When R_0 is reduced, as can be seen from Eqs. (2.3) and (2.2), the initial cooling rate is increased and the gas density falls more rapidly in time. This facilitates production of a higher "frozen" temperature T_3^*. However, due to the rapid reduction in the density the number of active CO_2 molecules at the moment of maximum inversion falls as R_0 is decreased. Competition among these factors also results in the existence of R_0^{opt} and in its shift rightward as the initial density is reduced. Another characteristic is the larger value of R_0^{opt} on going from the one-dimensional to the two-dimensional case; this is explained by the more rapid change in the gas density with time for $n = 2$.

Figure 5 shows ΔN_{max} and T_3^* as functions of the initial partial pressure of CO_2 (reduced to a temperature of 300°K) for a CO_2-N_2 mixture at a temperature of 1900°K and dimensions $R_0 = 0.1$ and 0.3 cm. As the pressure is increased the values of T_3^* decrease due to an increase in the relaxation rate of the energy of the asymmetric vibrational of CO_2, while the inversion increases at first (due to an increased density of CO_2 molecules), reaches a maximum, and then decreases since the increase in $W_{3,\Sigma}$ then begins to play a predominant role. The shift in $p_0^{opt}(CO_2)$ toward higher values as R_0 is reduced is explained by the more rapid fall in the gas density with time and, thus, by the smaller effect of the initial pressure on the relaxation rate of the 00^01 level.

The problem of producing an inversion during expansion of a gas is basically a nonstationary one since the time variation of the processes must be considered. However, by modeling free expansion with continuous escape of gas from a slit ($n = 1$ or 2), it is possible to obtain a stationary pattern. Then a time scan of the expansion is projected along the axis of the escaping flow. Figure 6 shows the variation of the inversion along the expansion axis for several values of R_0, $p_0(CO_2)$, and n (here the distance along the axis was taken as $l = v_0t$).

We note that the possible efficiency of a laser with the adiabatic gas cooling method considered here is low. Estimates yielded an efficiency of less than about 0.3% [8], which is much less than that for a CO_2 laser with electronic excitation of the vibrational levels. However, the thermal-pumping technique is of interest because of its uniqueness and the prospect of obtaining large gain coefficients. The maximum inversion $\Delta N_{max} \sim 2 \cdot 10^{16}$ cm^{-3} (cf. Figs. 2-6) considerably exceeds that for a cw CO_2 laser.†

These calculations may be used as the basis for specific recommendations on the formulation of experiments. For production of an inverted population in an expanding gas mixture of CO_2 and N_2 a mixture composition of $CO_2 : N_2 = 1:3-1:5$ with an initial gas temperature of roughly 1200-2000°K is optimal. The temperature is not a critical parameter and choosing it within these limits has little effect on the magnitude of the inversion. However, the inversion is very sensitive to the initial pressure and characteristic dimension of the gas. In the case of gas outflow through a long slit (n = 1) the optimum parameters are a CO_2 partial pressure p_0^{opt} of about 1-2 atm (at 300°K) and a slit size of $2R_0 \sim 2$ mm; for n = 2, the corresponding values are $p_0^0 \sim 1$ atm and $R_0 \sim 5$ mm.

The distance from the outflow plane at which the maximum inversion is achieved depends on the initial gas pressure and the size of the opening. For the conditions specified above it is roughly 4 and 2 cm, respectively.

As a conclusion to this paragraph we note that [8] stimulated a number of experiments [15, 21] in the Laboratory of Low-Temperature Plasma Optics at the Physical Institute of the Academy of Sciences of the USSR as a result of which a gas-dynamic laser was made for the first time in our country [15].

2. A Method for Calculating the Vibrational Relaxation of a $CO_2 - N_2 -$ He Mixture in a Supersonic Nozzle

Because of the large amount of experimental research on gas-dynamic lasers using supersonic nozzles the need has arisen for a detailed analysis of the kinetics of vibrational relaxation in such systems in order to optimize them. We shall consider a method for designing these lasers based on the correct relaxation equations and including the latest data on the probabilities of various processes. Within the framework of this computational scheme [86] it is possible to analyze vibrational relaxation in nozzles over a very wide temperature interval (up to about 4000°K).

A study of the kinetics of a moving and expanding gas requires joint consideration of gas-dynamics, chemical processes (for T > 2000°K), and vibrational relaxation. We shall analyze each type of equation in isolation in detail and establish its range of applicability. Let us begin with chemical kinetics.

In the ternary mixture CO_2-N_2-He only the CO_2 molecules are affected by any chemical reactions at initial temperatures of 2500-4000°K (N_2 is anywhere near noticeably dissociated only at higher temperatures.). Thus, the six (in general) principal components of the mixture will be O_2, CO, O, CO_2, N_2, and He. We shall denote their molar fractions by $\gamma_i = N_i/N$, where N_i is the concentration of the i-th component and N is the total number of particles per unit volume. Then we have four relations for finding the six unknown γ_i's:

$$\left.\begin{aligned} &\sum_i \gamma_i(t) = 1, \\ &\frac{\gamma_{CO}(t) + \gamma_{CO_2}(t)}{\gamma_{CO}(t) + \gamma_O(t) + 2\gamma_{O_2}(t) + 2\gamma_{CO_2}(t)} = b, \\ &\gamma_{N_2}(t) = \frac{\mu(t)}{\mu_{N_2}} a_{N_2}, \quad \gamma_{He}(t) = \frac{\mu(t)}{\mu_{He}} a_{He}, \end{aligned}\right\} \tag{2.5}$$

† Including the temperature and density inhomogeneity of the gas means that the volume averaged inversion is somewhat less than the calculated value.

where $\mu(t)$ is the molecular weight of the mixture and b, a_{N_2}, and a_{He} are constants depending on the initial conditions. The first of Eqs. (2.5) follows from the definition of γ_i; the second reflects the fact that the ratio of the number of carbon atoms to the number of oxygen atoms remains constant under all chemical transformations; and the third and fourth are related to the fact that N_2 and He do not participate in reactions (more precisely, they participate only as a third body) and the mass of each of these substances relative to the entire mass of gas must remain constant.

For a mixture with an equilibrium chemical composition, $b = {}^1/_2$, and specifying concrete values of a_{N_2} and a_{He} determines their mass fractions relative to the remaining components. If a disequilibrium in the composition is assumed from the start (for example, a chemical reaction that does not go to completion; mixing in some components; etc.) with the ratios of the molar fractions given for some time (e.g., the initial time), then for b, a_{N_2}, and a_{He} we find

$$b = \frac{\gamma_{CO}(0) + \gamma_{CO_2}(0)}{\gamma_{CO}(0) + \gamma_O(0) + 2\gamma_{O_2}(0) + 2\gamma_{CO_2}(0)},$$
$$a_{N_2} = \frac{\mu_{N_2}\gamma_{N_2}(0)}{\sum_i \gamma_i(0)\mu_i}, \qquad a_{He} = \frac{\mu_{He}\gamma_{He}(0)}{\sum_i \gamma_i(0)\mu_i}. \tag{2.6}$$

Thus, with four relations for finding six unknown molecular fractions, a system of two equations in all must be solved. For concreteness we shall assume that the variables in this system are γ_{CO_2} and γ_{O_2}. Substituting the given values of the molecular weights of the components in Eq. (2.5), we now express the remaining unknowns in terms of γ_{CO_2} and γ_{O_2} as

$$\gamma_{CO} = \frac{7b(1 - a_{N_2} - a_{He})(1 + \gamma_{O_2}) - \gamma_{CO_2}[7(1 - b) + a_{N_2}(10b - 3) + 7a_{He}(4b + 3)]}{7 + 3a_{N_2}(b - 1) + 21a_{He}(b + 1)},$$

$$\gamma_O = \gamma_{CO}\left(\frac{1}{b} - 1\right) + \gamma_{CO_2}\left(\frac{1}{b} - 2\right) - 2\gamma_{O_2}, \qquad \gamma_{N_2} = \frac{\Phi}{7}a_{N_2}, \qquad \gamma_{He} = \Phi a_{He}, \tag{2.7}$$

$$\Phi = \frac{1}{1 - a_{N_2} - a_{He}}(7\gamma_{CO} + 4\gamma_O + 11\gamma_{CO_2} + 8\gamma_{O_2}).$$

We next write the equations to determine γ_{CO_2} and γ_{O_2} due to gas-dynamic expansion in the presence of chemical reactions. In this case the concentration of the i-th component N_i is found from the equation

$$\frac{\partial N_i}{\partial t} + \operatorname{div} N_i\mathbf{V} = \left(\frac{dN_i}{dt}\right)_{chem}.$$

Substituting $N_i = \rho\gamma_i/\mu$, we obtain

$$\frac{\gamma_i}{\mu}\left(\frac{\partial\rho}{\partial t} + \operatorname{div}\rho\mathbf{V}\right) + \rho\left(\frac{\partial}{\partial t}\frac{\gamma_i}{\mu} + \mathbf{V}\operatorname{grad}\frac{\gamma_i}{\mu}\right) = \left(\frac{dN_i}{dt}\right)_{chem}, \tag{2.8}$$

where ρ is the density of the gas and $(dN_i/dt)_{chem}$ is the rate of change of the concentration of the i-th component due to a chemical reaction.

For a steady-state flow the continuity equation holds, so the first term on the left-hand side of Eq. (2.8) disappears, and the equation takes the form

$$\rho\frac{d}{dt}\frac{\gamma_i}{\mu} = \left(\frac{dN_i}{dt}\right)_{chem},$$

which yields

$$\frac{d\gamma_i}{dt}=\frac{1}{N}\left(\frac{dN_i}{dt}\right)_{\text{chem}}+\frac{\gamma_i}{\mu}\frac{d\mu}{dt}. \tag{2.9}$$

Using Eq. (2.7) we easily find that the molecular weight of the mixture is simply expressed in terms of the independent unknowns γ_{CO_2} and γ_{O_2}:

$$\mu=\frac{28\,(3b+4)}{7+3a_{N_2}(b-1)+21a_{He}(b+1)}(1+\gamma_{CO_2}+\gamma_{O_2}). \tag{2.10}$$

Then the desired system of two equations may be written in the form [87]

$$\begin{aligned}\frac{d\gamma_{CO_2}}{dt}&=\frac{\gamma_{CO_2}}{N}\left(\frac{dN_{O_2}}{dt}\right)_{\text{chem}}+\frac{\gamma_{CO_2}+1}{N}\left(\frac{dN_{CO_2}}{dt}\right)_{\text{chem}},\\ \frac{d\gamma_{O_2}}{dt}&=\frac{\gamma_{O_2}+1}{N}\left(\frac{dN_{O_2}}{dt}\right)_{\text{chem}}+\frac{\gamma_{O_2}}{N}\left(\frac{dN_{CO_2}}{dt}\right)_{\text{chem}}.\end{aligned} \tag{2.11}$$

To be more specific in Eq. (2.11) we must limit ourselves to a certain group of reactions responsible for changing the chemical composition. In our case we must include the following reactions as among the most important [87]:

$$CO_2+M\underset{K_{-1}^M}{\overset{K_1^M}{\rightleftarrows}}CO+O+M,\quad CO_2+O\underset{K_{-2}}{\overset{K_2}{\rightleftarrows}}CO+O_2,\quad O_2+M\underset{K_{-3}^M}{\overset{K_3^M}{\rightleftarrows}}O+O+M. \tag{2.12}$$

Here M is any of the six components of the mixture, and K_i^M, K_{-i}^M are the rate constants of the direct and reverse reactions, respectively. The particle (number) balance equations that determine the right-hand sides of Eqs. (2.11) then have the form

$$\begin{aligned}\frac{1}{N}\left(\frac{dN_{O_2}}{dt}\right)_{\text{chem}}&=\frac{p}{kT}\left(\frac{p}{kT}\gamma_O^2\sum_M K_{-3}^M\gamma_M+\gamma_{CO_2}\gamma_O K_2-\gamma_{O_2}\sum_M K_3^M\gamma_M-\gamma_{CO}\gamma_{O_2}K_{-2}\right),\\ \frac{1}{N}\left(\frac{dN_{CO_2}}{dt}\right)_{\text{chem}}&=\frac{p}{kT}\left(\frac{p}{kT}\gamma_{CO}\gamma_O\sum_M K_{-1}^M\gamma_M+\gamma_{CO}\gamma_{O_2}K_{-2}-\gamma_{CO_2}\sum_M K_1^M\gamma_M-\gamma_{CO_2}\gamma_O K_2\right),\end{aligned} \tag{2.13}$$

where $N = p/kT$ is the number of particles per unit volume, p is the pressure, and T is the gas temperature.

Therefore, Eqs. (2.7), (2.11), and (2.13) together with the equations of gas dynamics completely determine the chemical composition.

The variation of the gas-dynamic parameters is described by a system made up of the equations of motion, energy conservation, and continuity:

$$\frac{\partial \mathbf{v}}{\partial t}+(\mathbf{v}\nabla)\mathbf{v}=-\frac{1}{\rho}\nabla p,\quad h+\frac{v^2}{2}=\text{const},\quad \frac{\partial\rho}{\partial t}+\operatorname{div}\rho\mathbf{v}=0, \tag{2.14}$$

where h is the enthalpy per unit mass and $\mathbf{v} = \mathbf{v}(x, y, z, t)$ is the directed velocity of the gas.

Usually in discussions of the motion of a gas mixture in a supersonic nozzle it is assumed for simplicity that the flow is one-dimensional and all the gas-dynamic parameters depend on a single coordinate x directed along the axis of the nozzel. This assumption is justified in the great majority of cases since the principal variation in the flow velocity is indeed along the x axis (see also [32]). In this case the continuity equation simply corresponds to constancy of the mass flow rate through the nozzle, and for steady flow the system of Eqs. (2.14) takes the form

$$\rho v f = \rho^* v^* = \text{const}, \quad \rho v \frac{dv}{dx} + \frac{dp}{dx} = 0, \quad \frac{5}{2}\frac{\Re T}{\mu} + \sum_i (e_r^i + e_v^i) + \frac{v^2}{2} = \text{const}, \tag{2.15}$$

where $f = f(x)$ is the ratio of the area of the nozzle cross section to the area of the nozzle at its narrowest part (the critical cross section), ρ^* and v^* are the density and velocity at the critical cross section, $\Re$ is the gas constant, and e_r^i and e_v^i are the rotational and vibrational energies per unit mass of the i-th component, respectively. In Eq. (2.15) and all other equations in this section by a summation we shall mean summing over the molecular components.

We further assume that the rotational degrees of freedom are in equilibrium with the translational, and will use the equation of state $p = \rho \Re T/\mu$. Then the system of equations (2.15) takes the form

$$\begin{gathered} \frac{p\mu v f}{\Re T} = \text{const}, \\ \frac{dp}{dx} = -\frac{p\mu v}{\Re T}\frac{dv}{dx}, \\ \frac{\Re T}{\mu}\left(\frac{5}{2} + \sum_i \gamma_i\right) + \sum_i e_v^i + \frac{v^2}{2} = \text{const}, \end{gathered} \tag{2.16}$$

whence

$$\begin{gathered} \frac{dT}{dx} = -T\frac{\left\{(1-\lambda)\sum_i \frac{d\gamma_i}{dx} + \frac{1-\lambda}{\lambda v^2}\sum_i \frac{de_v^i}{dx} + \frac{1}{f}\frac{df}{dx} - \left[\left(\frac{5}{2} + \sum_i \gamma_i\right)(1-\lambda) - 1\right]\frac{1}{\mu}\frac{d\mu}{dx}\right\}}{\left(\frac{5}{2} + \sum_i \gamma_i\right)(1-\lambda) - 1}, \\ \frac{dv}{dx} = v\frac{\lambda \sum_i \frac{d\gamma_i}{dx} + \frac{1}{v^2}\sum_i \frac{de_v^i}{dx} + \lambda\left(\frac{5}{2} + \sum_i \gamma_i\right)\frac{1}{f}\frac{df}{dx}}{\left(\frac{5}{2} + \sum_i \gamma_i\right)(1-\lambda) - 1}, \\ \frac{dp}{dx} = -\frac{p}{\lambda v}\frac{dv}{dx}. \end{gathered} \tag{2.17}$$

Here $\lambda = \Re T/\mu v^2$, $\sum_i \gamma_i = 1 - \gamma_O - \gamma_{He}$, and μ is given by Eq. (2.10).

Using Eqs. (2.7), (2.10), and (2.11), we obtain

$$\begin{gathered} \frac{1}{\mu}\frac{d\mu}{dx} = \frac{1}{1 + \gamma_{CO_2} + \gamma_{O_2}}\left(\frac{d\gamma_{CO_2}}{dx} + \frac{d\gamma_{O_2}}{dx}\right), \\ \sum_i \frac{d\gamma_i}{dx} = \frac{d\gamma_{CO_2}}{dx}\frac{7b(1 - a_{He}) + 4a_{N_2}(1-b)}{7 + 3a_{N_2}(b-1) + 21a_{He}(b+1)} + \frac{d\gamma_{O_2}}{dx}\frac{7(1+b) + 7a_{He}(2b+3) + a_{N_2}(1-b)}{7 + 3a_{N_2}(b-1) + 21a_{He}(b+1)}. \end{gathered} \tag{2.18}$$

It can be seen from Eq. (2.17) that the shape of the nozzle may have a substantial effect on the rate of change of the gas-dynamic parameters of the flow, which in turn determines the rates of all relaxation processes. Not only is the shape beyond the critical cross section along the x axis important, but so is that before the critical cross section. This is mainly true in the case of nonequilibrium stagnation conditions.

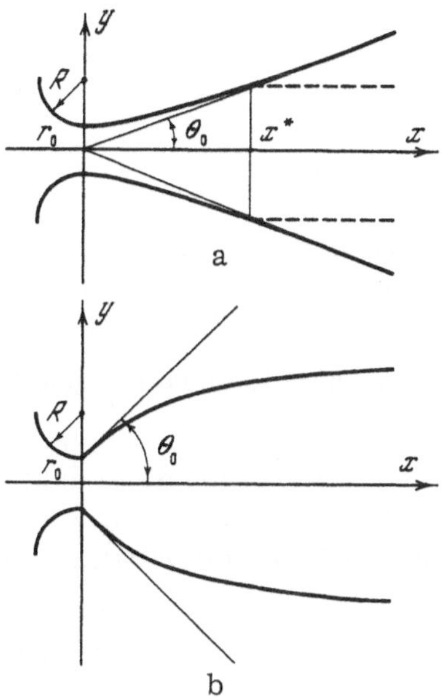

Fig. 7. The nozzle shapes.

In our calculations we considered two types of nozzle: hyperbolic and logarithmic (shape for $x \geq 0$; Fig. 7).

For a hyperbolic nozzle in coordinates (x, y) we have (Fig. 7a)

$$y = r_0 \sqrt{1 + \left(\frac{x}{r_0} \tan\theta_0\right)^2}, \tag{2.19}$$

where r_0 is the half-width (or radius) of the aperture at the critical cross section and θ_0 is the limiting aperture angle of the nozzle. Here, if the cross section is circular (axisymmetric Laval nozzle) or square, then

$$f = 1 + \left(\frac{x}{r_0} \tan\theta_0\right)^2. \tag{2.20}$$

If, on the other hand, the cross section is rectangular with parallel side walls, then

$$f = \sqrt{1 + \left(\frac{x}{r_0} \tan\theta_0\right)^2}. \tag{2.21}$$

For a rectangular nozzle generated by different hyperbolae in the y and z coordinates, we obtain

$$f = \sqrt{1 + \left(\frac{x}{r_{01}} \tan\theta_{01}\right)^2} \sqrt{1 + \left(\frac{x}{r_{02}} \tan\theta_{02}\right)^2}. \tag{2.22}$$

The last expression is a generalization of the preceding two since for $r_{01} = r_{02} = r_0$ and $\theta_{01} = \theta_{02} = \theta_0$ it transforms to Eq. (2.20), while for $\theta_{02} = 0$, and $r_{02} = \text{const}$ or $r_{02} \to \infty$ and $\theta_{02} = 0$, we obtain Eq. (2.21).

Also of interest is the case when beginning from some x a nozzle of the types in Eqs. (2.20)-(2.22) transforms to a channel of constant radius (that is, for $x \geq x^*$, $f = \text{const}$).

The shape of a type *b* (Fig. 7b) nozzle is most conveniently given by an expression of the form

$$y = q \ln(ax + c). \tag{2.23}$$

Here the constants a and c can be given in terms of q and the usual characteristics θ_0 and r_0:

$$a = c^{\frac{\tan\theta_0}{q}}, \quad c = e^{\frac{r_0}{q}}.$$

Thus, for given θ_0 and r_0 the parameter q determines the quantity f such that the greater q, the greater f (for fixed x). This type of nozzle for large (~45°) angles resembles a slit of half-width r_0, for which the calculation was done in [8].

The use of larger angles in our scheme is unacceptable since the quasiuniform flow condition may be destroyed.

As can be seen from Eqs. (2.20)-(2.23) type a and b nozzles differ fundamentally from one another in that for type a nozzles $(1/f)(df/dx)$ rises from zero (for x = 0) to some maximum value and again falls as $x \to \infty$. For type b nozzles $[(1/f)(df/dx)]_{x=0} = \max = \tan\theta_0/r_0$ and then falls as $x \to \infty$.

The shape of the nozzle inlet for both types of nozzle in our calculations was the same, a quarter circle with radius $R = 2r_0$ (see Fig. 7).

For the chosen composition there are only four molecular components, CO_2, CO, O_2, and N_2, during the chemical reactions (2.12) [although in general at higher temperatures (T ≳ 4000°K) a small amount of NO may be present]. Along with reactions (1.20) the principal channels for vibrational relaxation will be taken as the following:

$$\begin{aligned}
&CO_2(00^01) + CO(v=0) \rightleftarrows CO_2(00^00) + CO(v=1), && W_{3,5},\\
&CO_2(00^00) + CO(v=1) \rightleftarrows \begin{cases} CO_2(11^10) + CO(v=0), & W_{5,12},\\ CO_2(03^10) + CO(v=0), & W_{5,2},\end{cases}\\
&CO_2(00^00) + O_2(v=1) \rightleftarrows CO_2(10^00) + O_2(v=0), && W_{6,1}\\
&N_2(v=1) + CO(v=0) \rightleftarrows N_2(v=0) + CO(v=1), && W_{4,5},\\
&N_2(v=1) + O \rightleftarrows N_2(v=0) + O, && W_{4,0},\\
&O_2(v=1) + M \rightleftarrows O_2(v=0) + M, && W^M_{6,0}.
\end{aligned} \tag{2.24}$$

Here the corresponding probabilities of the direct elementary processes are shown on the right. The subscripts 1 to 3 designate symmetric, deformed, and asymmetric oscillations of CO_2, respectively; subscript 4 refers to the N_2 molecule; 5, to the CO molecule; and 6, to O_2 (see Fig. 1).

In the presence of chemical reactions the variation in the average amount ε_i of quanta stored in the i-th vibrational degree of freedom per particle is given by the equation [88]

$$\frac{d\varepsilon_i}{dt} = \left(\frac{d\varepsilon_i}{dt}\right)_v + \sum_j \left(\varepsilon^{(j)}_{i\,\text{chem}} - \varepsilon_i\right)\frac{1}{N_i}\left(\frac{dN_i}{dt}\right)^{(j)}_{\text{chem}}, \tag{2.25}$$

where by the summation we mean summation over all the chemical reactions in which a given molecule participates, and $\varepsilon^{(j)}_{\text{chem}}$ is the average energy, expressed as the number of quanta $h\nu_i$ going into the i-th oscillator as a result of a single event of reaction j. The first term on the right-hand side of Eq. (2.25) describes the usual collisional relaxation over vibrational levels. The relation between the average amount of quanta stored by a single particle ε_i and the vibrational energy per unit mass e^i_v of the i-th component is given by

$$e^i_v = \frac{\theta_i \mathfrak{R}}{\mu_i}\varepsilon_i, \tag{2.26}$$

where $\theta_i = h\nu_i/k$, and k is the Boltzmann constant. With Eq. (2.26), Eq. (2.25) takes the form

$$\frac{de_v^i}{dt}=\frac{\theta_i\Re}{\mu_i}\left[\left(\frac{d\varepsilon_i}{dt}\right)_v+\sum_j\left(\varepsilon^{(j)}_{i\,\mathrm{chem}}-\varepsilon_i\right)\frac{1}{N_i}\left(\frac{dN_i}{dt}\right)^{(j)}_{\mathrm{chem}}\right]. \tag{2.27}$$

We shall assume additionally that dissociation and recombination take place primarily through the upper vibrational levels. Then for the molecules participating in these processes the amount of vibrational energy lost or gained is roughly equal to the dissociation energy. For simplicity we assume for the exchange reaction (2.12) that $\varepsilon_{i_{\mathrm{chem}}} = 0$ for all the components, CO, O_2, and CO_2. Since the CO_2 molecule has four types of vibrations (the deformed type is doubly degenerate), it was assumed for it that the vibrational energy of the chemical reactions is equally distributed, that is, that $\varepsilon_{i_{\mathrm{chem}}} = D\alpha_i$, where D is the dissociation energy of CO_2, $\alpha_i = 0.25$ for the asymmetric and symmetric modes and $\alpha_2 = 0.5$ for the deformed mode.

The equations describing collisional relaxation of the vibrational energy of a harmonic oscillator in a mixture with other oscillators were derived for the general case in Chapter I. Using them, we now write the system of Eqs. (2.27) in the form

$$\left.\begin{aligned}
\frac{de_{12}}{dx}&=\frac{\theta_2\Re}{\mu_{CO_2}v}\biggl\{\biggl[3\,\frac{x_3-x_2^3e^{-\frac{500}{T}}}{(1-x_3)(1-x_2)^3}\sum_M W^M_{3,\Sigma}\gamma_M+3W_{4,\Sigma}\gamma_{N_2}\frac{x_4-x_2^3e^{-\frac{473}{T}}}{(1-x_4)(1-x_2)^3}+\\
&\quad+3W_{5,\Sigma}\gamma_{CO}\frac{x_5-x_2^3e^{-\frac{204}{T}}}{(1-x_5)(1-x_2)^3}+2W_{6,1}\gamma_{O_2}\frac{x_6-x_2^2e^{-\frac{320}{T}}}{(1-x_6)(1-x_2^2)}-\\
&\quad-2\,\frac{x_2-x_{02}}{1-x_2}\sum_M W^M_{2,0}\gamma_M\biggr]\frac{p}{kT}+\biggl[\frac{D_{CO_2}}{2h\nu_2}(\alpha_1+\alpha_2)-\frac{x_2(1+2x_2)}{1-x_2^2}\biggr]\times\\
&\quad\times\frac{2}{\gamma_{CO_2}}\frac{1}{N}\biggl[\left(\frac{dN_{CO_2}}{dt}\right)_{\mathrm{rec}}+\left(\frac{dN_{CO_2}}{dt}\right)_{\mathrm{diss}}\biggr]-\frac{x_2(1+2x_2)}{1-x_2^2}\frac{2}{\gamma_{CO_2}}\frac{1}{N}\left(\frac{dN_{CO_2}}{dt}\right)_{\mathrm{exch}}\biggr\},\\
\frac{de_3}{dx}&=\frac{\theta_3\Re}{\mu_{CO_2}v}\biggl\{-\biggl[W_{3,4}\gamma_{N_2}\frac{x_3-x_4e^{-\frac{27}{T}}}{(1-x_4)(1-x_3)}+W_{3,5}\gamma_{CO}\frac{x_3-x_5e^{-\frac{296}{T}}}{(1-x_5)(1-x_3)}+\\
&\quad+\frac{x_3-x_2^3e^{-\frac{500}{T}}}{(1-x_3)(1-x_2)^3}\sum_M W^M_{3,\Sigma}\gamma_M\biggr]\frac{p}{kT}+\left(\frac{D_{CO_2}}{h\nu_3}\alpha_3-\frac{x_3}{1-x_3}\right)\times\\
&\quad\times\frac{1}{\gamma_{CO_2}}\frac{1}{N}\biggl[\left(\frac{dN_{CO_2}}{dt}\right)_{\mathrm{rec}}+\left(\frac{dN_{CO_2}}{dt}\right)_{\mathrm{diss}}\biggr]-\frac{x_3}{1-x_3}\frac{1}{\gamma_{CO_2}}\frac{1}{N}\left(\frac{dN_{CO_2}}{dt}\right)_{\mathrm{exch}}\biggr\},\\
\frac{de_4}{dx}&=\frac{\theta_4\Re}{\mu_{N_2}v}\biggl[W_{3,4}\gamma_{CO_2}\frac{x_3-x_4e^{-\frac{27}{T}}}{(1-x_3)(1-x_4)}-W_{4,5}\gamma_{CO}\frac{x_4-x_5e^{-\frac{269}{T}}}{(1-x_4)(1-x_5)}-\\
&\quad-W_{4,\Sigma}\gamma_{CO_2}\frac{x_4-x_2^3e^{-\frac{473}{T}}}{(1-x_4)(1-x_2)^3}-W_{4,0}\gamma_O\frac{x_4-x_{04}}{1-x_4}\biggr]\frac{p}{kT},\\
\frac{de_5}{dx}&=\frac{\theta_5\Re}{\mu_{CO}v}\biggl\{\biggl[W_{3,5}\gamma_{CO_2}\frac{x_3-x_5e^{-\frac{296}{T}}}{(1-x_3)(1-x_5)}+W_{4,5}\gamma_{N_2}\frac{x_4-x_5e^{-\frac{269}{T}}}{(1-x_4)(1-x_5)}-\\
&\quad-W_{5,\Sigma}\gamma_{CO_2}\frac{x_5-x_2^3e^{-\frac{204}{T}}}{(1-x_5)(1-x_2)^3}\biggr]\frac{p}{kT}-\frac{x_5}{1-x_5}\frac{1}{\gamma_{CO}}\frac{1}{N}\left(\frac{dN_{CO}}{dt}\right)_{\mathrm{chem}}\biggr\},\\
\frac{de_6}{dx}&=\frac{\theta_6\Re}{\mu_{O_2}v}\biggl\{\biggl[-W_{6,1}\gamma_{CO_2}\frac{x_6-x_2^2e^{-\frac{320}{T}}}{(1-x_6)(1-x_2^2)}-\frac{x_6-x_{06}}{1-x_6}\sum_M W^M_{6,0}\gamma_M\biggr]\frac{p}{kT}+\\
&\quad+\left(\frac{D_{O_2}}{h\nu_6}-\frac{x_6}{1-x_6}\right)\frac{1}{\gamma_{O_2}}\frac{1}{N}\biggl[\left(\frac{dN_{O_2}}{dt}\right)_{\mathrm{rec}}+\left(\frac{dN_{O_2}}{dt}\right)_{\mathrm{diss}}\biggr]-\frac{x_6}{1-x_6}\frac{1}{\gamma_{O_2}}\frac{1}{N}\left(\frac{dN_{O_2}}{dt}\right)_{\mathrm{exch}}\biggr\}.
\end{aligned}\right\} \tag{2.28}$$

Here $\theta_2 = 960°$; $\theta_3 = 3380°$; $\theta_4 = 3353°$; $\theta_5 = 3084°$, $\theta_6 = 2240°$; $D_{CO_2}/k = 63{,}290°$; $D_{O_2}/k = 59{,}420°$; and $x_{0i} = \exp(-\theta_i/T)$.

In writing Eq. (2.28), as in Eq. (1.21a) it was assumed that there is an exact resonance, $h\nu_1 = 2h\nu_2$, and that $T_1 = T_2$; then $e_{12} = e_1 + e_2$ and the probabilities of several processes may be combined to yield

$$W^M_{3,\Sigma} = (1+\varepsilon_{02})^{-1} W^M_{3,12} + W_{3,2}, \qquad W_{4,\Sigma} = (1+\varepsilon_{02})^{-1} W_{4,12} + W_{4,2},$$
$$W_{5,\Sigma} = (1+\varepsilon_{02})^{-1} W_{5,12} + W_{5,2}.$$

The chemical terms on the right-hand sides of Eqs. (2.28) have the forms [see Eqs. (2.12) and (2.13)]

$$\frac{1}{N}\left[\left(\frac{dN_{CO_2}}{dt}\right)_{\rm rec} + \left(\frac{dN_{CO_2}}{dt}\right)_{\rm diss}\right] = \frac{p}{kT}\left(\frac{p}{kT}\gamma_{CO}\gamma_O \sum_M K^M_{-1}\gamma_M - \gamma_{CO_2}\sum_M K^M_1\gamma_M\right),$$
$$\frac{1}{N}\left(\frac{dN_{CO_2}}{dt}\right)_{\rm exch} = \frac{p}{kT}(\gamma_{CO}\gamma_{O_2}K_{-2} - \gamma_{CO_2}\gamma_O K_2),$$
$$\frac{1}{N}\left(\frac{dN_{CO}}{dt}\right)_{\rm chem} = -\left\{\frac{1}{N}\left[\left(\frac{dN_{CO_2}}{dt}\right)_{\rm rec} + \left(\frac{dN_{CO_2}}{dt}\right)_{\rm diss}\right] + \frac{1}{N}\left(\frac{dN_{CO_2}}{dt}\right)_{\rm exch}\right\}, \tag{2.29}$$
$$\frac{1}{N}\left[\left(\frac{dN_{O_2}}{dt}\right)_{\rm rec} + \left(\frac{dN_{O_2}}{dt}\right)_{\rm diss}\right] = \frac{p}{kT}\left(\frac{p}{kT}\gamma_O^2 \sum_M K^M_{-3}\gamma_M - \gamma_{O_2}\sum_M K^M_3\gamma_M\right),$$
$$\frac{1}{N}\left(\frac{dN_{O_2}}{dt}\right)_{\rm exch} = \frac{1}{N}\left(\frac{dN_{CO_2}}{dt}\right)_{\rm exch}.$$

The rate constants of the chemical reactions were taken from [89].

Therefore, for the model chosen here Eqs. (2.11), (2.17), and (2.28), together with Eqs. (2.7), (2.13), and (2.29), completely determine the vibrational relaxation in a chemically reacting gas expanding through a nozzle.

3. A Gas-Dynamic Laser with a Supersonic Nozzle. Effect of the Nozzle Shape on Laser Operation for $T_0 \leqslant 2500°K$

Substantial powers have been obtained from gas-dynamic lasers with supersonic nozzles at the present time [14], and there is great interest in optimizing them.

The dependence of the inversion in low-temperature lasers on the initial parameters (temperature and stagnation pressure, i.e., the conditions at the nozzle inlet) and on the composition of the mixture has been studied in comparative detail. On the other hand, the effect of the size and shape of the nozzle has not been sufficiently investigated yet. However, it is just these which are many times the controlling factors in the design of specific devices. In most papers, as already noted in the review, only one type of nozzle is considered, and studies in this area have a disconnected character. Even if assembled together they do not allow one to make a systematic study of gas-dynamic lasers with different nozzles.

Meanwhile calculations using the method in Section 2 for low-temperature mixtures have shown that the parameters of gas-dynamic lasers depend strongly on the nozzle shape. It appears that chemical reactions do not take place and their effect on vibrational relaxation may be neglected. Acceleration of the gas from the nozzle inlet to the critical cross section may be regarded (as shown in [11]) as taking place in complete thermodynamic equilibrium (isentropic process). Our calculations confirm this, since the maximum deviation of the vibrational temperature from the gas temperature did not exceed about 80°K for an absolute value of T ~ 2000°K. Thus, in this case the flow ahead of the critical cross section was regarded as independent of the shape of the nozzle inlet and the parameters at the critical cross section were found from the equations [90]

$$p = p_0\left(1+\frac{\gamma-1}{2}M^2\right)^{-\frac{\gamma}{\gamma-1}}, \quad T = T_0\left(1+\frac{\gamma-1}{2}M^2\right)^{-1}. \tag{2.30}$$

Here p_0 and T_0 are the stagnation parameters, $M = v/c$ is the local Mach number, v and $c = \sqrt{\gamma \Re T/\mu}$ are the local flow and sound velocities, and γ is the adiabatic index defined in our case as

$$\gamma = \frac{5/2 + \sum_i \gamma_i}{3/2 + \sum_i \gamma_i}$$

(summation is over the molecular components). Putting M = 1 at the critical cross section, we find†

$$p_{cr} = p_0\left(\frac{2}{\gamma+1}\right)^{\frac{\gamma}{\gamma-1}}, \quad T_{cr} = T_0\frac{2}{\gamma+1}. \tag{2.31}$$

As can be seen from Eq. (2.17) the numerator and denominator on the right-hand sides of the equations for T and v change sign as the gas moves, beginning at the nozzle entrance some place before the critical cross section. For the supersonic nozzle regime both the numerator and denominator should only change sign simultaneously. The numerator goes to zero at a critical point x_1 located just before the critical (where $df/dx = 0$) cross section for type *a* nozzles (Fig. 7a) and, as a rule, precisely at the critical cross section for type *b* nozzles. The denominator is equal to zero only when the flow velocity equals the local sound speed. It should also be noted that for equilibrium initial stagnation conditions the distance from the critical cross section to the critical point is very small; thus, in the low-temperature mixture calculations the origin was located at the critical cross section. Taking concrete values of T_{cr} and p_{cr} [the corresponding T_0 and p_0 are found using Eq. (2.31)], the equilibrium composition was computed for certain values of a_{N_2} and a_{He} using Eq. (2.11) with the derivatives set equal to zero. Knowing the composition and temperature we find the local sound speed c_{cr}. Then, in order to find the critical point a small (not more than 0.001 of that quantity, depending on the nozzle shape) increment was added to c_{cr} and the resulting value was taken for the initial value of the flow velocity.

The probabilities in Eqs. (1.20) and (2.24) which enter into the system of Eqs. (2.28) were taken from [82, 91] subject to the comments in Section 4 of Chapter I.

In this systematic study of the dependence of the gas-dynamic laser parameters on the nozzle shape each nozzle was examined in terms of its individual elements: the region up to critical, the shape of the critical cross section (circular, rectangular), its half-width (radius) r_0, the aperture angle of the widening part of the nozzle θ_0 (the parameter $\tan\theta_0/r_0$), the nature of the transition to expansion (nozzle of type *a* or *b*), the shape of the nozzle in the region

† Equations (2.30) and (2.31) are true only for an ideal gas. In real cases, γ varies as the gas mixture moves from the beginning of the subsonic part of the nozzle up to the critical cross section. However, in the temperature range T_0 = 3000-2000°K and for the chemical compositions typical for gas-dynamic lasers this change does not exceed about 5%.

far from critical and lying beyond the expanding part (a constant-cross-section channel or slight narrowing), and the length of the expansion channel x*.

It is clear from Eqs. (2.20)-(2.23) that each of these elements has its specific effect on the characteristics of a gas-dynamic laser: the maximum value of the inversion, ΔN_{max}, the location of this maximum on the nozzle axis, the width of the region where ΔN occurs, and the way the inversion depends on pressure and temperature. The purpose of the calculations was to study the effect of each of these closely interrelated factors under different stagnation conditions so as to permit evaluation of the prospects for specific nozzles as a whole on this basis. The results of the calculations are given in Figs. 8-16. In them the coordinate origin is located at the critical cross section. A type *a* nozzle (Fig. 7a) was used first in our analysis of the effect of the character of the transition from critical to expansion.

Figure 8 shows typical distributions of the inversion $\Delta N = N_{00^01} - N_{10^00}$ and of the vibrational and gas temperatures along the axis of *a* nozzle for complete thermodynamic equilibrium at the critical cross section. It is clear (dashed curves) that at some distance from the critical cross section (here $x \approx 7$ cm) the vibrational temperatures of N_2 and the asymmetric mode of CO_2 become "frozen," while T_2 falls further until $x \approx 12$ cm. The inversion, which appears when a certain difference is achieved between T_3 and T_2 (the ratio T_3/T_2 must be about 1.73), passes through a maximum, and then falls. The existence of a maximum is due to an increase in the difference between T_3 and T_2 and also to a reduction in the density during expansion.

Under the same conditions but with a nozzle which becomes a cylindrical channel of constant cross section (smooth curves) at some distance x* from critical (x* = 10 cm in Fig. 8) the pattern is somewhat different. Beginning at that time the gas temperature T increases due

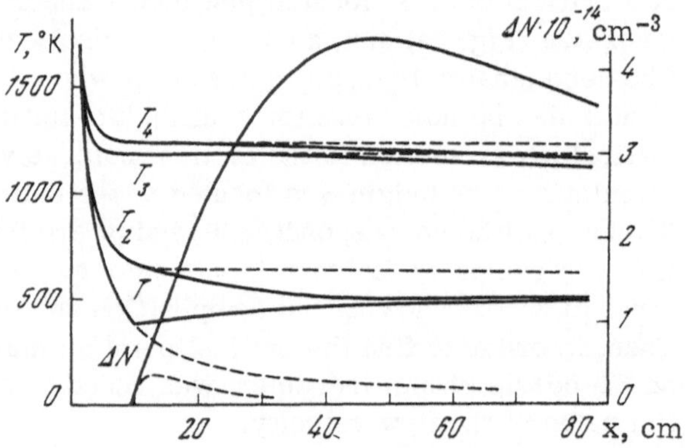

Fig. 8. Distributions of the inversion ΔN and vibrational and gas temperatures along the axis of an axially symmetric type *a* nozzle. $T_{cr} = 2000$°K; $p_{cr} = 15$ atm; $\tan\theta_0/r_0 = 1.25$; $a_{N_2} = 0.6$; $a_{He} = 0.085$, x* = 10 cm.

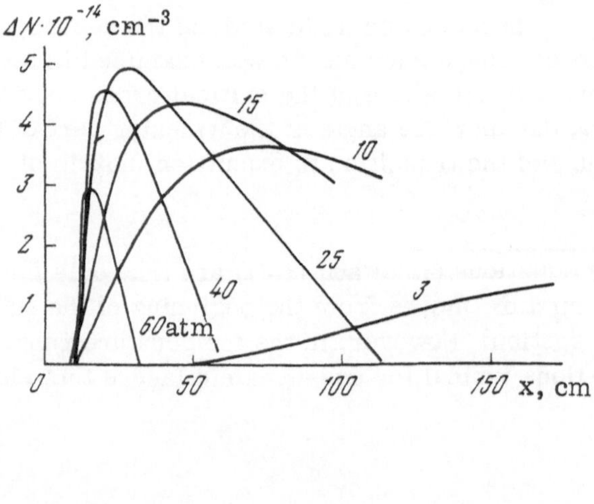

Fig. 9. Dependence of the inverted populations on the distance along the axis of an axially symmetric type *a* nozzle for various pressures p_{cr} at the critical cross section. $T_{cr} = 2000$°K; $\tan\theta_0/r_0 = 1.25$; $a_{N_2} = 0.6$; $a_{He} = 0.085$; x* = 10 cm.

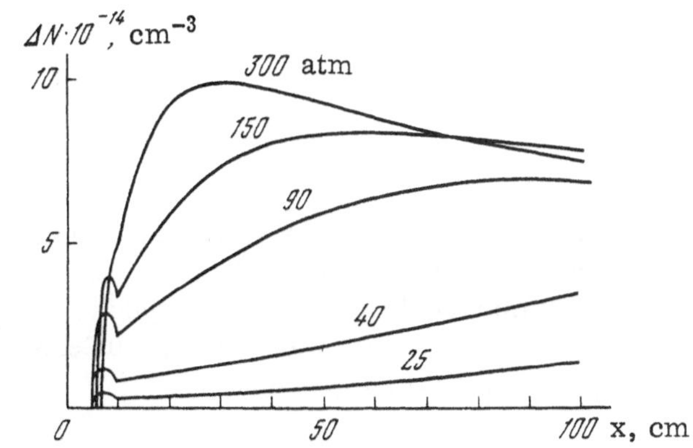

Fig. 10. Dependence of the inverted populations on the distance along the axis of an axially symmetric type a nozzle for various pressures p_{cr} at the critical cross section. T_{cr} = 2000°K; $\tan\theta_0/r_0$ = 2.5; a_{N_2} = 0.6; a_{He} = 0.085; x* = 10 cm.

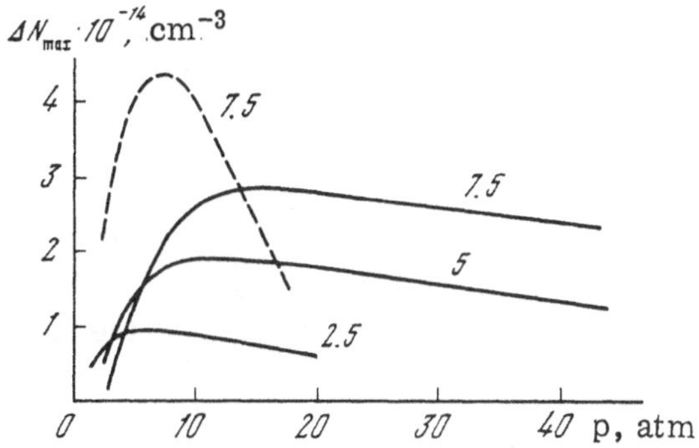

Fig. 11. The maximum inversion ΔN_{max} as a function of p_{cr} for a plane type a nozzle for various values of the expansion parameter $\tan\theta_0/r_0$ (continuous curves). T_{cr} = 2000°K; a_{N_2} = 0.6; a_{He} = 0.085. The dashed curve corresponds to the same conditions for a nozzle that becomes a constant-cross-section channel at x* = 10 cm.

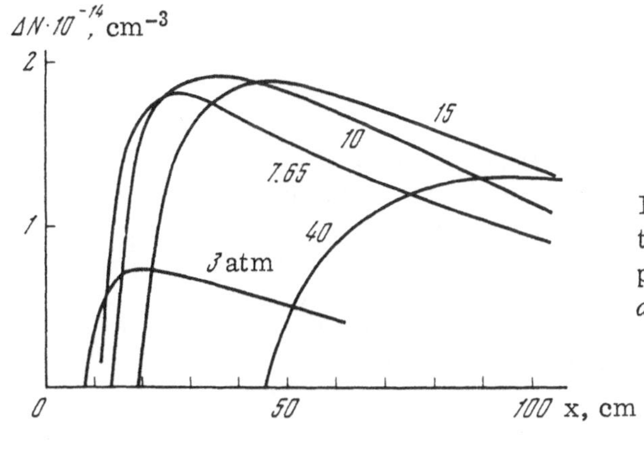

Fig. 12. Population inversion distribution along the axis of a plane type a nozzle for different pressures p_{cr}. T_{cr} = 2000°K; $\tan\theta_0/r_0$ = 5; a_{N_2} = 0.6; a_{He} = 0.085.

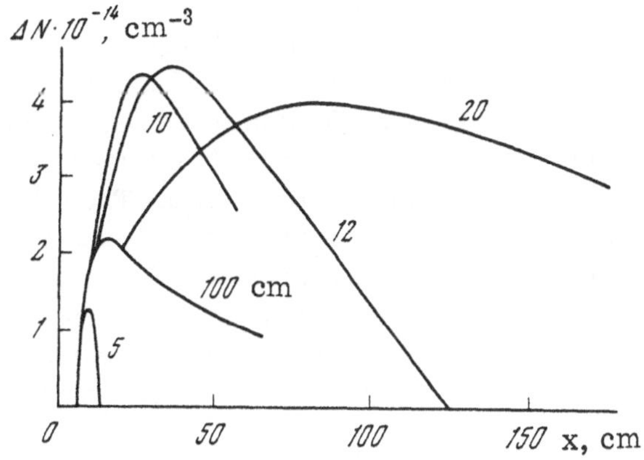

Fig. 13. Population inversion distribution along the axis of a plane type a nozzle that becomes a constant-cross-section channel at x = x* for different values of x*. T_{cr} = 2000°K; p_{cr} = 7.65 atm, $\tan\theta_0/r_0$ = 7.5; a_{N_2} = 0.6; a_{He} = 0.085.

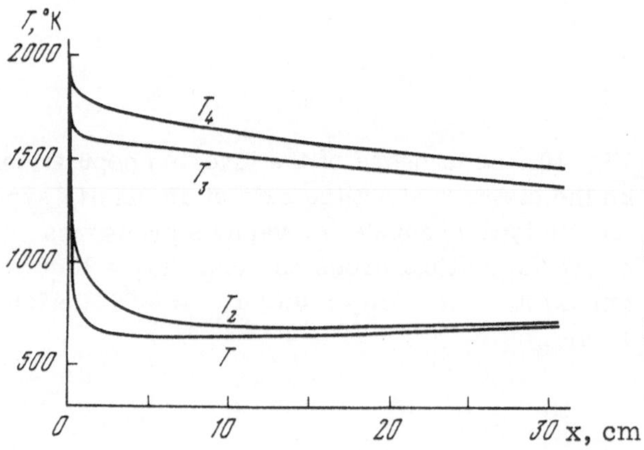

Fig. 14. Dependences of the vibrational and gas temperatures on the distance along the axis of a logarithmic type *b* nozzle. $T_{cr} = 2000°K$; $p_{cr} = 15$ atm, $\tan\theta_0/q = 2.663$; $r_0/q = 0.0525$ ($q = 1/3$); $a_{N_2} = 0.6$; $a_{He} = 0.085$.

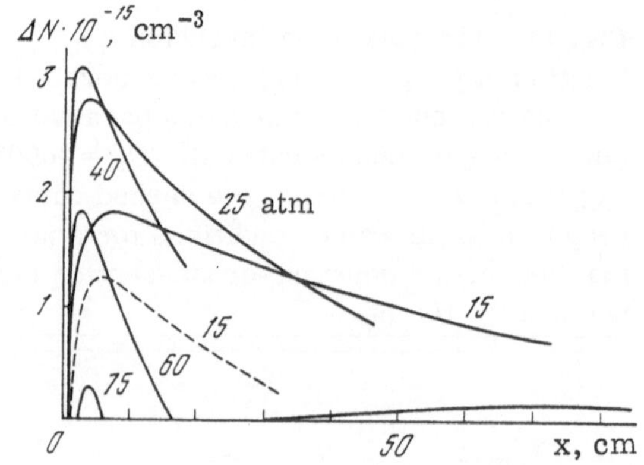

Fig. 15. Distributions of ΔN along the axis of a type *b* nozzle for various pressures at the critical cross section. $T_{cr} = 2000°K$; $\tan\theta_0/q = 2.663$; $r_0/q = 0.0525$ ($q = 1/3$); $a_{N_2} = 0.631$; $a_{He} = 0.1122$ (corresponds to the mixture CO_2:N_2:He = 1:4:5). The dashed curve corresponds to $a_{N_2} = 0.6$; $a_{He} = 0.085$ (composition 1:3:3).

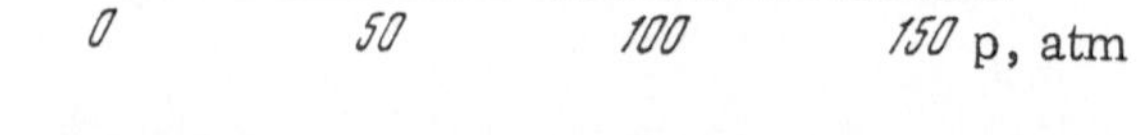

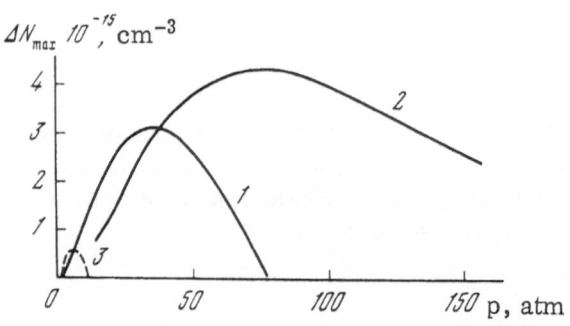

Fig. 16. Dependence of ΔN_{max} along the axis of a logarithmic nozzle on the pressure at the critical cross section. $T_{cr} = 2000°K$; $a_{N_2} = 0.631$; $a_{He} = 0.1122$. Curve 1) $\tan\theta_0/q = 2.633$; $r_0/q = 0.0525$ ($q = 1/3$); curve 2) $\tan\theta_0/q = 0.888$; $r_0/q = 0.0175$ ($q = 1$); curve 3) $\tan\theta_0/q = 0.662$; $r_0/q = 0.065$ ($q = 1$).

to curtailment of expansion and transfer of energy from vibrational degrees of freedom to translational. This growth in T promotes a continuation of the rather intense drop in T_2 and subsequently results in faster relaxation of T_3 and T_4. Here the portion over which the divergence between T_3 and T_2 increases may be considerably extended and, since the density of particles is constant for $x \geq x^*$, the maximum inversion (other conditions being equal) is much higher than in the first case. The fall in the inversion is due not to a reduction in density, but to a slowing down of the fall and a subsequent increase in T_2 at the same time T_3 and T_4 are relaxing. The approach of T_3 and T_2 to one another is comparatively slow; thus, the region with a high inversion may be large (sometimes more than 1-2 m). The values of a_{N_2} and a_{He} chosen here correspond (for our conditions) to the composition $CO_2 : N_2 : He \approx 1:3:3$ (the remaining components make up less than 1% of the whole).

Calculations were also done for nozzles with slightly narrowing cross sections after some distance $x' > x^*$. At supersonic velocities this makes it possible [see Eq. (2.17)] to accelerate somewhat the rise in T and more sharply (at small x) increase the divergence between T_3 and T_2. As this takes place the total number of particles per unit volume also increases. As a whole these factors should increase ΔN_{max}. The calculation showed, however, that the increase is only about 40% of the value of ΔN_{max} for a nozzle of constant (without narrowing) cross section even in the optimal variant. This is explained by the fact that the upper level relaxes so much more rapidly when the temperature and density increase. These calculations neglected the possible appearance of shock waves reflected from the narrowing section of the nozzle which might affect the character of the flow.

Figure 9 shows the distributions of the population inversions along the axis of a nozzle with $x^* = 10$ cm for different pressures p_{cr} at the critical cross section. It is apparent that for these system parameters the optimum value is $p_{cr} \approx 27$ atm. This is due to an increase in the number of working particles at higher pressures and the resulting acceleration of vibrational relaxation. The increased relaxation rate also leads to a more rapid rise in the gas temperature, so the inversion region becomes narrower as the pressure is increased.

The role of the expansion parameter ($\tan\theta_0/r_0$) is clear from Fig. 10, which shows the analogous dependences for a nozzle with a larger value of $\tan\theta_0/r_0$. Here the optimum inversion is already achieved in the expanding part of the nozzle; however, the transition into a constant-cross-section channel makes it possible to extend the relazation of the deformed oscillations of CO_2 and to increase the divergence between T_3 and T_2. At constant density this causes increased inversion and attainment of a second, higher maximum. The maximum value of ΔN increases up to pressures $p_{cr} \sim 300$ atm and is located closer to x^* at higher pressures while the inversion region, as in the case of Fig. 9, becomes narrower.

To examine the effect of the shape of the critical cross section we have computed the inversion distributions along the axis of a plane type *a* nozzle. The vibrational temperature and inversion distributions in this case are analogous to those given in Fig. 8; however, the range of optimal densities (Fig. 11) is found at much smaller values of p_{cr} (cf. Fig. 10 in the region of the first maxima) due to the lower expansion velocity [see Eqs. (2.20) and (2.21)]. It is clear from Fig. 11 that the higher the expansion parameter, $\tan\theta_0/r_0$, the more the optimum inversion occurs at higher pressures and the higher its absolute value (continuous curves).

The inversion distributions along the axis of a plane type *a* nozzle are shown in Fig. 12. The maximum in ΔN appears earlier at lower pressures. This is explained by an increase in the relaxation rates as the pressure increases so that the necessary divergence between T_3 and T_2 is achieved later for large p_{cr}. The existence of an optimum pressure at $p_{cr} \approx 10$ atm is due, as in the case of an axisymmetric nozzle, to two competing factors, a growth in the total number of working particles and more rapid relaxation as the pressure is increased.

As a plane type *a* nozzle changes to a constant-cross-section channel, a picture similar to that shown in Fig. 10 is observed, with the difference that, as pointed out above, the region of optimal pressures is shifted toward smaller values. Figures 13 shows the inverted population distributions along the axis for different lengths x^* of the expanding part of a nozzle. The dependence of ΔN_{max} on pressure for $x^* = 10$ cm is shown in Fig. 11 (dashed curve). For the conditions of Fig. 13 the optimal value of x^* is about 14 cm and is determined, on the one hand, by the fact that for $x^* < x^*_{opt}$ the contribution of relaxation is large, and on the other, by the fact that for $x^* > x^*_{opt}$ the number of working particles is small due to the considerable expansion. As in the cases of Figs. 8 and 10 the size of the maximum inversion may be higher than for a nozzle with $x^* \to \infty$; however, here this difference is not so large, even in the optimum variant (see Fig. 11).

Larger values are obtained for type *b* nozzles (see Fig. 7b) since with them it is possible to realize substantial initial cooling rates. As can be seen from Fig. 14, T_3 and T_4 do not "freeze out." However, the divergence between T_3 and T_2 attained for $x \sim 1.5$ cm in this case is sufficient to produce an inversion, and the further increase in this divergence due to the difference in the relaxation times of the vibrational degrees of freedom of CO_2, even with a small change in the density, results in an increase in ΔN (Fig. 15). As the (geometric) rate of expansion of the nozzle df/dx is reduced, the gas temperature not only ceases to fall, but rises due to accelerated relaxation of T_3 and T_4 (here T_2 is already close to T); this together with the slight drop in density leads to a reduction in ΔN. For comparison the inversion distribution in a 1 : 3 : 3 mixture is plotted in Fig. 15 (dashed curve), the other parameters being the same. It is clear that despite the small absolute concentration of CO_2 in a 1 : 4 : 5 mixture the inversion is somewhat greater.

Figure 16 illustrates the dependence of the maximum inversion on the pressure at the critical cross section for various nozzle characteristics. The parameter q in curves 1 and 2 differs by a factor of 3 and the higher q is, the more the region of optimal values of ΔN_{max} is shifted toward higher pressures. The calculations show that the greater the ratio $\tan\theta_0/r_0$ is, the more effectively the gas is cooled (compare, for example, with curve 3). Here higher initial temperatures require larger values of q (to achieve sufficiently intense cooling). The absolute values of the optimal ΔN are roughly four to eight times higher than in the best variants of hyperbolic nozzles (to within small differences in the mixture compositions).

Therefore, it has been shown that the nozzle shape may have a significant effect on the operation of a gas-dynamic laser. On the whole, with these data on the dependence of the inversion on the individual elements of a nozzle in the low-temperature case, one can derive the characteristics of a laser with various specific nozzles. Naturally, in doing this it is necessary to keep in view the possible differences between real and calculated flows (due to a number of factors neglected here, such as the effect of the boundary layer, replacement of three-dimensional flow by one-dimensional motion, formation of turbulence in real cases, etc.); however, the semiquantitative picture can be illustrated by this computational scheme.

These results are in satisfactory agreement with comparable data obtained by a number of other authors [7, 9, 11, 19] for the temperature range up to about 2500°K.

CHAPTER III

CHEMICAL AND NONEQUILIBRIUM VIBRATIONAL PROCESSES IN THE MIXTURES USED IN GAS-DYNAMIC LASERS. CHEMICAL GAS-DYNAMIC LASERS

1. Gas-Dynamic Lasers Using Combustion Products

In the previous chapter we examined gas-dynamic lasers with equilibrium initial conditions and unchanged chemical composition of a CO_2-N_2-He mixture. In actual mixtures for gas-dynamic lasers there may be unavoidable impurities and chemical processes may take place. Thus, in the great majority of cases H_2O, molecular and atomic oxygen, and a number of other materials are contained in the mixtures in amounts that depend on the means of producing the lasing medium and the design of the device. On the other hand, the following may be used to optimize the operation of a gas-dynamic laser: selection of special mixtures, combustible fuels, mixing of gases, and stimulation of certain chemical reactions.

Two types of phenomena, chemical processes and nonresonance vibrational exchange, are important in an analysis of relaxtion of mixtures in gas-dynamic flows. Both processes produce additional deviations from equilibrium and may have a dominant role in the operation of a gas-dynamic laser. We shall first consider chemical processes.

The idea of the chemical gas-dynamic laser was advanced by Biryukov and Shelepin [17] and is based on the fact that vibrationally excited molecules are formed in nonequilibrium chemical combustion reactions and may be used as the active medium of a gas-dynamic laser. By creating the conditions for rapid depopulation of the lower levels it is possible to obtain an inversion. The latter is achieved either by rapid cooling of the combustion products or by introducing suitable impurities into the gaseous mixture to speed up relaxation of these levels. A number of multiatomic molecules produced by combustion of appropriate materials may be used as the working gas.

We shall examine the operating principle of such lasers with the CO_2 molecule as an example. In accordance with [41] we shall characterize the populations of the vibrational levels by the vibrational temperatures T_i (symmetric, T_1; deformed, T_2; and asymmetric, T_3). When CO is burned in air vibrationally excited CO_2 molecules are formed with T_1, T_2, and T_3 greater than the gas temperature T. The efficiency of vibrational excitation is indicated by an experiment [92] in which cold CO_2 was irradiated by the light from a CO flame. The result was lasing at a wavelength of 10.6 μ in the volume of cold CO_2 with a power of about 1 mW. According to [92] more than 20% of the heat of combustion of CO or hydrocarbons may be radiated in the infrared. A substantial portion of this radiation is in the vibrational–rotational band at 4.4 μ corresponding to the asymmetric vibrational modes of CO_2 with which pumping to the upper laser level is achieved.

In CO_2 as a combustion product the vibrational temperature is high and may be estimated at several thousand degrees. However, an inversion does not immediately occur among CO_2 molecules formed by combustion of CO since the vibrational excitation is distributed over all the vibrational modes ($T_1 \approx T_2 \approx T_3$). To produce an inversion among the level populations in the combustion products it is necessary to ensure rapid relaxation of the energy of the symmetric and deformed vibrational modes over a time t less than the energy relaxation time of the asymmetric vibrational modes, τ. Radiative decay of the levels may be neglected at pressures above 1 torr. The characteristic collisional relaxation times for asymmetric vibrational modes at pressures of about 1 torr and T ≈ 1000°K are on the order of 1-5 · 10^{-3} sec for CO_2-CO_2, CO_2-N_2, and CO_2–He collisions.

Along with excited CO_2 molecules in a CO–air flame there are also vibrationally excited CO and N_2 which, because of resonance energy transfer in collisions with CO_2, serve as a reservoir that feeds energy to the asymmetric vibrational modes of CO_2, thereby increasing the relaxation time of these modes. With this the time τ is about 5 · 10^{-4} sec for pressures of several torr. The conditions for rapid relaxation of the vibrational energy from the symmetric and deformed vibrational modes are realized when vibrationally excited CO_2 is abruptly cooled. The cooling rate achieved during expansion of a gas through a slit or supersonic nozzle is about 10^9°K (see [4]), which exceeds that required for this kind of laser using the 00^01-10^00 transition.

Figure 17a shows a schematic diagram of a laser with cooling of combustion products as they pass through a nozzle. In this type of apparatus hydrocarbons (e.g., C_2H_2) or ordinary coal may be burned instead of CO and the gaseous combustion products directed through the nozzle into the resonator. A variant in which the reactions

$$C_2N_2 + O_2 \rightarrow 2CO + N_2, \quad 2CO + O_2 \rightarrow 2CO_2$$

take place in succession is shown in Fig. 17b. Among the combustion products are excited CO and N_2 which slow down relaxation of the 00^01 level.

Depopulation of the lower laser levels and production of an inversion in the $00^01 \rightarrow 10^00$, 02^00 transitions are possible as well during rapid mixing of CO_2 formed during combustion with

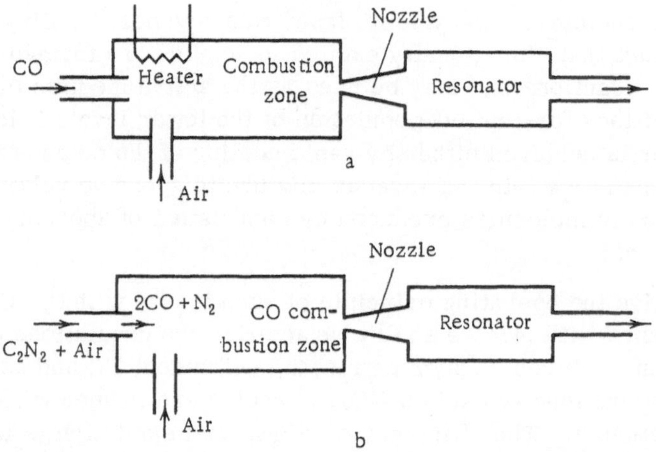

Fig. 17. Schematic diagram of gas-dynamic lasers using combustion products.

helium and water vapor. Because of the closeness of the 01^10 levels of the CO_2 molecule resonant transfer of vibrational energy takes place from the lower working levels of CO_2 to the unexcited H_2O molecule which has small relaxation times (cf. [43]). Helium effectively destroys the deformed vibrational modes and, as it has good thermal conductivity, reduces the gas temperature T, thus creating favorable conditions for enhancement of the inverted population. A diffusion calculation shows that over times of about $5 \cdot 10^{-4}$ sec the gases are mixed over distances of 3-4 mm at pressures of several torr.

A schematic diagram of a laser with gas mixing is shown in Fig. 18a. The flame in the combustion chamber is initiated and maintained by supplementary heating. The apparatus works in a flow-through regime. For more complete mixing of the entering He and H_2O with the combustion products a device of the type shown in Fig. 18b may be used. Here H_2O and He are let into the inside of the disc, the points denote through holes by which the combustion products pass, and the circles denote holes in one side of the disc through which helium and water vapor pass into the resonator. In this case relaxation of the lower levels is sped up due to cooling during expansion because the gas passes through a hole.

This method of producing a population inversion with its comparative simplicity is promising. The possibility of using combustion products of materials in the gaseous as well as in the liquid and solid phases should be especially emphasized.

In addition to CO_2, other molecules, such as N_2O and COS, may be used as a working gas.

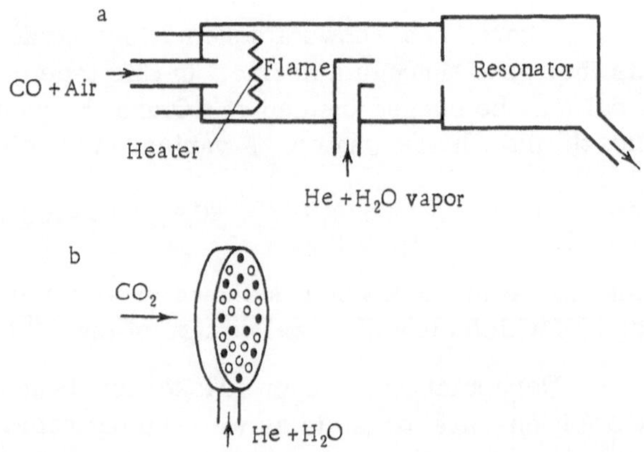

Fig. 18. Schematic diagram of a laser with gas mixing.

2. Recombination Processes in Molecular Lasers

Chemical recombination reactions may play an important role in gas-dynamic flows. One of the problems is therefore to examine the possibility of using recombination as a chemical reaction for producing additional deviations from equilibrium in gas-dynamic lasers and for obtaining inversion during mixing of reacting gas flows.

In this analysis we have used a simplified variant of the scheme for calculating an expanding flow in which chemical reactions may take place. Here the nozzle was modeled by adiabatic expansion through a slit with half-width R_0 (see Chapter II, Section 1) and the relaxation equations were written with the assumption that the deformed and symmetric vibrational modes of CO_2 as well as the vibrations of O_2 have vibrational temperatures equal to the gas temperature T (this is satisfied if the appropriate amount of He is added to increase the relaxation rate of these types of vibrations). It was further assumed that the nearly resonant exchange between CO and the 00^01 level of the asymmetric mode of CO_2 is the dominant process, so that equilibrium (quasiequilibrium) is rapidly established among these oscillators. Then the vibrational energy relaxation equation for CO_2 in the truncated harmonic oscillator model may be written in the form [see also Eq. (2.25)]

$$\frac{d\varepsilon_3}{dt}=\frac{1}{\beta}\left[-(\varepsilon_3-\varepsilon_{03})\sum_M W^M_{3,\Sigma}N_M+\left(\frac{D\alpha_3}{h\nu_3}-\varepsilon_3 e^{\frac{296}{T}}\right)\frac{1}{N_{CO_2}}\left(\frac{dN_{CO_2}}{dt}\right)_{\text{chem}}\right]. \tag{3.1}$$

Here ε_{03} is the value of ε_3 at temperature T, D is the dissociation energy of CO_2, M = CO_2, CO, O, O_2, and He; $N_M = N_M(0)R_0/(R_0 + v_0t)$.

Since a reliable value of α_3 is not available at present, we assume, as in Eq. (2.28), that $\alpha_3 = 0.25$. The equations used to find the concentrations of the chemical components of the mixture have the same form as in Chapter II. The factor β in Eq. (3.1) effectively takes exchange between CO_2 and CO into account and has the form (for details, see Chapter III, Section 3)

$$\beta=1+\frac{N_{CO}}{N_{CO_2}}\exp\left(-\frac{\theta_5-\theta_3}{T}\right). \tag{3.2}$$

This equation together with the chemical kinetic equations were solved on an electronic computer with the magnitudes and temperature dependences of $W^M_{3,\Sigma}$ and the rate constants of reactions (2.12) taken from experiment and the remaining unavailable values calculated by the SSH method [62, 63].

Two cases of initial conditions were examined:

a. Complete thermodynamic equilibrium at t = 0;

b. Initial nonequilibrium particle concentrations; here an excess† of CO molecules was assumed.

The calculated results showed that for expansion from a state of complete thermodynamic equilibrium the pumping of energy into asymmetric vibrational modes of CO_2 due to recombination is inefficient. For $T_0 \leqslant 3500$°K, this is due to rapid chemical "quenching" because of a fall in the density of particles; for $T_0 \geqslant 4000$°K, to the high rate of change of the vibrational temperature T_3 which prevents development of the divergence between T_3 and T needed for production of an inversion. The effect of the initial degree of equilibrium dissociation on the relaxation kinetics and population inversion amounts to some reduction in the number of CO_2

† This may be obtained by mixing flows of partially dissociated CO_2 molecules and CO molecules.

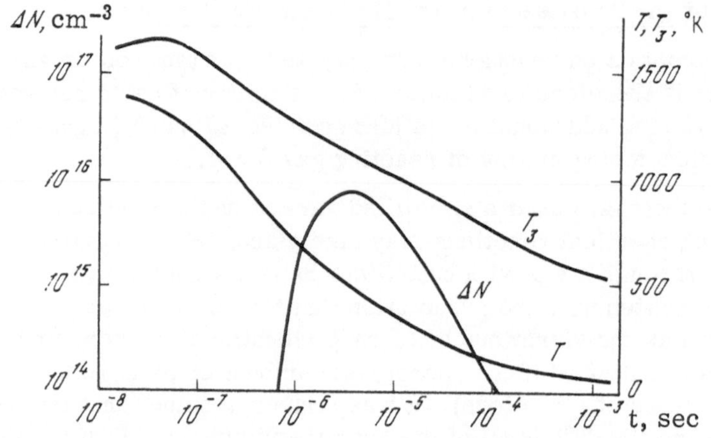

Fig. 19. Time dependences of the temperatures T and T_3 and of the population inversion ΔN of the 00^0-10^00 levels of CO_2. $T_0 = 1500°K$, $R_0 = 0.1$ cm; $\gamma = 1.3$; $\alpha_3 = 0.25$; $[N_{CO}/(N_{CO} + N_{CO_2})]_0 = 0.9$; $(N_{CO}/N_O)_0 = 9$; $(N_{He}/N_{CO_2})_0 = 91$; $p^0_{tot} = 162.8$ atm.

molecules and to a change in the effective relaxation time τ_1 of the asymmetric mode due to exchange with CO [the factor β in Eq. (3.1)] and due to direct destruction of the 00^01 level in collisions with CO and O. Thus, for $T_3 = 1000°$ and $T = 300°$ the relaxation time in dissociated CO_2 is reduced by n times, where

$$n = \left[\left(W^{He}_{3,\Sigma} N_{He} + W^{CO_2}_{3,\Sigma} N_{CO_2}\right)\left(1 + 3\frac{N_{CO}}{N_{CO_2}}\right)\right]^{-1} \sum_M W^M_{3,\Sigma} N_M.$$

For $N_O/N_{CO_2} \approx N_{CO}/N_{CO_2} \approx 1$ and $N_{He}/N_{CO_2} \approx 10$, we obtain $n \approx 3$.

For the case of an initial concentration disequilibrium when the amount of CO exceeds the equilibrium value corresponding to temperature T_0, there is a disequilibrium both in temperature and concentration in the course of the expansion. This may shift the equilibrium for the reaction $CO + O + M = CO_2 + M$ strongly to the right since recombination will provide substantial pumping of energy into the asymmetric vibration modes of CO_2. Figure 19 illustrates the effect of the initial concentration disequilibrium on the relaxation of T_3. Clearly, even in the initial stage of the expansion, when the concentrations of the particles are still large, the divergence between T and T_3 may be substantial. When the initial temperature is lowered this divergence increases. It is very important that in such a system an inversion may be obtained at high pressures, thus making possible increased efficiency in gas-dynamic lasers. We note that the gas-dynamic problems become nontrivial here. Choice of the optimum conditions for expansion requires further study.

The gas-dynamic laser design with a supplementary concentration disequilibrium discussed above may be called a chemical gas-dynamic laser. Its limiting case is very interesting, a purely chemical recombination laser (nonequilibrium concentration). Two features of recombination as a chemical reaction should be noted: First, almost all the reaction energy, equal to the dissociation energy [for many molecules, a large quantity ($\gtrsim 5$ eV)], is released in the form of vibrational energy, and second, the reaction is termolecular with a rate that depends strongly on the gas pressure.

We shall examine the possibilities that follow from these features of using recombination to produce a population inversion by obtaining high vibrational temperatures at high gas pressures. Let the molecule XY be formed in the reaction

$$X+Y+M \underset{K_-}{\overset{K_+}{\rightleftarrows}} XY+M. \tag{3.3}$$

To analyze the qualitative features we shall assume for simplicity that the gas temperature is constant and the initial concentration obeys $[X]_0 \gg [Y]_0$, so that during the reaction $[X] \approx \mathrm{const} = [X]_0$. Then for the initial concentrations [XY] and [Y] we have

$$\frac{d[XY]}{dt}=-\frac{d[Y]}{dt}=K_+[X]_0[M][Y]. \tag{3.4}$$

In this equation dissociation may be neglected since for $[X]_0 \gg [Y]_0$ and comparatively low temperatures the equilibrium in reaction (3.3) is strongly shifted to the right. Let $[XY]_0 = 0$ at the initial moment; then from Eq. (3.4) we find

$$[XY]=[Y]_0\left(1-e^{-\frac{t}{\tau_r}}\right); \quad \tau_r^{-1}=K_+[X]_0[M]. \tag{3.5}$$

The equation for vibrational relaxation is similar to Eq. (3.1) and has the form

$$\frac{d\varepsilon}{dt}=-\frac{\varepsilon-\varepsilon_0}{\tau_v}+\left(\frac{D}{h\nu}-\varepsilon\right)\frac{1}{[XY]}\frac{d[XY]}{dt}. \tag{3.6}$$

The vibrational relaxation time τ_v is determined by the concentration $[X]_0$ (or [M]) and does not change in time; that is, $\tau_v^{-1} = K_v[X]_0$ (since $[X]_0 \gg [Y]$, [XY] and T = const).

Using Eqs. (3.4) and (3.5) we find from Eq. (3.6) that

$$\varepsilon=\varepsilon_0\frac{1-e^{-\frac{t}{\tau_v}}}{1-e^{-\frac{t}{\tau_r}}}+\frac{\frac{D}{h\nu}\tau_v-\varepsilon_0\tau_r}{\tau_r-\tau_v}\,\frac{e^{-\frac{t}{\tau_r}}-e^{-\frac{t}{\tau_v}}}{1-e^{-\frac{t}{\tau_r}}}, \tag{3.7}$$

where ν is the vibration frequency of the XY molecule. The value of ε determines the vibrational temperature in XY and, therefore, the possibility of creating a population inversion. The amount of vibrational energy E per unit volume which is responsible for the absolute magnitude of this inversion and the output power is equal to

$$E=\varepsilon h\nu[XY]=h\nu[Y]_0\left[\varepsilon_0\left(1-e^{-\frac{t}{\tau_v}}\right)+\frac{\frac{D}{h\nu}\tau_v-\varepsilon_0\tau_r}{\tau_r-\tau_v}\left(e^{-\frac{t}{\tau_r}}-e^{-\frac{t}{\tau_v}}\right)\right]; \tag{3.8}$$

E is a nonmonotonic function of time. Its maximum occurs at the time

$$t_1=\frac{\tau_r\tau_v}{\tau_r-\tau_v}\ln\left[\frac{\tau_r}{\tau_v}\left(1+\frac{\tau_r-\tau_v}{\frac{D}{h\nu}\tau_v-\varepsilon_0\tau_r}\varepsilon_0\right)\right]. \tag{3.9}$$

If $\varepsilon_0 \ll 1$, then

$$E_{\max}=D[Y]_0\frac{1}{a-1}\left(a^{\frac{1}{1-a}}-a^{\frac{a}{1-a}}\right); \quad a=\frac{\tau_r}{\tau_v}=\frac{K_v}{K_+[M]}. \tag{3.10}$$

For $a \lessdot 1$, $E_{max} \sim D[Y]_0$, and for $a \gg 1$, $E_{max} \sim D[Y]_0 a^{-1}$.

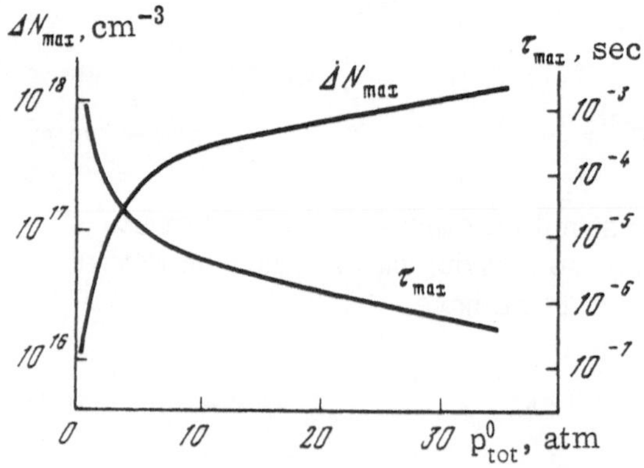

Fig. 20. Dependence of the maximum inversion ΔN_{max} of the 00^01-10^00 levels of CO_2 and of the time to achieve it on the total pressure p^0_{tot} of the mixture. T = 300°K; $\alpha_3 = 0.25$; $[N_{CO}/(N_{CO} + N_{CO_2})]_0 = 0.9$; $(N_{He}/N_{CO_2})_0 \approx 100$; $(N_{CO}/N_O) \approx 9$.

By increasing the total concentration of particles, that is, by reducing a, it is possible in principle to obtain large values of E and, thus, large specific laser powers. It should be noted, however, that when this is done the time the nonequilibrium energy E exists is reduced; thus, the requirements on the rate of merging of the flows are increased. In the case of a bimolecular reaction the parameter a is practically independent of the pressure and E_{max} is determined only by the temperature. Thus, for the reaction $F + H_2 \rightarrow HF^* + H$ with excess H_2, at $T \approx 500$°K, $a \approx 10^5 \cdot P_{10}$, where P_{10} is the probability (per collision) of relaxation of HF^* in a collision with H_2. For $P_{10} \sim 10^{-3}$, $a \approx 10^2$, $E_{max} \lesssim 10^{-2} E_{chem}[F]_0$ (where E_{chem} is the reaction energy released in the form of vibrational energy). On the other hand, for the recombination reaction $CO + O + M \rightarrow CO_2 + M$ with excess CO, at $T \approx 500°$, $a = 2 \cdot 10^{20}/[CO]$. With [CO] $\sim 10^{19}$ cm^{-3} we obtain $E_{max} \sim 5 \cdot 10^{-2}\ \alpha_3 D[O]_0$ for the energy stored in CO and CO_2 (assuming rapid vibrational energy exchange).

The equations analogous to Eq. (3.1) for recombination of CO into CO_2 were solved on a computer but with the assumption of a constant volume of reacting particles and T = const. The dependence of the maximum inverted population in CO_2 on the total pressure is illustrated in Fig. 20. It is clear that the resulting population inversions may be large. The times required to achieve the maximum inversion for $p^0_{tot} \lesssim 5$ atm are not inflexible. A CO_2 recombination chemical laser may be realized in practice by merging flows of cold CO gas with a smaller quantity of hot dissociated oxygen. In doing so it is important to ensure the absence of water vapor. Here the inverted population may reach more than 10^{17} cm^{-3}; however, to specify it further will require a refinement in the value of α_3.

In gas-dynamic lasers the existence of an additional disequilibrium (besides the nonequilibrium temperature) makes it possible to increase the inversion with the density. Along with the creation of a disequilibrium by combining a combustion reaction with gas-dynamics, an initial concentration nonequilibrium is also of interest. In that case the choice of specific optimal designs for the nonequilibrium gas flows requires further study.

3. The Mechanism of Nonresonance Vibrational Exchange in Gas-Dynamic Lasers

In [67, 93, 94] it was shown that nonresonance vibrational exchange plays an important role in a number of chemical and electrical discharge lasers, as well as in the process of nonequilibrium dissociation. It is important to include it in gas-dynamic lasers as well. During expansion of gaseous mixtures, nonresonance exchange processes lead to a noticeable redistribution of the vibrational level populations and to a change in the effective relaxation times. Thus, according to [67] in a mixture of two harmonic oscillators with different vibrational quanta E_A and E_B, when the gas is cooled rapidly the effective energy relaxation time of the oscillator with the smaller quantum E_B may increase greatly due to nonresonance exchange depending on the ratio of the rates of cooling and vibrational–translational (V–T) relaxation.

In this section we consider the prospects for gas-dynamic lasers if nonresonance exchange [95] is adequately accounted for. To analyze the kinetics we shall use the system of balance equations for the vibrational energies of a system of two harmonic oscillators. Assuming that molecules of type A may be formed as a result of chemical reactions and that the temperature T and density N of the gas vary as during expansion into a vacuum from a slit of half-width R_0 (i.e., according to the model of a plane nozzle discussed in Chapter II, Section 1, for n = 1), the system of equations has the form

$$\frac{d\varepsilon_A}{dt}=-\frac{\varepsilon_A-\varepsilon_{0A}}{\tau_{VT}^A}-W_{10}^{01}N_B\left[\varepsilon_A(\varepsilon_B+1)\exp\left(\frac{E_A-E_B}{T}\right)-\varepsilon_B(\varepsilon_A+1)\right]+(\chi-\varepsilon_A)\frac{1}{N_A}\left(\frac{dN_A}{dt}\right)_{\text{chem}},$$
$$\frac{d\varepsilon_B}{dt}=-\frac{\varepsilon_B-\varepsilon_{0B}}{\tau_{VT}^B}+W_{10}^{01}N_A\left[\varepsilon_A(\varepsilon_B+1)\exp\left(\frac{E_A-E_B}{T}\right)-\varepsilon_B(\varepsilon_A+1)\right], \tag{3.11}$$
$$\frac{dT}{dt}=-(\gamma-1)\left(\frac{v_0T}{R_0+v_0t}-\frac{Q}{k}\right).$$

Here $N_i(t) = N_i(0)R_0/(R_0 + v_0t)$, W_{10}^{01} is the rate constant for single-quantum exchange processes between type A and B oscillators, $[\tau_{VT}^{(i)}]^{-1} = (W_{10}^{iB}N_B + W_{10}^{iA}N_A)(1 - x_{0i})$ is the vibrational-translational relaxation time of the oscillator, χ is the fraction of the energy of the chemical reactions that goes into vibrational degrees of freedom, expressed as the number of E_A quanta per A molecule, γ is the adiabatic index, and Q determines the effect of vibrational relaxation and chemical reactions on the variation of the gas temperature.

The rate of exchange of quanta between A and B (V−V' exchange) usually exceeds the rate of any processes controlling the level populations. Then a qausistationary distribution is established in the system and "follows" V−T and other processes that are slower than V−V' processes, while the vibrational energies are related by [67]

$$\varepsilon_A=\frac{\varepsilon_B}{(\varepsilon_B+1)\exp(\Delta E/T)-\varepsilon_B}. \tag{3.12}$$

If $T_B \lessgtr E_B$ and $E_A > E_B$, then by Eqs. (3.11) and (3.12) we obtain an equation for the change in ε_B of the form

$$\frac{d\varepsilon_B}{dt}=-\frac{\varepsilon_B-\varepsilon_{0B}}{1+k_A\exp(-\Delta E/T)}\left[\frac{1}{\tau_{VT}^B}+\frac{k_A\exp(-\Delta E/T)}{\tau_{VT}^A}\right]-$$
$$-\frac{\varepsilon_B}{1+k_A\exp(-\Delta E/T)}\left[\frac{k_A\exp(-\Delta E/T)\,\Delta E}{T^2}\frac{dT}{dt}-\left(\frac{\chi}{\varepsilon_B}-e^{-\frac{\Delta E}{T}}\right)\left(\frac{dk_A}{dt}\right)_{\text{chem}}\right], \tag{3.13}$$

where $k_A = N_A/N_B$.

In the range of applicability of Eq. (3.13) both cooling of the gas when $\Delta E > 0$ and chemical reactions resulting in formation of vibrationally excited A molecules increase the effective relazation time τ. Here the condition $\chi > \varepsilon_B e^{-\Delta E/T} \sim 10^{-1}$ (for $E_B \sim 3000$°K, $\Delta E \sim 500$°K, $T \sim 1500$°K) must be satisfied. If unexcited molecules are also formed ($\chi = 0$) or they are not excited enough, then the energy delivered to each particle decreases and is redistributed among newly formed molecules, and the vibrational temperature T_B decreases.

We shall examine the conditions under which nonresonance exchange plays a significant role in the operation of a gas-dynamic laser. It follows from Eq. (3.11) that the quasistationary distribution of Eq. (3.12) occurs when

$$\frac{1}{\varepsilon_i}\left|\frac{d\varepsilon_i}{dt}\right|+\frac{1}{\tau_{VT}^{(i)}}+\left|\frac{1}{\tau_1}\right|\ll\frac{1}{\tau_{VV'}}, \tag{3.14}$$

where $i = B, A$, $\tau_1^{-1}=\frac{1}{N_A}\left(\frac{dN_A}{dt}\right)_{\text{chem}}$, and $\tau_{VV'}^{-1} = W_{10}^{01}\min(N_A, N_B)$.

Usually, in order to increase the nonresonance pumping of a type B molecule the condition $k_A > 1$ is used; hence $\tau_{VV'}^{-1} = W_{10}^{01}N_B$. For most real mixtures, as a rule, the condition $(\tau_{VV'}/\tau_{VT}) \ll 1$ is satisfied over a wide range of temperatures. We now transform the first term on the left-hand side of Eq. (3.14) as follows:

$$\frac{1}{\varepsilon_i}\left|\frac{d\varepsilon_i}{dt}\right| = \frac{1}{\varepsilon_i}\left|\frac{T}{\tau_2}\frac{d\varepsilon_i}{dT}\right| = \left|\frac{1+\varepsilon_i}{\tau_2}\frac{E_i}{T_i}\frac{T}{T_i}\frac{dT_i}{dT}\right| \sim \left|\frac{1}{\tau_2}\right|.$$

Here $\tau_2 = \left(\frac{1}{T}\frac{dT}{dt}\right)^{-1}$ is the characteristic time for cooling of the gas. Equation (3.14) now takes the form

$$\frac{1}{T}\left|\frac{dT}{dt}\right| + \frac{1}{N_A}\left(\frac{dN_A}{dt}\right)_{\text{chem}} \ll W_{10}^{01}N_B. \tag{3.15}$$

For adiabatic expansion of a gas into a vacuum, neglecting the term Q/k in Eq. (3.11) and in the case of a chemical reaction following the scheme

$$C + M \xrightarrow{K_+} A + F,$$

we find

$$\frac{v_0(\gamma-1)}{R_0} + \frac{K_+}{k_A}\frac{N_{0C}N_{0M}}{N_{0B}} \ll W_{10}^{01}N_{0B},$$

where K_+ is the rate constant of the chemical reaction; and N_{0C} and N_{0M} are the initial densities of the original materials. Thus, in our case Eq. (3.13) is applicable in that temperature and density range over which

$$N_{0B} \gg \max\left\{\frac{v_0(\gamma-1)}{R_0W_{10}^{01}};\ \sqrt{\frac{K_+N_{0C}N_{0M}}{k_AW_{10}^{01}}}\right\}. \tag{3.16}$$

On the other hand, for effective utilization of nonresonance exchange the corresponding inflow of energy to a B molecule must be no smaller than the outflow from it due to V–T relaxation. Then, from Eq. (3.13) we find

$$\frac{1}{\tau_{VT}^B} + \frac{k_Ae^{-\frac{\Delta E}{T}}}{\tau_{VT}^A} \leqslant \frac{k_A\Delta E}{T^2}\left|\frac{dT}{dt}\right| + \left(\frac{dk_A}{dt}\right)_{\text{chem}} \tag{3.17}$$

or

$$\max\{N_AW_{10}^A,\ N_BW_{10}^B\} \leqslant \frac{1}{T}\left|\frac{dT}{dt}\right|\frac{\Delta E}{T} + \frac{1}{k_A}\left(\frac{dk_A}{dt}\right)_{\text{chem}}$$

Let us suppose that $\{N_AW_{10}^A,\ N_BW_{10}^B\} = N_BW_{10}^B$. Then in our model for the variation in the gas-dynamic parameters we find for

$$N_{0B} \leqslant \min\left\{\frac{v_0(\gamma-1)}{R_0W_{10}^B}\frac{\Delta E}{T};\ \sqrt{\frac{K_+N_{0C}N_{0M}}{k_AW_{10}^B}}\right\} \tag{3.18}$$

that the condition (3.17) is always satisfied.

Analogous equations may also be obtained for termolecular reactions.

We now examine the case in which there are no chemical reactions. Then

$$\frac{1}{W_{10}^{01}} \ll \frac{N_{0B}R_0}{v_0(\gamma-1)} \lesssim \frac{\Delta E}{T}\frac{1}{W_{10}^{B}}. \tag{3.19}$$

If $\Delta E \approx 0$ then the usual V−T relaxation of the A + B system takes place with a characteristic time of

$$\tau = \left(1 + k_A e^{-\frac{\Delta E}{T}}\right)\left(\frac{1}{\tau_{VT}^{B}} + \frac{k_A e^{-\frac{\Delta E}{T}}}{\tau_{VT}^{A}}\right)^{-1}.$$

Under the conditions (3.19) and for $\Delta E \sim E_i/2$ relaxation is so inhibited that T_B may increase, with its value determined by the rate of V−T transitions.

When nonresonance exchange is included, the relaxation processes and the efficiency of a gas-dynamic laser depend strongly on the exact mixtures and specific conditions. The characteristic features will be examined below using the example of two mixtures, $CO-N_2$ and CO_2-HCl.

We shall first discuss a CO_2-N_2 mixture and use Eq. (3.13) to describe the variation of the vibrational temperature in CO. It follows from [96] that the rate of V−V' exchange between CO and N_2 is not very high; nevertheless, it is higher than the V−T rates. Thus, up to some stage of the cooling process Eq. (3.13) is applicable [see Eq. (3.16)].

To solve Eq. (3.13) an analytic expression is needed for the probabilities W_{10} in τ_{VT}. It is shown in [97] that an expression of the form

$$W_{10} = ATe^{-BT^{-1/3}} \tag{3.20}$$

gives fair agreement with experiment for diatomic molecules. Here A and B are constants which depend on the energy defect ΔE and the reduced mass μ for a given pair of colliding molecules. Since in our case μ is the same for all three pairs N_2-N_2, N_2-CO, and $CO-CO$, $W_{10}^{N_2-N_2} = W_{10}^{N_2-CO}$ and $W_{10}^{CO-CO} = W_{10}^{CO-N_2}$. The relaxation times τ_{VT} therefore have the form

$$\begin{aligned}(\tau_{VT}^{CO})^{-1} &= T_0 N_{CO}^{0}\left(\frac{T}{T_0}\right)^{\frac{\gamma}{\gamma-1}}(1+k_{N_2})A_1\exp(-B_1T^{-1/3}),\\ (\tau_{VT}^{N_2})^{-1} &= T_0 N_{CO}^{0}\left(\frac{T}{T_0}\right)^{\frac{\gamma}{\gamma-1}}(1+k_{N_2})A_2\exp(-B_2T^{-1/3}).\end{aligned} \tag{3.21}$$

In this cooling model Eq. (3.13) may be integrated assuming that $k_{N_2}/(k_{N_2}+e^{\Delta E/T})$ depends weakly on the temperature and using Eq. (3.21). The result of the integration is shown in Fig. 21, from which it is clear that the stored vibrational energy (thus, also T_{CO}) increases as the gas temperature falls, while in pure CO this effect is absent. This increase is due to nonresonance exchange, which for weak relaxation of a $CO + N_2$ system leads to a substantial breakaway of the vibrational temperature from the gas temperature. In the real case T_{CO} will not increase indefinitely since beyond a certain gas temperature (in our case about 800°K) the condition (3.16) is no longer satisfied and Eq. (3.13) cannot be used. At such temperatures the role of collisions is so small that the vibrational energies of both molecules no longer change and take on their "frozen" values (dashed curves in Fig. 21).

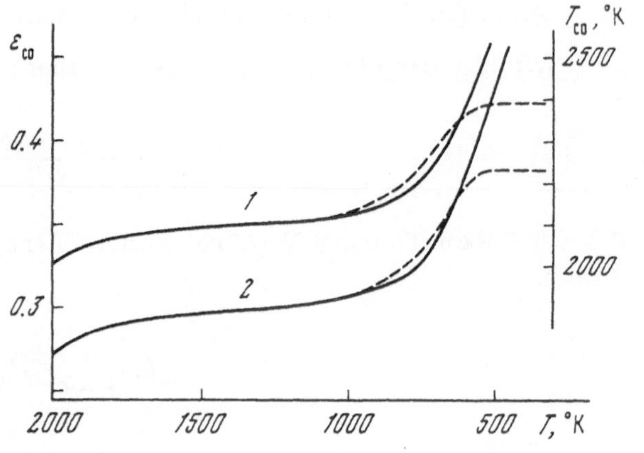

Fig. 21. $T_{CO}(T)$ (curve 1) and $\varepsilon_{CO}(T)$ (curve 2) for the mixture $CO:N_2 = 1:10$, $T_0 = 2000$°K; $R_0N^0_{CO} = 5 \cdot 10^{19}$ cm^{-2}, $A_1 = 4 \cdot 10^{-12}$, $A_2 = 7.8 \cdot 10^{-12}$, $B_1 = 195.2$, $B_2 = 217.8$.

By increasing N^0_{CO} up to values† of the order of 10^{22} cm^{-3}, we substantially expand the domain of applicability of Eq. (3.13) to lower temperatures and may obtain large values of "frozen" T^*_{CO} (cf. Fig. 21). Such large initial densities are possible for a $CO-N_2$ mixture in view of the very low V−T relaxation rate [see Eq. (3.19)]. The results obtained agree with the experimental data of [98, 99]. Thus, in [98] Eq. (3.12) was derived and experimental dependences of the vibrational temperatures in N_2 and CO on the initial gas-dynamic parameters were presented for the $CO-N_2$ mixture. From this it is clear that for mixtures with large k_{N_2}, T^*_{CO} may greatly exceed T_0. Lasing from CO in a $CO-N_2$ mixture was observed in [99]. To do this temperatures $T_0 \sim 2000$°K and very large pressures $p_0 \sim 100$-250 atm were used, in agreement with Eq. (3.19).

We now apply the same approach to CO_2-HCl mixtures. We shall assume that the temperatures of the deformed and symmetric vibrational modes of CO_2 are close to the gas temperature, as is certainly true for high T when the relaxation rates are large. Thus, as before, two oscillators will be examined: the asymmetric vibrations of CO_2 and the vibrations of the HCl molecule. HCl relaxes to the translational degrees of freedom rather rapidly [100] but V−V' exchange takes place more rapidly in this case [101]. Thus, Eq. (3.13) may be used with the time $\tau_{VT}^{CO_2}$ corresponding to the time for release of the energy of the asymmetric mode to all other vibrations of CO_2. We obtain analytic expressions for $W_{10}^{HCl-HCl}$ and $W_{10}^{CO_2-CO_2}$ on the basis of experimental data [82, 100] in the form of Eq. (3.20) and assume that $W_{10}^{CO_2-CO_2} \approx W_{10}^{CO_2-HCl}$ and $W_{10}^{HCl-CO_2} \approx 4W_{10}^{HCl-HCl}$ [101]. Over the temperature range 700-2000°K we have [100]

$$W_{10}^{HCl-HCl} \approx 7.2 \cdot 10^{-14} T \exp(-58.3 T^{-1/3}), \text{ cm}^3/\text{sec},$$

and for CO_2 [82],

$$W_{10}^{CO_2-CO_2} \approx 1.2 \cdot 10^{-14} T \exp(-38.3 T^{-1/3}), \text{ cm}^3/\text{sec}.$$

Substituting these equations in Eq. (3.13) and integrating we find ε_{CO_2} as a function of the gas temperature (Fig. 22, curves 1 and 1'). A slow rise in T_{CO_2} can be seen (which is not seen in the pure gas). However, in view of the relatively high V−T relaxation rate the growth in the vibrational temperature is very small and when the parameter $R_0N^0_{CO_2}$ decreases to $5 \cdot 10^{17}$ cm^{-2} there is no increase and T_{CO_2} decreases to $T^*_{CO_2}$, which is determined by the specific value of $R_0N^0_{CO_2}$.

† At such densities, relaxation due to three-body collisions may set in; however, at this time data on the cross sections for these processes are lacking.

Nonresonance exchange is more efficient when vibrationally excited HCl molecules are formed in a chemical reaction, among the more important steps of which are

$$H_2 + Cl \rightleftarrows HCl + H \tag{3.22}$$

and

$$H + Cl_2 \rightleftarrows HCl^* + Cl. \tag{3.23}$$

We shall assume that the conditions for efficient production of atomic chlorine exist (for example, by ultraviolet radiation which breaks up Cl_2). Then reaction (3.22) takes place rapidly with formation of H needed in reaction (3.23), so that the concentrations of H and Cl_2 may be regarded as weakly dependent on the temperature. In this case the equilibrium for reaction (3.23) will be shifted to the right and the equation for the change in the density of HCl^* will take the simple form

$$\left(\frac{dN_{HCl}}{dt}\right)_{chem} = N_H^0 N_{Cl_2}^0 K_+ \left(\frac{R_0}{R_0 + v_0 t}\right)^2. \tag{3.24}$$

For temperatures of 700-2000°K the rate constant for reaction (3.23) is well approximated by a constant [102], $K_+ = 2.8 \cdot 10^{-10}$ cm^3/sec. With these assumptions we obtain

$$N_{HCl}(t) = \left[N_{HCl}^0 + \frac{N_H^0 N_{Cl_2}^0 K_+ R_0}{v_0}\left(1 - \frac{R_0}{R_0 + v_0 t}\right)\right]\frac{R_0}{R_0 + v_0 t} \tag{3.25}$$

and

$$k_{HCl}(t) = k_{HCl}^0 + \frac{GR_0}{v_0}\left(1 - \frac{R_0}{R_0 + v_0 t}\right), \quad \left(\frac{dk_{HCl}}{dt}\right)_{chem} = G\left(\frac{R_0}{R_0 + v_0 t}\right)^2, \tag{3.26}$$

where $G = (N_H^0 N_{Cl_2}^0 K_+ / N_{CO_2}^0)$. Substituting Eqs. (3.25) and (3.26) in Eq. (3.13) and integrating over the temperature, we find the result shown in Fig. 22 (curves 2 and 2'). Here, as in the case of a $CO-N_2$ mixture, the growth in the vibrational temperature due to nonresonance exchange is limited by the increase in $\tau_{VV'}$ (see the dashed curves in Fig. 22).

An increase in the initial density $N_{CO_2}^0$ leads to a drop in $T_{CO_2}^*$. However, in this case this density may be 5-10 times higher than in the absence of a chemical reaction. Thus, for $R_0 N_{CO_2}^0 \sim 5 \cdot 10^{18}$ cm^{-2} there is no growth in the vibrational temperature, but the value of $T_{CO_2}^* \sim T_0$ is sufficiently high to obtain an inverted population among various vibrational modes. For

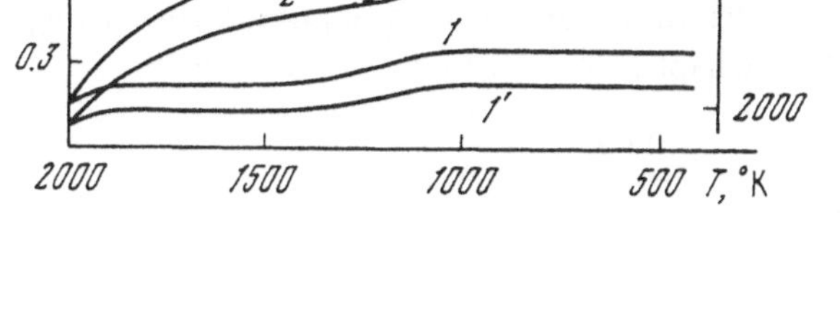

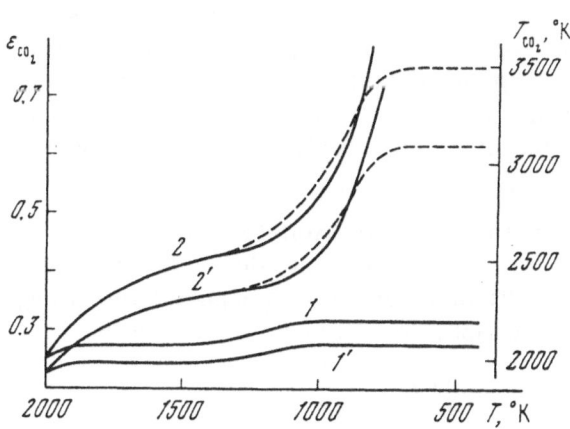

Fig. 22. $T_{CO_2}(T)$ and $\varepsilon_{CO_2}(T)$ for the mixture CO_2:HCl = 1:10 (curves 1 and 1') and for a CO_2-HCl mixture in which a chemical reaction takes place (curves 2 and 2'). T_0 = 2000°K; $k_{HCl}^0 = 10$; $R_0 N_{CO_2}^0 = 10^{16}$ cm^{-2}; R_0 = 0.1 cm; χ = 0.2; $G = 2 \cdot 10^7$.

$R_0N^0_{CO_2} \gtrsim 10^{19}$ cm^{-2} there is still no inversion and to produce one it is necessary to increase the intensity of the chemical reaction (for example, by increasing the parameter G). It should be noted that in this case, when the reaction involves atomic and molecular hydrogen (which can speed up the vibrational relaxation) the value of $T^*_{CO_2}$ will be somewhat lower than shown in Fig. 22. However, the quasiquilibrium values of the densities of these components are small because they take part in the chain reaction, and the corresponding corrections are small.

Thus, we have established the limits of applicability of Eq. (3.13) and pointed out the range of variation of parameters over which nonresonance exchange operates effectively (in a number of cases it includes very high initial pressures). Use of chemical reactions in expanding gas flows makes it possible to further increase the density. With the equations obtained here one can choose a mixture of gases for which the principal mechanism determining the vibrational temperature is nonresonance exchange (of course, the probabilities W_{10} and W^{01}_{10} must be known).

CHAPTER IV

ELECTRICAL AND GAS-DYNAMIC PROCESSES. ELECTRICAL GAS-DYNAMIC LASERS

1. Electrical Methods of Producing an Inversion, Their Limitations, and the Role of Gas Dynamics

At present a combination of electrical and gas-dynamic methods of producing an inversion seems very promising. To analyze this approach to the development of lasers we shall briefly discuss the physical kinetics of the processes in electrical discharge lasers. The mechanism for inversion formation in cw lasers has been fairly well studied [40-43]. It has been shown that the principal factor limiting the growth of an inversion is increased heat release as the discharge current (i.e., the electron density N_e) rises.

The transition to a pulsed discharge makes it possible [8], as will be shown below, to avoid many of the limitations of operation in the steady state since the thermal factor almost ceases to affect the inversion. We shall examine the basic features of inversion formation by pulsed electronic excitation of vibrational levels. Besides having intrinsic interest, the results may serve as the basis of an analysis of electrical gas-dynamic lasers (Section 2, Chapter IV).

A study of relaxation in a number of multiatomic molecules [43] showed that CO_2 has advantages over other molecules for producing an inversion. Thus, in the following we shall dwell on the CO_2 laser.

Let some CO_2 gas (or a mixture of it with N_2 and/or He) be placed in a tube of radius R. In this case vibrational relaxation must be considered together with changes in the state of the free electrons. We shall assume that as the electrical pulse, of duration τ, acts on the gas, the electron density N_e and temperature T_e remain constant during the entire pulse. After the pulse ends, N_e and T_e begin to fall, with their time variation determined by volume recombination, ambipolar diffusion (for N_e), and the cooling law (for T_e) [8].

The equations describing vibrational relaxation with electronic excitation are similar to Eqs. (2.1), but must also include electron impact pumping of energy into different vibrational modes as well as diffusive decay of vibrationally excited molecules. It was assumed that along with the relaxation processes described in Section 1 of Chapter II, the system includes

$$N_2(v=0)+e \rightleftarrows N_2(v=1-8)+e, \quad W_{e,4},$$
$$CO_2(00^\circ 0)+e \rightleftarrows CO_2(00^\circ 1)+e, \quad W_{e,3},$$
$$CO_2(00^\circ 0)+e \rightleftarrows CO_2(10^\circ 0)+e, \quad W_{e,1},$$

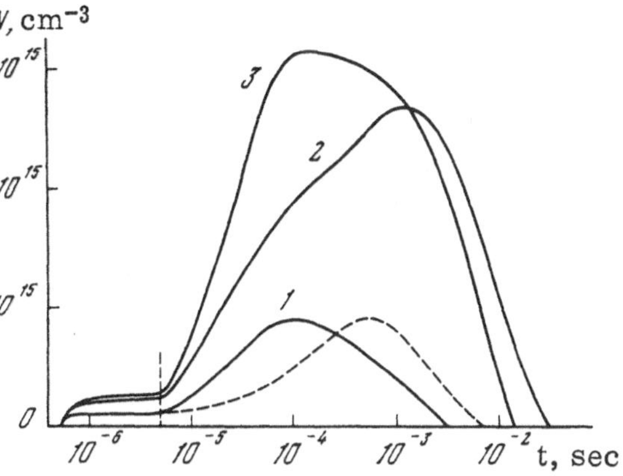

Fig. 23. Time variation of the inverted population for pulsed electrical excitation of vibrational levels in CO_2:He = 1:7 (curve 1), $CO_2:N_2$ = 1:3 (curve 2), and $CO_2:N_2$:He = 1:3:7 (curve 3) mixtures. $N_e = 2 \cdot 10^{12}$ cm^{-3}; $\tau = 5 \cdot 10^{-6}$ sec; R = 1.7 cm; T_0 = 300°K. The dashed curve is drawn using experimental data [47]. The vertical line denotes the time of the pulse ends.

where $W_{e,i}$ is the probability (per electron) of excitation of the i-th vibrational mode by electron impact.

As already noted, the thermal regime and gas temperature play an important role in molecular lasers. Thus, along with the relaxation equations and the equations for T_e and N_e we must consider the nonstationary heat conduction equation [8] which determines the time variation of the gas temperature during pulsed electrical excitation. The joint solution of these equations with specific probabilities for the elementary processes involved yields the time dependences of the vibrational temperatures (T_i), the vibrational level populations, the electron and gas temperatures, and the electron density. Schulz's [103] measurements of the cross sections for electron impact excitation of the vibrational levels of N_2 and CO_2 molecules averaged over a Maxwellian distribution were used for $W_{e,i}$. We also note that in computing $W_{e,i}$ we used the cross section found by Schulz only for the 10^00 level. However, a comparison of the results of the relaxation calculations with experiment [47] made it possible to determine the effective probability $\omega_{12} W_{e,12}$ and to take into account the role of the remaining levels $v_1 0^0 0$ and $0 v_2^l 0$ in the total excitation cross section [104] (here ω_{12} is the average number of quanta excited per electron impact).

The results of a solution of this system of equations with initial conditions $T_e^0 = 3.3 \cdot 10^4$°K and $T_0 = T_w$ = 300°K (where T_0 is the gas temperature before the current pulse acts on it and T_w is the temperature of the discharge tube walls) for a tube of radius R = 1.7 cm are shown in Figs. 23-28.

Figure 23 shows the dependence of the population inversion of the 00^01-10^00 levels of the CO_2 molecule when a short current pulse acts on the gas mixture.† The parameters are the same as in the experiment of [47]. Since the pulse duration τ taken in the calculation is less than the relaxation time of the 10^00 level, the inversion produced during the pulse increases after the pulse ends, and only after a certain time does it fall due to relaxation of the asymmetric vibrational mode. For the mixture 1 torr CO_2 + 7 torr N_2 there appears to be good quantitative agreement between theory and experiment [47] as regards the relative increase in the inversion. The disagreement in the time of the inversion maxima is not fundamental and is explained by the choice of the probability $W_{2,0}^{He}$. The growth in the inversion during the interval $5 \cdot 10^{-6}$ to $3 \cdot 10^{-4}$ sec for a 1 torr CO_2 + 3 torr N_2 mixture (curve 2) is due not to relaxation of the 10^00 level, but to transfer of energy from N_2 to the asymmetric vibrational modes

† For equations such as Eqs. (1.14b) and (2.1) to be applicable it is necessary that the current pulse duration τ and the characteristic electron pumping times $(\omega_i W_e, N_e)^{-1}$ be greater than the times for establishing a Boltzmann distribution in each vibrational mode and establishing equilibrium between the symmetric and deformed vibrational modes. For $\tau \geqslant 5 \cdot 10^{-6}$ sec, N $\leqslant 2 \cdot 10^{13}$ cm^{-3}, and pressures p > 1 torr, these conditions are satisfied.

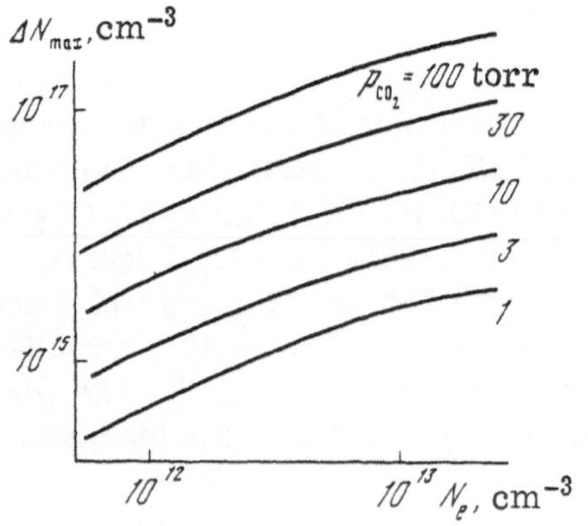

Fig. 24. Dependence of the maximum inversion on the electron density N_e for a constant current pulse duration of $\tau = 5 \cdot 10^{-6}$ sec and various partial pressures p_{CO_2}. CO_2:He = 1:3; R = 1.7 cm; T_0 = 300°K.

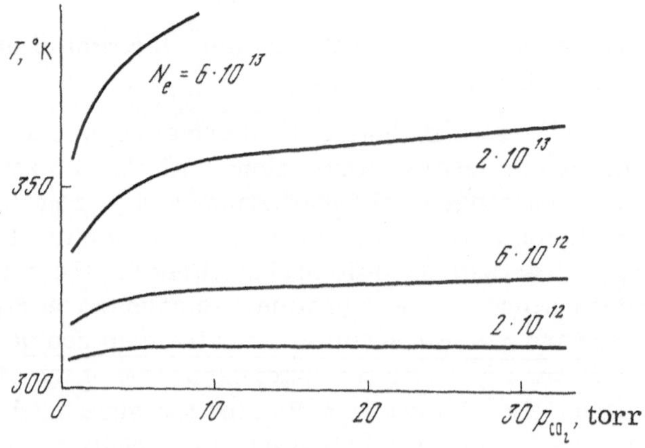

Fig. 25. Maximum heating of the gas as a function of the partial pressure p_{CO_2} for electrical pulses with different N_e. $\tau = 5 \cdot 10^{-6}$ sec; T_0 = 300°K, R = 1.7 cm, mixture CO_2:He = 1:3.

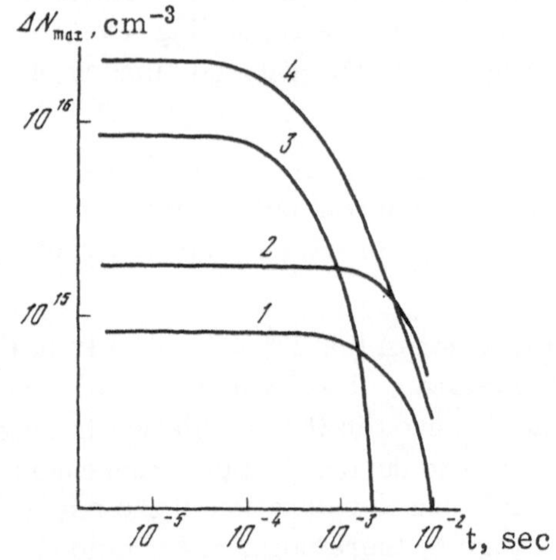

Fig. 26. Dependence of the maximum inversion on the current pulse duration τ for $N_e\tau$ = const and various pressures p_{CO_2}. $N_e\tau = 10^7$ sec · cm^{-3} (1, 3) and $3 \cdot 10^7$ sec · cm^{-3} (2, 4); p_{CO_2} = 1 torr (1, 2) and 10 torr (3, 4). Mixture CO_2:He = 1:3, R = 1.7 cm, T_0 = 300°K.

of CO_2, and only later, in the interval $3 \cdot 10^{-4}$ to 10^{-3} sec, is the further growth in the inversion due to relaxation of this level. We note that the computed increase in the inversion in the afterglow for mixtures containing nitrogen does not agree with the experimental value. This is because the ratio of the probabilities $\omega_3 W_{e,3}/\omega_4 W_{e,4}$ used in the calculation differs somewhat from experiment.

Figures 24 and 25 give the maximum inversion and maximum gas temperatures as functions of the electron density N_e (current density) and partial pressure of CO_2 for a CO_2 : He mixture. In the interval $N_e = 6 \cdot 10^{11}$ to $2 \cdot 10^{13}$ cm^{-3} and P_{CO_2} = 1-100 torr the inversion increases roughly linearly with N_e and P_{CO_2}. A further increase in the inversion with N_e would be limited by dissociation of CO_2. Estimates showed that 10% of the CO_2 molecules may dissociate for $N_e \sim 4$-$20 \cdot 10^{12}$ cm^{-3} due to a single pulse of duration $\tau = 5 \cdot 10^{-6}$ sec.

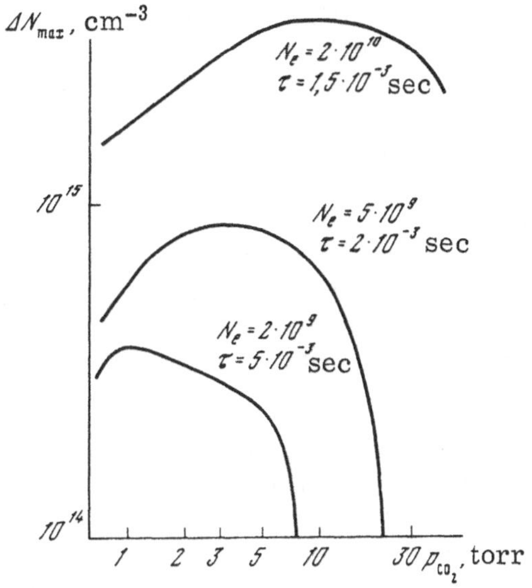

Fig. 27. Dependence of the maximum inversion of the partial pressure p_{CO_2} for various $N_e \tau$ and different current pulse durations τ. Mixture CO_2:He = 1:3; R = 1.7 cm; T_0 = 300°K.

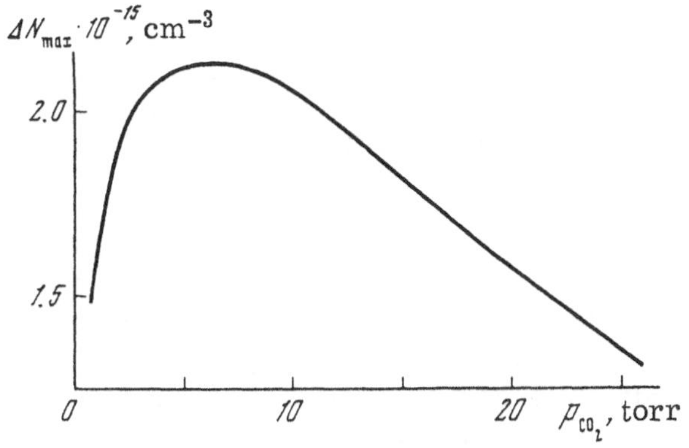

Fig. 28. Dependence of the maximum inversion on the partial pressure p_{CO_2}. $\tau = 10^{-2}$ sec; $N_e = 6 \cdot 10^9$ cm^{-3}, R = 2.5 cm; T_0 = 300°K.

One important feature of operation with current pulses of duration less than the relaxation time of the upper laser level 00^01 should be noted. In this case the maximum population of the 00^01 level and, therefore, the maximum inversion are independent of the relaxation rate and are determined by the total vibrational energy $\mathcal{E}_3$ obtained from the electrons. For just this reason when $\tau = 5 \cdot 10^{-6}$ sec the inversion increases linearly with pressure up to $p_{CO_2} \sim$ 100 torr. In addition, temperature effects play a considerably smaller role here and any reduction in the inversion with increasing temperature may occur only due to an increase in the equilibrium population of the lower laser level 10^00.

For pulsed electronic excitation the product $N_e \tau$ is the characteristic parameter which determines the stored molecular vibrational energy, the magnitude of the maximum inversion, the gas temperature, and the degree of dissociation. Figure 26 shows the dependence of ΔN_{max} on the pulse duration τ for constant $N_e \tau$ and various pressures p_{CO_2}. It is clear that for sufficiently small τ the inversion is independent of τ and is determined by $N_e \tau$. However, if $\tau > \tau_1$, where τ_1 is the relaxation time of the 00^01 level, then the maximum inversion begins to fall with increasing current pulse duration. This is explained by the fact that for $\tau > \tau_1$ the upper laser level begins to relax even while the current is acting, thus limiting the inversion. The increased relaxation rate of this level with rising pressure leads to the existence of an optimum CO_2 pressure (as opposed to small $\tau \sim 10^{-5}$ sec, when the inversion increases linearly with the pressure; see Fig. 24). The pressure dependence of the inversion for large pulse durations $\tau > \tau_1$ is illustrated in Figs. 27 and 28. The shift in $p_{CO_2}^{opt}$ toward larger values with decreasing τ (for constant $N_e \tau$) and with increasing $N_e \tau$ is due to the small effect of relaxation of the 00^01 level on its population.

When studying the action of a series of pulses one after another at equal time intervals, rather than a single pulse, one must include the possible dissociation of CO_2 and heating of the gas. The effect of these factors on the inverted populations and output power will be greater the greater the pulse repetition rate. We have computed the variation in the maximum inversion ΔN_{max} over a sequence of pulses due to heating of the gas. It turns out that a thermal regime and constant value of ΔN_{max} for a repetition rate of 50 Hz are established after several (two to four) pulses. The reduction in the inversion due to heating is especially important for current pulses of duration $\tau > \tau_1$. Due to an increase in the relaxation rate of the 00^01 level the optimum CO_2 pressures in this case are less than for a single pulse. In addition, heating of the gas leads to a limitation on the inversion as the current rises and to the appearance of optimum values of N_e. Typical dependences of the inverted populations on N_e and time in the repeated pulse regime are given in [8].

As can be seen from the above analysis, in the case of short initiating pulses ($\tau < \tau_1$) with high current densities ($N_e \sim 10^{12}$-10^{13} cm^{-3}) at large partial pressures ($p_{CO_2} \sim 10^2$ torr) it is possible to obtain an inversion of the 00^01-10^00 transition by $\gtrsim 10^{17}$ cm^{-3}, i.e., two orders of magnitude greater than for a dc discharge or a pulsed discharge at low pressures. A similar gain in the power of a pulsed laser in both the conventional mode and in the Q-switched mode might be expected as well.

The basic experimental data of [44-50] agree qualitatively with the theoretical analysis above. In particular, the conclusions of this section are in agreement with the results of [50] where a high output power was obtained and no limitations on its growth were found up to pressures $\gtrsim 60$ torr for current pulse lengths $\tau = 5$-50 μsec.

Thus, the output power of a CO_2 laser with pulsed electronic excitation may in principle be increased by two orders of magnitude over a conventional dc discharge. However, to maintain the required $N_e \sim 10^{12}$-10^{13} cm^{-3} as the pressure increased, older laser designs needed very high voltages on the discharge tube (for a length of about 1 m, this voltage was 0.5-1 MV). Later this difficulty was avoided by replacing a longitudinal discharge by a transverse one. (The discharge gap is then reduced by about 10^2 times and the required voltage is greatly reduced.) This made it possible to build the so-called TEA lasers with working pressures $\gtrsim 1$ atm [51, 52]. A further reduction in the applied voltage is ensured by auxiliary initiation of the discharge (by ultraviolet radiation [105], electron beam [106], etc.), and the pressures may then be higher. In fact, a steady-state regime in sealed-off systems at these pressures is impossible; indeed, the repetition rate of a pulsed laser is limited by the temperature rise. This situation changes when electronic excitation and gas-dynamic flows are combined in specific devices.

2. Electrical Gas-Dynamic Lasers

In this section we shall analyze the prospects for using gas dynamics together with electrical excitation. The goal is to rapidly replace the gas in the discharge zone (with its higher temperature) and to increase the repetition rate (reaching a continuous regime in the limit) thereby increasing the average laser power. In the literature on this topic (see, for example, [51-56]) gas is pumped both across [53] and along [54-56] the discharge at speeds of no more than about 50-100 m/sec as a rule. Such speeds do not permit cw operation at high pressures (~1 atm). This can be done only at supersonic pumpthrough speeds, that is, in electrical gas-dynamic lasers.

In the following we shall examine two tyes of transverse discharge lasers, with subsonic and with supersonic pumpthrough speeds [57]. We shall begin with low speeds ($\lesssim 100$ m/sec). To analyze the population kinetics of the vibrational levels in this case it is necessary to use the system of relaxation equations with the condition N_i = const. For simplicity we shall assume that the temperature of the gas changes only due to transfer of energy from vibrational degrees of freedom to translational (i.e., neglect thermal conductivity, etc.).

An analysis [8] showed that the front of the current pulse is very steep for a step-function voltage pulse and so, to good accuracy, we may assume T_e = const and N_e = const for $0 \le t \le \tau$ and $T_e = 0$ and $N_e = 0$ for $t > \tau$.

With these assumptions the system of equations describing relaxation in CO_2-N_2-He mixtures takes the form [the difference from Eqs. (2.1) and [8] lies in the use of the exact equations (1.14b) and (1.17) and current data on the probabilities]

$$\left.\begin{aligned}
&\frac{d\varepsilon_3}{dt}=N_{CO_2}\left[W_{3,4}k_{N_2}\frac{x_4-x_3}{(1-x_4)(1-x_3)}-\frac{x_3-x_2^3e^{-\frac{500}{T}}}{(1-x_3)(1-x_2)^3}\sum_M W_{3,\Sigma}^M k_M\right]+\\
&\quad+W_{e,3}\omega_3N_eS\left(1-x_3e^{\frac{3380}{T_e}}\right),\\
&\frac{d(\varepsilon_2+2\varepsilon_1)}{dt}=N_{CO_2}\left[3\frac{x_3-x_2^3e^{-\frac{500}{T}}}{(1-x_3)(1-x_2)^3}\sum_M W_{3,\Sigma}^M k_M+3W_{4,\Sigma}k_{N_2}\times\right.\\
&\quad\left.\times\frac{x_4-x_2^3e^{-\frac{500}{T}}}{(1-x_4)(1-x_2)^3}-2\frac{x_2-x_{02}}{1-x_2}\sum_M W_{2,0}^M k_M\right]+2W_{e,12}\omega_{12}N_eS\left(1-x_2^2e^{\frac{1920}{T_e}}\right),\\
&\frac{d\varepsilon_4}{dt}=N_{CO_2}\left[-W_{3,4}\frac{x_4-x_3}{(1-x_4)(1-x_3)}-W_{4,\Sigma}\frac{x_4-x_2^3e^{-\frac{500}{T}}}{(1-x_4)(1-x_2)^3}\right]+\\
&\quad+W_{e,4}\omega_4N_e(1-x_4)\left(1-x_4e^{\frac{3380}{T_e}}\right),\\
&\frac{dT}{dt}=(\gamma-1)\frac{Q}{k},
\end{aligned}\right\}\quad(4.1)$$

where Q is given by [cf. Eq. (2.4)]

$$\frac{Q}{k}=N_{CO_2}\left[500\frac{x_3-x_2^3e^{-\frac{500}{T}}}{(1-x_3)(1-x_2)^3}\sum_M W_{3,\Sigma}^M k_M+500W_{4,\Sigma}k_{N_2}\frac{x_4-x_2^3e^{-\frac{500}{T}}}{(1-x_4)(1-x_2)^3}+\right.$$
$$\left.+1920\frac{x_2-x_{02}}{1-x_2}\sum_M W_{2,0}^M k_M\right]\frac{1}{1+k_{N_2}+k_{He}}. \quad (4.2)$$

The probabilities of various processes were taken from experimental data in [79, 82, 103]. Typical discharge parameters for transversely excited pulsed lasers were chosen: $N_e = 2\cdot10^{12}$ cm^{-3}, $T_e = 3$ eV, $T_0 = 300$°K, $p_{tot} = 250$ torr in a $CO_2 : N_2 = 1:1$ mixture.

The vibrational temperatures and inverted populations are plotted as functions of time in Fig. 29 for two current pulse lengths. The value of ΔN is determined by the difference between T_3 and T_2 since the total number of CO_2 molecules remains constant. It is clear that as the pulse duration is increased by an order of magnitude the maximum inversion increases by roughly four times; however, the lifetime of the inversion was shortened by a factor of 3. For a longer current pulse the energy transferred from the discharge to vibrational degrees of freedom becomes substantial, the gas temperature increases, relaxation speeds up, and, at the same time, the lifetime of the inversion decreases. For the same reason a further increase in τ is inappropriate, as illustrated by Fig. 30 which shows the same functions assuming $\tau \to \infty$ (dc discharge). Increasing the pulse length to about 10^{-5} sec yields little gain in the maximum inversion and shortens the lifetime of the inversion still more. The growth in the gas temperature and in the divergence between T_3 and T_2 cause T_3 and T_4 to decrease even while the discharge is acting and, later (after about $5\cdot10^{-5}$ sec) they already "follow" the

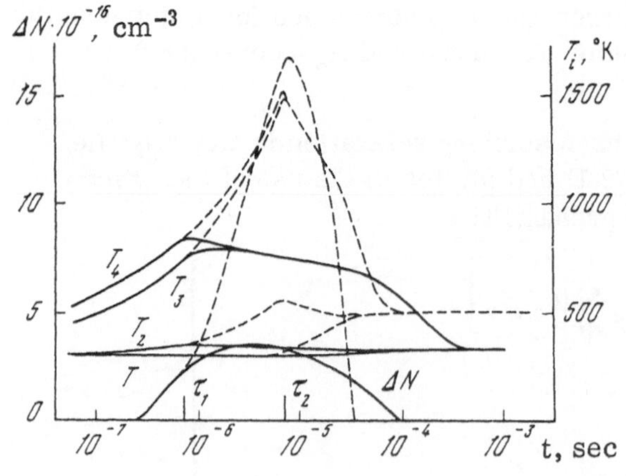

Fig. 29. The inverted population ΔN and vibrational and translational temperatures as functions of time for transverse discharge excitation of a CO_2:N_2 = 1:1 mixture. T_0 = 300°K; p_{tot} = 250 torr; $N_e = 2 \cdot 10^{12}$ cm^{-3}; T_e= 3 eV. The continuous curves refer to pulse durations of $\tau_1 = 7 \cdot 10^{-7}$ sec; the dashed curves, to $\tau_2 = 7 \cdot 10^{-6}$ sec.

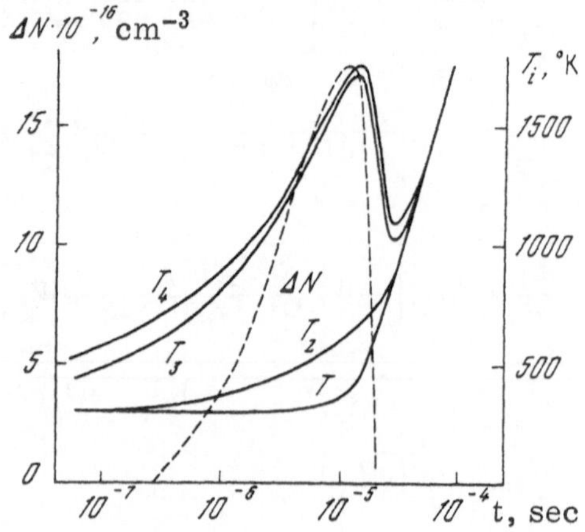

Fig. 30. The inverted population ΔN and the vibrational and gas temperatures as functions of time for the conditions of Fig. 29 with a dc discharge ($\tau = \infty$).

gas temperature; that is, all the energy (for constant N_e) leaving the discharge goes into heating the gas.

The inversion lifetime, which determines the possibility of producing a continuous pumpthrough laser with transverse excitation, depends on the total pressure and the gas temperature. A cw regime is possible when the gas in the discharge gap can be completely replaced over the lifetime of the inversion. The length of the discharge gap along the pumpthrough flow in real cases is of order 1-5 cm. At flow speeds of about 10^2 m/sec this corresponds to a time of about 10^{-4} to $5 \cdot 10^{-4}$ sec for complete exchange of the gas. Such inversion lifetimes are possible for small N_e or total pressures (compared to the conditions in Fig. 30). In fact, in this case the maximum value of ΔN will be less than in the pulsed regime. Thus, for example, with $N_e = 3 \cdot 10^{10}$ cm^{-3} and p_{tot} = 50 torr in a $CO_2 : N_2$ = 1 : 1 mixture a cw regime is possible since the lifetime of the inversion is approximately $4 \cdot 10^{-4}$ sec and for a discharge gap of 2 cm the gas must be pumped through at a speed of $\gtrsim$50 m/sec. Accordingly, an increase in the discharge gap requires a higher pumpthrough speed. The maximum inversion in such a regime is about $2 \cdot 10^{16}$ cm^{-3}.

We now discuss electrical gas-dynamic lasers. For simplicity in the calculations we shall assume, as in Section 1 of Chapter II, that a supersonic gas-dynamic flow is formed as a result of adiabatic escape of gas from a long thin slit into a vacuum (plane nozzle model). In this model the density N(t) of a gas heated initially to temperature T_0 and expanding through a slit of half-width R_0 varies as

$$N(t) = N(0) \frac{R_0}{R_0 + v_0 t},$$

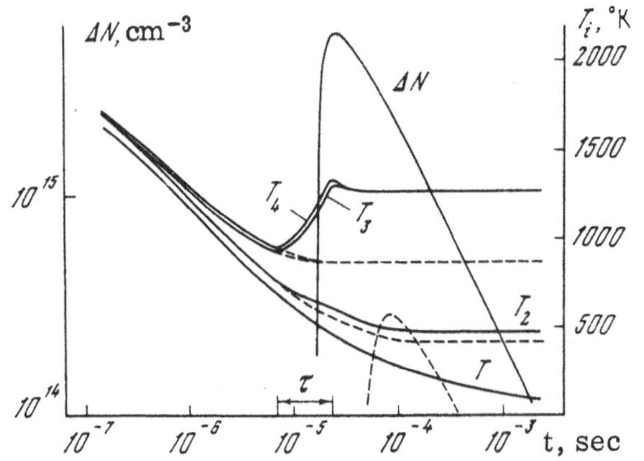

Fig. 31. Dependences of the inverted population ΔN and the vibrational and gas temperatures on the time for a $CO_2:N_2 = 1:5$ mixture expanding through a slit. $T_0 = 1900°K$; $p_0(CO_2) = 1$ atm; $R_0 = 0.1$ cm; $N_e = 3 \cdot 10^{11}$ cm^{-3}, $T_e = 2$ eV; $l_1 = 3$ cm; $l_2 = 10$ cm. The pressure is reduced to that at 300°K. The dashed curves represent the corresponding results without a discharge.

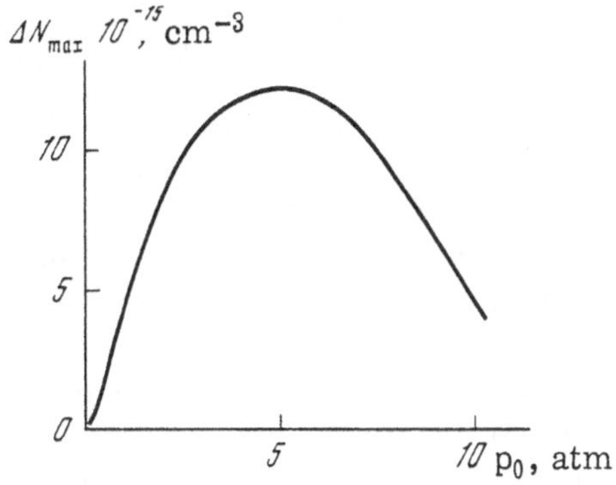

Fig. 32. The maximum inversion as a function of the initial CO_2 pressure. Pressure reduced to 300°K value; $T_0 = 1900°K$, $R_0 = 0.1$ cm; $N_e = 3 \cdot 10^{11}$ cm^{-3}, $T_e = 2$ eV, $l_1 = 3$ cm; $l_2 = 10$ cm; $CO_2:N_2 = 1:5$ mixture.

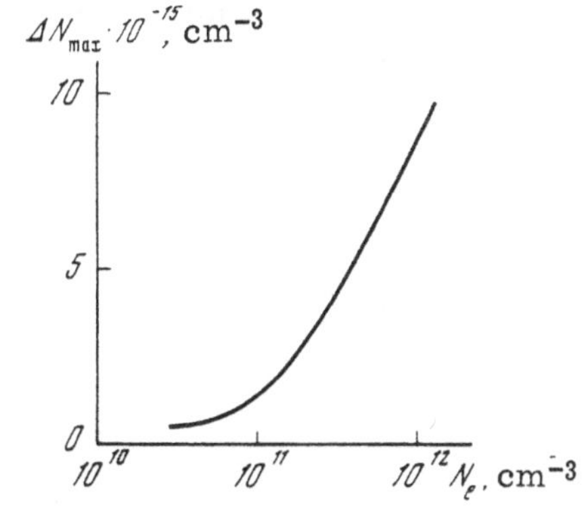

Fig. 33. The maximum inversion as a function of the electron density for a $CO_2:N_2 = 1:5$ mixture. $p_0(CO_2) = 1$ atm; $T_0 = 1900°K$; $R_0 = 0.1$ cm; $T_e = 2$ eV; $l_1 = 3$ cm; $l_2 = 10$ cm.

and the gas temperature is given by

$$\frac{dT}{dt} = (\gamma - 1)\left[\frac{Q}{k} - \frac{v_0 T}{R_0}\left(\frac{R_0}{R_0 + v_0 t}\right)\right]. \tag{4.3}$$

According to this scheme an expanding CO_2-N_2-He gas mixture with initial pressure p_0 and temperature T_0 reaches a discharge gap at some distance l_1 from the slit along the axis of the expanding flow. In a coordinate system attached to the moving gas this corresponds to an electrical pulse of duration $\tau = (l_2 - l_1)/v_0$ acting on the gas, where $l_2 - l_1$ is the extent of the discharge gap along the axis of the flow. With good accuracy it can be assumed that the electron density N_e and temperature T_e are constant over the entire length of the discharge gap. Outside it $N_e = 0$ and $T_e = 0$.

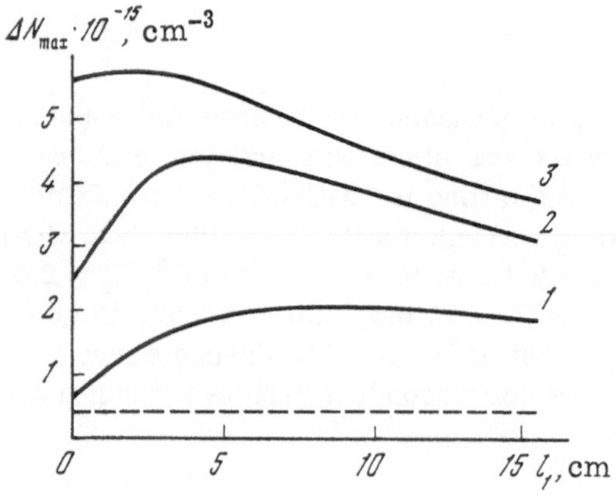

Fig. 34. Dependence of the maximum inversion on the location of the discharge gap for various τ in a CO_2:N_2 = 1:5 mixture. p_0 (CO_2) = 1 atm; T_0 = 1900°K; R_0 = 0.1 cm; $N_e = 3 \cdot 10^{11}$ cm^{-3}; T_e = 2 eV. The dashed curve corresponds to the absence of a discharge. 1) $\tau = 7.1 \cdot 10^{-6}$ sec ($l_2 - l_1$ = 3 cm); 2) $\tau = 1.65 \cdot 10^{-5}$ sec ($l_2 - l_1$ = 7 cm); 3) $\tau = 3.53 \cdot 10^{-5}$ sec ($l_2 - l_1$ = 15 cm).

Equations (4.1) together with Eq. (4.3) and including Eq. (4.2) were solved on a computer. The results of the calculations are shown in Figs. 31-37.

Figure 31 shows typical time dependences of the inverted population and vibrational temperatures. The same functions without a discharge are shown there for comparison. It is clear that electronic excitation makes it possible to raise the maximum inversion by more than an order of magnitude. This is because, first, a divergence between T_3 and T_2 sufficient to produce an inversion may be attained sooner than with the same arrangement without a discharge (here the total number of particles is greater) and, second, this divergence is greater in absolute magnitude. In Fig. 31 it is clear that from the moment the gas enters the discharge gap the vibrational temperatures T_3 and T_4 begin to rise noticeably while T_2 continues to fall despite the fact that the total probability of excitation of the lower laser states ($v_1v_2^l0$) is somewhat greater than the probability of electronic pumping into the asymmetric vibrational mode ($\omega_3 W_{e,3}$) and into $N_2(\omega_4 W_{e,4})$. This is explained by the high relaxation rate of the defomed vibrational mode. The inversion increases rapidly with the growth in the difference $x_3 - x_2^2$, passes through a maximum, and then falls as does the density since at that moment T_3 and T_2

Fig. 35. Dependence of the maximum inversion on the half-width of the slit R_0. Mixture CO_2:N_2 = 1:5; p_0(CO_2) = 1 atm; T_0 = 1900°K; T_e = 2 eV; $N_e = 3 \cdot 10^{11}$ cm^{-3}; l_1 = 3 cm; l_2 = 10 cm.

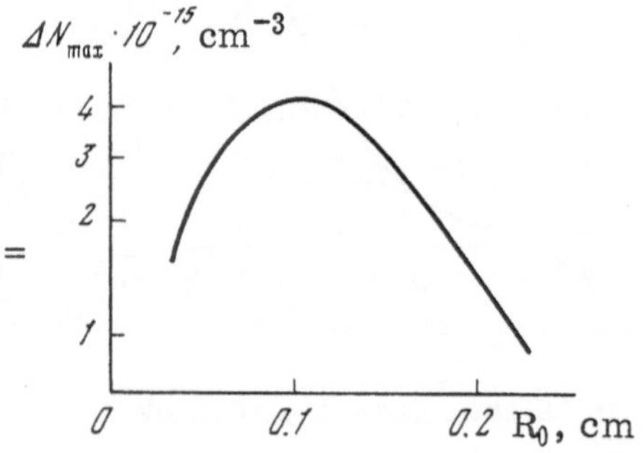

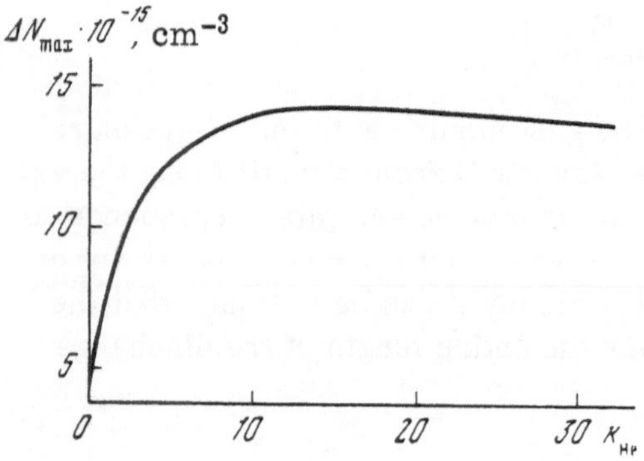

Fig. 36. Dependence of ΔN_{max} on the amount of He. k_{N_2} = 5; p_0(CO_2) = 1 atm; T_0 = 1900°K; T_e = 2 eV; $N_e = 3 \cdot 10^{11}$ cm^{-3}; l_1 = 3 cm; l_2 = 10 cm; R_0 = 0.1 cm.

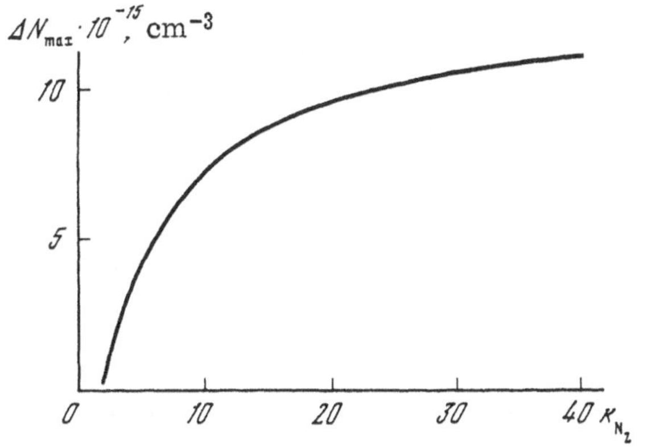

Fig. 37. Dependence of ΔN_{max} on the amount of N_2 for a CO_2-N_2 mixture. $T_0 = 1900°K$; $T_e = 2$ eV; $p_0(CO_2) = 1$ atm; $N_e = 3 \cdot 10^{11}$ cm^{-3}; $l_1 = 3$ cm; $l_2 = 10$ cm; $R_0 = 0.1$ cm.

have already reached their "frozen" values. For adiabatic expansion without a discharge there is a slight divergence from the results of Chapter II, Section 1. This is due to including the process $W_{4,\Sigma}$ in the relaxation scheme and to the use of exact equations. Since there are no experimental data on this probability it was assumed that $W_{4,\Sigma} \approx W_{3,\Sigma}^{N_2}$ in accordance with a calculation using the method of [62, 84].

The dependences shown in Fig. 31 are typical but not optimal. Variation of N_e, T_0, p_0, τ, and the mixture composition makes it possible to obtain fairly large ΔN ($\sim 10^{17}$ cm^{-3}). Thus, the dependence of the maximum inversion on the initial partial pressure of CO_2 is shown in Fig. 32. The existence of an optimum in the curve, as in [8], is explained by the influence of two factors: a growth in the total number of active particles and an increase in the rate of relaxation of the 00^01 level due to collisions. Depending on N_e, τ, the location of the discharge gap, the composition of the mixture, and T_0, this optimum may be shifted slightly to one side or another and vary in absolute magnitude.

Increasing N_e promotes a growth in ΔN_{max} since the amount of energy pumped into vibrational degrees of freedom is increased. Even at other than optimal conditions (as regards p_0, τ, and the mixture composition), for $N_e = 10^{12}$ cm^{-3} we have ΔN_{max} $\sim 10^{16}$ cm^{-3} (Fig. 33). For the same conditions, but $p_0 = 5$ atm, we already find $\Delta N_{max} = 2 \cdot 10^{16}$ cm^{-3}. Increasing N_e above about 10^{13} cm^{-3} leads to noticeable dissociation of CO_2, and using this scheme to analyze the kinetics yields inaccurate results.

Figure 34 illustrates the dependence of ΔN_{max} on the location of the discharge gap and its length. Also shown there are the value of ΔN_{max} without a discharge. It is clear that as τ (or, equivalently, l_2) is increased for a given value of l_1 the maximum inversion increases and for $l_2 - l_1 = 15$ cm reaches about $6 \cdot 10^{15}$ cm^{-3} ($l_1 \approx 3$ cm). The existence of optima in the curves is due, on the one hand (for small l_1) to rapid relaxation which reduces the efficiency of electronic pumping and, on the other hand, to the fact that at large l_1 relaxation is slowed down (T and the total pressure decrease) and the lower level is no longer depopulated sufficiently rapidly causing the fall in T_2 to be slowed down (sometimes even to the point of rising) so that the deformed and symmetric vibrational modes are also highly excited by the electrons. For the same reason, further increasing τ is inappropriate since this no longer leads to an increase in ΔN_{max} but only complicates the technical task of achieving a discharge of greater extent.

The maximum inversion is plotted as a function of the slit half-width R_0 in Fig. 35. It is clear that for these parameters the optimum is $R_0 \sim 0.1$ cm. This optimum is due to the fact that for small R_0 the rate of change of the density (and, therefore, the temperature) is very high and by the time a sufficient invergence has developed between T_3 and T_2 to produce an inversion the number of active CO_2 particles per volume is small. At large R_0 expansion and cooling take place much more slowly so that due to rapid relaxation the divergence between T_3 and T_2 sets in much later. Thus, for $R_0 \gtrsim 0.3$ cm the relaxation is so strong that with these parameters an inversion does not occur before the end of the discharge gap. Only when the density and temperature have fallen and the number of damping collisions becomes small does an inversion appear. Thus, large values of R_0 ($\gtrsim 0.3$ cm) substantially reduce the efficiency of electronic pumping and the situation is similar to a conventional gas-dynamic laser without a discharge.

The results of varying the composition of the mixture are shown in Figs. 36 and 37. Adding He leads to an increase in ΔN_{max} (Fig. 36); however, the expansion velocity is thereby increased, and so τ falls. Increasing k_{He} from 10 to about 30 no longer causes a noticeable increase in the inversion since the relaxation rate of the upper level becomes large. For further increases in k_{He} the density falls so rapidly that the first inversion maximum is reached even before the discharge gap (a pure gas-dynamic laser) and its magnitude is reduced.

The dependence of ΔN_{max} on the amount of nitrogen in the mixture is shown in Fig. 37. For $k_{N_2} \lesssim 2$ there is no inversion at all without helium. Then it appears and increases all the way up to $k_{N_2} \sim 30$ despite the fact that collisional relaxation is very strong by then. The vibrational states of N_2 are highly excited by electrons and due to intense transfer of excitation to CO_2, relaxation of the latter is considerably slowed down. The increase in ΔN_{max} when N_2 is added becomes fairly weak after $k_{N_2} \sim 15$; thus, further increasing the amount of nitrogen only leads to technical difficulties.

On the whole, devices with transverse electrical excitation in expanding supersonic flows are promising. They make it possible to obtain substantially larger inversions than in ordinary gas-dynamic lasers. These lasers also have advantages over transverse discharges with subsonic pumpthrough gas speeds of $v \lesssim 10^2$ m/sec.

In conclusion we note that along with molecular mixtures certain atomic gases may also be used in electrical gas-dynamic lasers. With them the inversion lifetime may be 10^{-7}–10^{-6} sec; thus, to obtain effective cw lasers with transverse excitation the pumpthrough speeds must be supersonic. An example of a possible working gas is argon. The s and p states of argon are efficiently excited in a discharge while the radiative lifetime of the p state is more than an order of magnitude larger than that of the s state. With radiative relaxation this leads to an inversion between the s and p states (also between the p and d states [107]). The lifetime of this inversion is determined mainly by the lifetime of the p levels (which may exceed 10^{-6} sec for the comparatively high-lying levels). At currently attainable supersonic flow-through speeds a continuous regime may be realized in this manner.

Therefore, the electrical gas-dynamic method of creating an inverted population is of interest for a whole series of active media and is promising for various applications.

CHAPTER V

HIGH-TEMPERATURE GAS-DYNAMIC LASERS

1. The Role of Chemical Reactions in the Operation of a Gas-Dynamic Laser

Initial temperatures $T_0 \sim 1800$°K are optimal for gas-dynamic lasers. Higher temperatures are regarded as inefficient because of the large relaxation rate. However, a number of physical processes (for example, for $T \gtrsim 3000$°K carbon dioxide gas is more than half dissociated and thermal ionization becomes significant) have not been taken into account in the theoretical analysis. Each of these neglected factors requires a detailed study of the kinetics.

The technique for calculating the relaxation processes in a supersonic nozzle in the presence of chemical reactions was discussed in Chapter II. In view of the specific features of high-temperature gas-dynamic lasers their operation has been analyzed in a separate chapter. The results obtained using this method for initial temperatures $\geqslant 3200°K$ are given in this chapter. Here the calculation has been done both for equilibrium and for nonequilibrium (with respect to the composition of the mixture) slowing down (stagnation) conditions. In both cases a type *b* nozzle shape was used (see Fig. 7b).

Since for $T \geqslant 3000°K$ intense chemical reactions may take place and change the composition (especially under nonequilibrium conditions) the choice of the initial values of the velocity in Eq. (2.17) is not trivial. The most widespread method of selecting this value consists of choosing from several tests (computations with the entire system of equations) a value of the initial velocity v_0 that increases to the local sound speed at the critical cross section. The process is begun with that velocity which yields the sound speed at the critical cross section in the absence of chemical reactions (equilibrium flow). If this value of v_0 is higher than the required value, the solution becomes blocked and (the flow) does not reach the critical cross section. The v_0 quantity is reduced until the solution approximates a point with a coordinate differing from that of the critical cross section ($x = 2r_0$) by only 2-3%. In these cases all the instantaneous variables were carried unchanged to the point $x = 2r_0$, while the velocity had an increment such that the denominator in Eqs. (2.17) changed sign and the calculation was continued. The velocity increment was insignificant, as in the calculations for low-temperature gas-dynamic lasers.

The inversion was calculated including chemical kinetics for stagnation temperatures $T_0 = 3200\text{-}4000°K$. In both cases (equilibrium and clearly nonequilibrium initial chemical composition) the velocity was selected for $r_0 = 0.025$ and 0.05 cm. Figure 38 shows typical results of this choice for one variant of equilibrium stagnation conditions. It is clear that for both values of r_0 the vibrational temperatures are very close to the gas temperature (a difference of about 90-120°K) at $x = 2r_0$ and the chemical compositions correspond with good accuracy to equilibrium for p_{cr} and T_{cr}. This means that reactions (2.12) "succeed" in holding the system in chemical equilibrium under conditions of changing gas-dynamic parameters over the entire

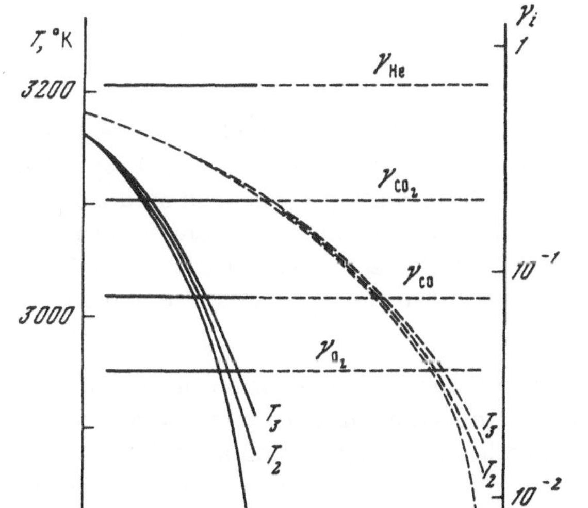

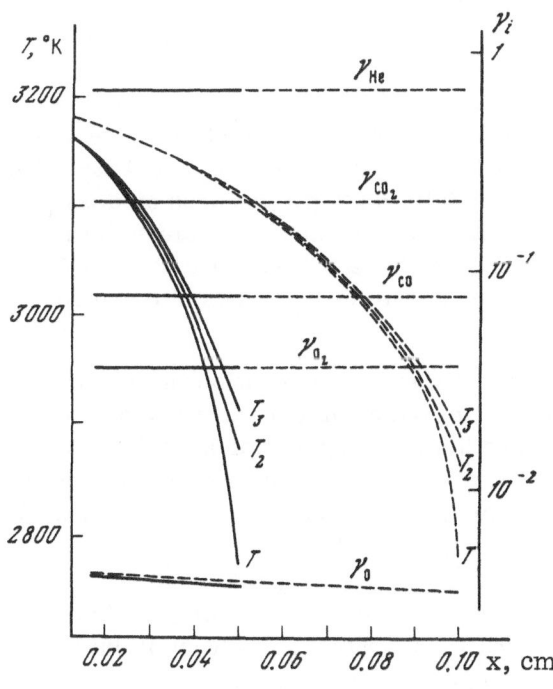

Fig. 38. The variation in the molar fractions and in the vibrational and gas temperatures along the nozzle entrance axis from the cutoff to the critical cross section in the case of equilibrium stagnation conditions. Continuous curves refer to $r_0 = 0.025$ cm; dashed, to $r_0 = 0.05$ cm. $T_0 = 3200°K$; $p_0 = 90$ atm, $a_{N_2} = 0$; $a_{He} = 0.178$; mixture $CO_2{:}CO{:}O{:}O_2{:}N_2{:}He$ = 1:0.37:0.02:0.17:0:3.26.

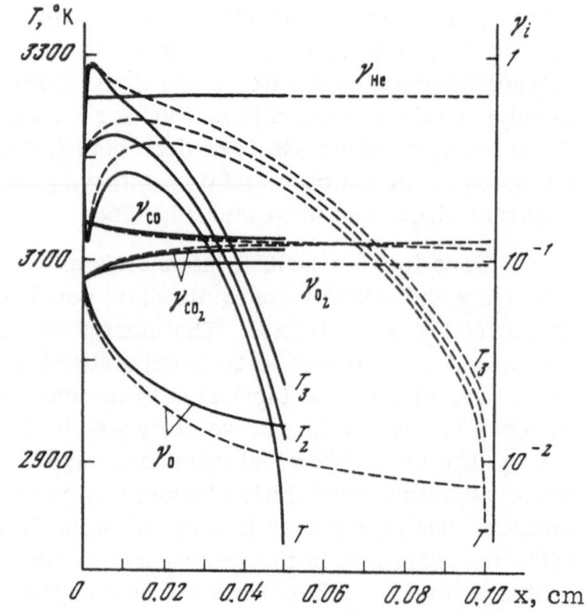

Fig. 39. The variation in the molar fractions and in the vibrational and gas temperatures along the nozzle entrance axis up to the critical cross section in the case of nonequilibrium chemical composition at stagnation. Continuous curves refer to $r_0 = 0.025$ cm; dashed, to $r_0 = 0.05$ cm. $T_0 = 3200$°K; $p_0 = 90$ atm, mixture CO_2:CO:O:O_2:N_2:He = 1:2:1:1:0:8.

length from x = 0 to x = $2r_0$. Here, as can be seen in Fig. 38, there is practically no change in the composition due to chemical reactions as the temperature falls from T_0 to T_{cr}.

The case of nonequilibrium chemical composition at p_0 and T_0 is illustrated in Fig. 39. The vibrational temperatures at the critical cross section are close to the gas temperature here; however, the reactions are much more intense and cause a noticeable nonmonotonicity in the behavior of the vibrational temperatures at the beginning of the nozzle entrance. Then the total particle density decreases due to recombination reactions, which leads to a rapid reduction in T. At the same time, it follows from Eqs. (2.27) and (2.28) that T_3 and T_2 increase and a divergence develops between them (about 80°K), which, however, is far from sufficient for an inversion. Then as the divergence between T_3 and T_2 increases at high gas temperatures, such intense relaxation of the energy in the asymmetric mode sets in that T begins to rise despite the fact the gas is accelerated (this is facilitated by the rise in T_2 as well). Later, depending on how closely the chemical composition approaches equilibrium, the vibrational temperatures also approach the gas temperature due to a weakening of the energy flux into the vibrational degrees of freedom and to the high relaxation rate. At the critical cross section the composition is already close to equilibrium. Increasing the pressure to $p_0 = 150$ atm with the conditions in Fig. 39 does not change anything substantially (only p_{cr} changes as does the composition, slightly, in accordance with the higher pressure). The absolute values of the temperatures at the critical cross section are ~2800-2900°K (for a temperature $T_0 = 4000$°K we have $T_{cr} = 3500$-3700°K), which are high, and, as noted above, large values of q are required to obtain an inversion during the expansion.

The subsequent motion of a mixture with almost equilibrium composition through the nozzle is accompanied (for $x \gtrsim 1$ cm) by rapid "freezing" of the chemical reactions due to the sharp drop in p and T so that in both cases (Figs. 38 and 39) the composition at distances $x \sim 1$ cm differs from that at the critical cross section by no more than 2% (in the amount of CO_2). Thus, we may assume with good accuracy that a mixture with constant chemical composition in almost complete thermodynamic equilibrium is expanding from the critical cross section. This problem was discussed previously in Chapter II, with the only difference here being the higher initial temperatures T and somewhat different proportions of the components. In this case the use of the same type *b* nozzles as in the low-temperature gas-dynamic laser calculations does not lead to an inversion since, despite the roughly identical initial rates of tem-

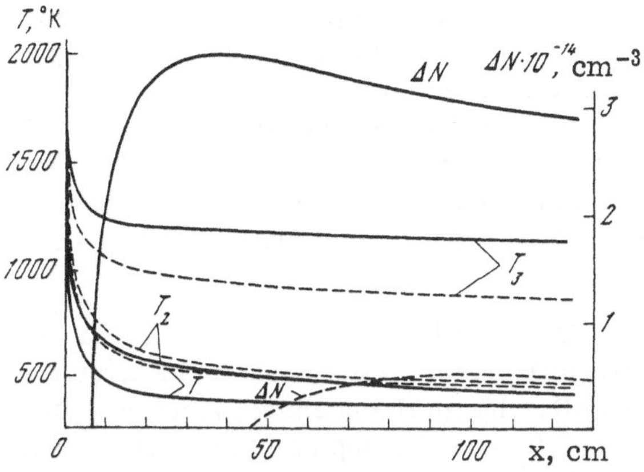

Fig. 40. Distributions of the inverted populations and the vibrational and gas temperatures along the axis of a logarithmic nozzle for the conditions of Fig. 38. Continuous curves refer to $r_0 = 0.025$ cm; dashed, to $r_0 = 0.05$ cm; $\tan\theta_0/q = 0.1$; $q = 10$.

perature drop in both cases $[(1/T)(dT/dx)]_{x=0} \sim (1/f)(df/dx) = \tan\theta_0/r_0$, the nozzle widening, $(1/f)(df/dx)$, at large distances ($x \geqslant 5$-10 cm) still cannot compete with the rapid relaxation because of the high gas temperature and since the divergence between T_3 and T_2 produced in the first stage of the expansion becomes smaller with the retarded drop in T. Increasing q increases $(1/f)(df/dx)$ at large distances from the critical cross section and makes it possible to cool the gas more thoroughly and obtain an inversion.

Figure 40 shows the distributions of the inversion and the vibrational and gas temperatures along the nozzle axis beyond the critical cross section for the conditions of Fig. 38. Clearly, for a nozzle with a smaller half-width r_0 the "frozen" value of T_3 is somewhat higher while the temperatures of the lower states in both cases are close to one another. In accordance with this the inversion for such a nozzle (as in the case of low stagnation temperatures) considerably exceeds ΔN for a nozzle with a larger critical dimension. However, the optimal values of the inversion for high T_0 is much less than at $T_0 \sim 2000$°K. The times that ΔN appears and ΔN_{max} is achieved correspond to larger distances (~10-40 cm in the first and ~50-100 cm in the second case) from the critical cross section where the total number of active particles is already small. Reducing r_0 within reasonable limits hastens the appearance of ΔN; however, this only makes it possible to approach the values of ΔN_{max} obtained at $T_0 \sim 2000$°K with nozzles having the customary half-width r_0.

Figure 41 illustrates the effect of the parameter q on the maximum values of the inversion for high initial gas temperatures. This effect is large for nozzles with small r_0 and the best results are obtained for $q \sim 5$-7. Increased r_0 requires larger q, and vice versa.

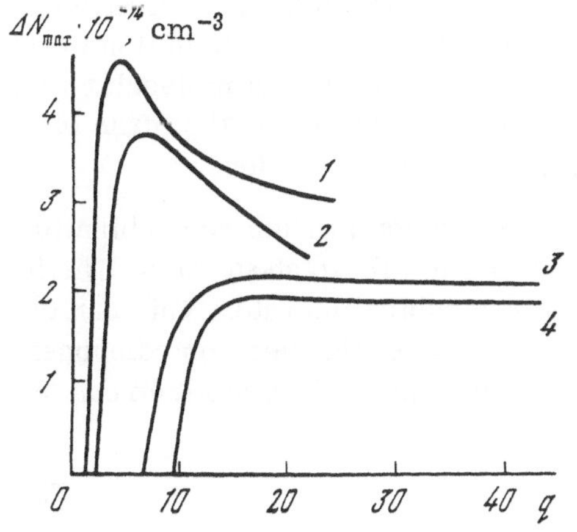

Fig. 41. Dependence of ΔN_{max} on q for a logarithmic nozzle. Curves 1 and 3 correspond to the conditions of Fig. 39; curves 2 and 4, to Fig. 38. $\tan\theta_0/r_0 = 40$ (curves 1 and 2); $\tan\theta_0/r_0 = 20$ (curves 3 and 4).

Thus, a joint examination of the chemical kinetics according to Eq. (2.12) with vibrational relaxation in nozzles has shown that accounting for the chemical reactions is important only in the neighborhood of the nozzle entrance up to the critical cross section (since here the greatest changes in the composition occur) and is not important in the expanding part where the chemistry does not produce any additional disequilibrium in the vibrational levels beyond that which existed at the critical cross section. This is because at the critical cross section the gas is almost in equilibrium with respect to composition and the reactions are nearly complete.

These calculations, which included the reactions (2.12), still do not make it possible to completely evaluate the role of chemistry. In fact, the mixtures actually used in gas-dynamic lasers contain some water vapor which may fundamentally change the chemical kinetics. Oxidation of CO catalyzed with water is a chain reaction and follows a completely different scheme from (2.12) (see, for example, [108]). The reaction is then exothermic and a high energy input into the vibrational degrees of freedom of the CO_2 molecules formed as a result of the reactions is very probable. However, a detailed discussion of the kinetics of this reaction and a study of its effect on vibrational relaxation are made difficult by the facts that it is not known what part of the energy of the reactions goes into vibrations, that the rate constants for several branches of the reaction are not known, and that the role of boundary surfaces on which active centers may be formed is not clear. Under these circumstances only qualitative estimates are possible, and these have been made using the simplest model for the chemical kinetics. It included the system of reactions (2.12) with water catalysis replaced by increased rate constants for all the reactions resulting in formation of CO_2. Here it was also assumed that part of the energy of the chemical reactions goes into vibrational degrees of freedom and the other part into translational, so there is no drop in the gas temperature due to the decrease in the total number of particles. Estimates using this model showed that the inversion can be about an order of magnitude greater than with the usual CO oxidation reaction.

2. Thermal Ionization and High-Temperature Gas-Dynamic Lasers

In gas-dynamic lasers thermal ionization and the free electrons in the expanding gas may be important in addition to the nonequilibrium chemical reactions when $T_0 > 3000°K$.

A calculation of the equilibrium degree of ionization i according to the Saha formula

$$i^2 = \frac{g_+ g_e}{g_0} \frac{1}{N_0} \left(\frac{2\pi m_e kT}{h^2}\right)^{3/2} e^{-\frac{I}{kT}} \tag{5.1}$$

for a CO_2-N_2-He mixture half made up of molecules at $T_0 = 4000°K$ and a pressure of $p_0 \approx 10^2$ atm yields $i \approx 10^{-6}$ if dissociation is taken into account. Here g_e, g_+, and g_0 are the statistical weights of an electron, a molecular ion, and a neutral molecule, respectively, N_0 is the density of heavy particles, m_e is the mass of an electron, I is the ionization potential, and $i = N_e/N_e$ is the degree of ionization.

Due to equilibrium recombination at the critical cross section of the nozzle the degree of ionization falls to about 10^{-7}. The further expansion of the mixture leads to slowing down of the recombination rate. This can easily be shown by making simple estimates of the drop in the density of the electron component in time as the gas expands into the vaccum. The equation describing the loss of electrons as the gas expands has the form

$$\frac{\partial N_e}{\partial t} + \operatorname{div} N_e \mathbf{v} = \left(\frac{dN_e}{dt}\right)_*, \tag{5.2}$$

where the right-hand side describes electron losses due to diffusion, attachment, and dissociative recombination. Substituting $N_e = iN_0 = i(\rho/\mu)$ into Eq. (52), using the continuity equation $\partial\rho/\partial t + \mathrm{div}\,\rho v = 0$, and assuming that the chemical composition of the mixture does not change after the critical cross section (μ = const), we obtain an equation for the degree of ionization:

$$di/dt = -(K_{\mathrm{dif}} + K_{\mathrm{at}})\,i - K_{\mathrm{rec}} N_0 i^2, \tag{5.3}$$

where K_{diff}, K_{at}, and K_{rec} are the diffusion, attachment, and recombination coefficients, respectively. Using approximate experimental temperature and pressure dependences [109] for these coefficients, we can find to good accuracy that for $T_{\mathrm{cr}} \approx 3500$°K, $p_{\mathrm{cr}} \approx 10^2$, and $i_0 \approx 10^{-8}$ at the critical cross section the principal electron loss process at distances from the critical cross section of up to about 10 cm is volume recombination. Equation (5.3) is easily solved in this approximation, and it follows from the solution that for these initial parameters the degree of ionization at distances of 3-5 cm from the critical cross section changes little with distance along the expansion axis and equals roughly 10^{-9}. Then the density of heavy particles is $N \sim (1\text{-}5) \cdot 10^{18}$ cm^{-3}. This value of the degree of ionization exceeds the equilibrium value by many orders of magnitude. Thus, the presence of a large number of electrons ($\sim 10^{12}$-10^{13} cm^{-3} at the critical cross section and $\sim 10^9$-10^{10} cm^{-3} in the working zone of the laser) may greatly ease the problems of initiating a discharge transverse to the flow and building a high-temperature electrical gas-dynamic laser without an external preionization source. The inversion may then be greatly increased as in low-temperature gas-dynamic lasers [57]. A gas-dynamic laser with excitation by a thermally ionized gas is studied in detail in the next article of this volume [110].

CHAPTER VI

NONEQUILIBRIUM PROCESSES AND INFRARED RADIATION IN THE WAKE OF AN OBJECT FLYING THROUGH THE UPPER ATMOSPHERE

The radiation of the upper atmosphere may be conditionally divided into three groups according to its cause:

1. Radiation from a quiet, unperturbed atmosphere (for example, night and twilight glow)
2. Radiation produced by global or large local perturbations of the atmosphere (for example, aurora, radiation caused by magnetic storms, etc.)
3. Radiation caused by comparatively small local perturbations as objects fly through the atmosphere, by artificial clouds in the upper atmosphere, and so on.

The infrared emission from an unperturbed, quiet atmosphere over the wavelength range $\lambda \sim 3$-$6\ \mu$ produced by vibrational-rotational transitions in molecules and molecular ions at altitudes $H \gtrsim 120$ km was studied in [111].

This section is devoted to an analysis of infrared radiation belonging to the third group, specifically that with $\lambda \sim 6\ \mu$ behind the shock front formed as a blunt object passes through the atmosphere with a supersonic velocity at altitudes $H = 20$-100 km.

Most articles on the flight of bodies examine supersonic flow around them, that is, the portion behind the shock front formed ahead of the body, while phenomena in the wake have been the subject of little research [112, 113]. It should be noted that a rigorous and systematic examination of the postshock emission from an object passing through the atmosphere is a very complex and difficult problem.

To solve it one must examine jointly the equations of gas dynamics, chemical kinetics, and vibrational relaxation. At the present time this problem has been stated only for that portion ahead of the body [114] and an illustrative solution has been presented only for a gas of oxygen molecules [114]. Usually for both shock waves in shock tubes [87] and supersonic flow around blunt objects [115-117] the gas dynamics are examined together with either the chemical kinetics (here it is assumed that the vibrational degrees of freedom are in equilibrium with the translational) or vibrational relaxation (assuming the absence of chemical processes).

Examining the phenomena in the wake of a body is still more complicated. Only the gas-dynamic problem was solved for this case in [112, 113] by numerical methods with various simplifying assumptions (the chemical kinetics and vibrational relaxation were neglected). The IR emission due to vibrational–rotational transitions in the NO molecule is estimated in [112] assuming complete thermodynamic equilibrium. However, at relatively high altitudes, $H \gtrsim 40$ km, the relaxation zone where thermodynamic equilibrium does not exist for either the vibrational temperatures or the concentrations may be large. This must lead to substantial deviation from equilibrium in the intensity of IR radiation as well.

In the present article we compute the basic parameters (molecular fractions and vibrational temperatures) determining the intensity of IR radiation from molecular vibrational–rotational transitions using a joint solution of the equations of chemical kinetics and vibrational relaxation without assuming thermodynamic equilibrium. Since we are examining conditions in which the concentration of NO molecules formed in hot air due to various chemical processes greatly exceeds the concentrations of the radiating molecules CO_2 and CO (i.e., N_{CO_2}/N, $N_{CO}/N \lesssim 3 \cdot 10^{-4}$, where N is the overall particle concentration), the entire analysis is made for only one radiating molecule, NO. Most attention is paid to molecular kinetics. The gas-dynamic problem is not solved, and to evaluate the gas-dynamic parameters (temperature and pressure) we have used a detonation analogy [112, 113, 118] and a numerical solution of the gas-dynamic equations [112, 113]. This approach to the problem simplifies it greatly and makes it possible to obtain numerical values for the parameters of interest to us over a wide range of altitudes and flight speeds. The effect of chemical kinetics and vibrational relaxation on the gas temperature may be accounted for approximately by varying the adiabatic index γ with the chemical composition and vibrational temperatures as an independent parameter.

We now turn to a determination of the gas-dynamic parameters and examine the equations of chemical kinetics and vibrational relaxation. Let an object fly at velocity v parallel to and at height H above the earth's surface. We shall assume the body is spherical with radius R_0. We choose a system of coordinates r and z attached to the body and arrange the origin and axes as shown in Fig. 42.† It is convenient to discuss the problem in terms of the dimensionless parameters $y = z/R_0$ and $x = r/R_0$. As the body moves through the atmosphere at a supersonic velocity a shock wave is formed. The pressure p_F at the front of the shock is much greater than atmospheric pressure p_0 at first and gradually decreases along the wake, approaching atmospheric in the limit, while the shock degenerates into an ordinary sound wave. In the coordinate system attached to the body all parameters are time independent and are functions of the variables of x and y. In the following we shall examine the shock zone with values of y such that the pressure $p_F(y)$ satisfies the conditions

$$p_F(y) \gg p_0. \quad (6.1)$$

The values of y for which this condition holds depend on the velocity of the body. For $v/c = M \approx 30$ it is of order 50. Here c is the speed of sound in the unperturbed atmosphere and M is the Mach number.

† The third coordinate is not introduced because of the axial symmetry of the problem.

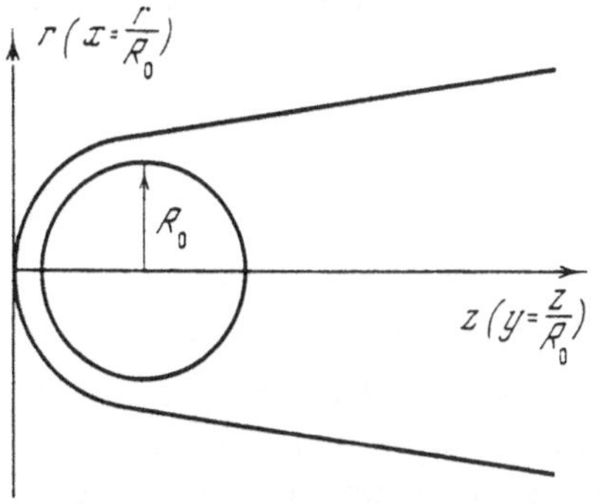

Fig. 42. Schematic diagram of the shock wave produced during supersonic motion of a spherical object through the atmosphere.

When Eq. (6.1) is satisfied, i.e., at high pressures, if we use a model of a cylindrical explosion for the pressure, $p_F(y)$, at the shock front [113, 118] we have

$$p_F(y) = p_0\left(k_1 \frac{M^2}{y} + k_2\right). \tag{6.2}$$

The constants k_1 and k_2 depend on the adiabatic index γ and are given in Table 1 for various γ.

From the numerical calculations by Sedov [119] of the radial dependences of the postshock pressure we can obtain an approximate analytic expression for the pressure at an arbitrary point x, y behind the shock front:

$$p(x,\ y) = p_F(y)\, f(x,\ x_F). \tag{6.3}$$

The function $f(x, x_F)$ is given by

$$f(x,\ x_F) = 0.627\,(x/x_F)^{9.5} + 0.373, \tag{6.4}$$

where x_F is the coordinate of the shock front and determines its form. The shape of the front may be regarded as parabolic [113, 118] to good accuracy. In that case

$$x_F \approx B y^{1/2}. \tag{6.5}$$

The constant B is of order unity and depends on γ [113]. The value of B for various γ is given in Table 1.

Thus, using Eqs. (6.2)-(6.5) p(x, y) is given by

$$p(x,\ y) = p_0\left(k_1 \frac{M^2}{y} + k_2\right)\left[0.627\left(\frac{x}{By^{1/2}}\right)^{9.5} + 0.373\right], \qquad x \geqslant 1, \quad y \geqslant \left(\frac{x}{B}\right)^2. \tag{6.6}$$

TABLE 1. The Parameters k_1, k_2, B, ξ, and β for Various Values of the Adiabatic Index γ

Adiabatic index γ	k_1	k_2	B	ξ	β
1.2	0.307	1.345	0.92	1.55	0.84
1.4	0.3564	1.085	1.104	0.848	0.77
1.667	0.388	0.887	1.273	0.577	0.745

The density ρ_F at the front of the shock may be found from the relation [113, 118]

$$\rho_F(y) = \rho_0\left(1 + \frac{2}{\gamma - 1}\cdot\frac{1.4}{1.4 + \sqrt{y}}\right). \tag{6.7}$$

Here ρ_0 is the density of the unperturbed atmosphere at the flight altitude H.

To find the gas temperature T(x, y) behind the shock front we shall use the numerical calculations of Feldman [112, 113], which imply that the temperature profile over x has roughly a Gaussian form. Also, using the relative values of the gas temperature computed numerically in [113] for a given y at $x = x_F$ and x = 1, the following approximate analytic expression may be found for the gas temperature:

$$T(x, y) = T_F(y)\exp[\xi x^2(B^{2\beta}x_F^{-2\beta} - y^{-2\beta})]. \tag{6.8}$$

Here ξ and β are constants depending on the chosen value of γ and given in Table 1, $T_F(y)$ is the temperature of the gas at the shock front determined by the gas law from the values of p_F and ρ_F:

$$T_F(y) = \frac{\mu_0}{\Re}\frac{p_F}{\rho_F} = \frac{\mu_0}{\Re}\frac{p_0}{\rho_0}\cdot\frac{k_1\frac{M^2}{y} + k_2}{1 + \frac{2}{\gamma - 1}\frac{1.4}{1.4 + \sqrt{y}}}. \tag{6.9}$$

Here $\Re$ is the gas constant and μ_0 is the molecular weight of air at the front. Since at the shock front chemical reactions and vibrational relaxation have just begun, it can be assumed that μ_0 in Eq. (6.9) is the molecular weight of the unperturbed air.

Therefore, Eqs. (6.6), (6.8), and (6.9) determine the values of the basic gas-dynamic parameters (pressure and temperature) behind the front of the shock in the wake left behind by a body flying through the atmosphere for $x \geq 1$, and $y \geq (1/B)^2$. In the central portion of the wake (of a spherical body) for $x < 1$ these expressions can also be used, but there may be significant error in p and T. In this region the gas-dynamic regime is complicated and the motion of the gas may be turbulent [112].

We now turn to a discussion of the chemical kinetics and vibrational relaxation behind the shock front in this region. The basic chemical reactions which take place in air at high temperatures are the following [87]:

$$\begin{aligned}
&1)\quad O_2 + M + 5.1\,\text{eV} \underset{K_{-1}}{\overset{K_1}{\rightleftarrows}} O + O + M, &&4)\quad O + N_2 + 3.3\,\text{eV} \underset{K_{-4}}{\overset{K_4}{\rightleftarrows}} NO + N,\\
&2)\quad N_2 + M + 9.8\,\text{eV} \underset{K_{-2}}{\overset{K_2}{\rightleftarrows}} N + N + M, &&5)\quad N + O_2 - 1.4\,\text{eV} \underset{K_{-5}}{\overset{K_5}{\rightleftarrows}} NO + O,\\
&3)\quad NO + M + 6.5\,\text{eV} \underset{K_{-3}}{\overset{K_3}{\rightleftarrows}} N + O + M, &&6)\quad N_2 + O_2 + 1.9\,\text{eV} \underset{K_{-6}}{\overset{K_6}{\rightleftarrows}} NO + NO.
\end{aligned} \tag{6.10}$$

The rate constants for a number of these reactions are presently known only to within an order of magnitude. Using the data of various authors [87, 89] we have chosen the most reliable (in our opinion) rate constants, and these are given in Table 2.

The chemical processes (6.10) begin in the shock front and evolve along the coordinates x and y toward the interior of the wake. In the range of applicability of the law of plane cross sections it is possible, by assuming the lines of flow to be parallel to the z(y) axis, to study the

TABLE 2. The Rate Constants for Reactions (6.10) Used in Our Calculations (temperatures in degrees Kelvin)

Reaction rate		Value of the constants	Third particle M
K_1, cm^3/sec	$K_1^{(1)}$	$1.8 \cdot 10^1 T^{-2.5} \exp\left(-\frac{59,415}{T_{O_2}}\right)$	O_2, NO, N
	$K_1^{(2)}$	$1.5 \cdot 10^{-4} T^{-1} \exp\left(-\frac{59,415}{T_{O_2}}\right)$	O
	$K_1^{(3)}$	$K_1^{(2)} \cdot 4 \cdot 10^{-2}$	N_2
K_{-1}, cm^6/sec	$K_{-1}^{(1)}$	$2.7 \cdot 10^{-32}\ T^{-0.41}$	O_2, NO, N
	$K_{-1}^{(2)}$	$1.3 \cdot 10^{-29}\ T^{-1}$	O
	$K_{-1}^{(3)}$	$K_{-1}^{(2)} \cdot 4 \cdot 10^{-2}$	N_2
K_2, cm^3/sec	$K_2^{(1)}$	$9.3 \cdot 10^{-2} T^{-1.7} \exp\left(-\frac{113,300}{T_{N_2}}\right)$	O_2, NO, N_2, O
	$K_2^{(2)}$	$1.2 \cdot 10^{-4} T^{-1} \exp\left(-\frac{113,300}{T_{N_2}}\right)$	N
K_{-2}, cm^6/sec	$K_{-2}^{(1)}$	$3.0 \cdot 10^{-32} T^{-0.5}$	O_2, NO, N_2, O
	$K_{-2}^{(2)}$	$6.6 \cdot 10^{-27} T^{-1.5}$	N
K_3, cm^3/sec	$K_3^{(1)}$	$6.6 \cdot 10^{-4} T^{-1.5} \exp\left(-\frac{75,530}{T_{NO}}\right)$	O_2, N_2
	$K_3^{(2)}$	$K_3^{(1)} \cdot 20$	NO, O, N
K_{-3}, cm^6/sec	$K_{-3}^{(1)}$	$2.8 \cdot 10^{-28} T^{-0.5}$	O_2, N_2
	$K_{-3}^{(2)}$	$K_{-3}^{(1)} \cdot 20$	NO, O, N
K_4, cm^3/sec		$1.2 \cdot 10^{-10} \exp\left(-\frac{38,000}{T}\right)$	—
K_{-4}, cm^3/sec		$2.6 \cdot 10^{-11}$	—
K_5, cm^3/sec		$2.2 \cdot 10^{-14}\ T \exp\left(-\frac{3562}{T}\right)$	—
K_{-5}, cm^3/sec		$5.2 \cdot 10^{-15}\ T \exp\left(-\frac{19,690}{T}\right)$	—
K_6, cm^3/sec		$1.5 \cdot 10^1 T^{-1.5} \exp\left(-\frac{64,700}{T}\right)$	—
K_{-6}, cm^3/sec		$2.2 \cdot 10^{-10} \exp\left(-\frac{38,080}{T}\right)$	—

chemical kinetics and vibrational relaxation along the y axis moving from a point at the shock front with a corresponding fixed value of the coordinate $x = y_F$. We introduce the notation

$$\gamma_i(x, y) = \frac{N_i(x, y)}{\alpha(x, y)}, \qquad \alpha(x, y) = \frac{\rho(x, y)}{\rho_0} N_0,$$

where N_0 is the concentration of air molecules ahead of the front, $N_i(x, y)$ is the instantaneous concentration of the i-th type of particle, and $\gamma_i(x, y)$ is the concentration of these particles normalized to the equivalent number of molecules, $\alpha(x, y)$. Then the chemical kinetics behind the shock front obey the equations (cf. [87])

$$\begin{aligned} d\gamma_{O_2}/dy &= -A_1^{(1)} - A_1^{(2)} + A_{-1}^{(1)} + A_{-1}^{(2)} - A_5 + A_{-5} - A_6 + A_{-6}, \\ d\gamma_{N_2}/dy &= -A_2^{(1)} - A_2^{(2)} + A_{-2}^{(1)} + A_{-2}^{(2)} - A_6 + A_{-6}, \\ d\gamma_{NO}/dy &= -A_3^{(1)} - A_{-3}^{(1)} + A_4 - A_{-4} + A_5 - A_{-5} + 2A_6 - 2A_{-6}, \end{aligned} \qquad (6.11)$$

where the following notation is introduced:

$$A_1^{(1)} = K_1^{(1)}(\gamma_{O_2} + \gamma_{NO} + \gamma_N)\gamma_{O_2}\alpha\frac{R_0}{v}, \qquad A_1^{(2)} = K_1^{(2)}(\gamma_O + 4 \cdot 10^{-2}\gamma_{N_2})\gamma_{O_2}\alpha\frac{R_0}{v},$$

$$A_{-1}^{(1)} = K_{-1}^{(1)}(\gamma_{O_2} + \gamma_{NO} + \gamma_N)\gamma_O^2\alpha^2\frac{R_0}{v}, \qquad A_{-1}^{(2)} = K_{-1}^{(2)}(\gamma_O + 4 \cdot 10^{-2}\gamma_{N_2})\gamma_O^2\alpha^2\frac{R_0}{v},$$

$$\left.\begin{aligned}
&A_2^{(1)} = K_2^{(1)}(\gamma_{N_2} + \gamma_{O_2} + \gamma_{NO} + \gamma_O)\gamma_{N_2}\alpha\frac{R_0}{v}, \qquad A_2^{(2)} = K_2^{(2)}\gamma_N\gamma_{N_2}\alpha\frac{R_0}{v},\\
&A_{-2}^{(1)} = K_{-2}^{(1)}(\gamma_{N_2} + \gamma_{O_2} + \gamma_{NO} + \gamma_O)\gamma_N^2\alpha^2\frac{R_0}{v}, \qquad A_{-2}^{(2)} = K_{-2}^{(2)}\gamma_N^3\alpha^2\frac{R_0}{v},\\
&A_3^{(1)} = K_3^{(1)}(\gamma_{O_2} + \gamma_{N_2} + 20\gamma_{NO} + 20\gamma_O + 20\gamma_N)\gamma_{NO}\alpha\frac{R_0}{v},\\
&A_{-3}^{(1)} = K_{-3}^{(1)}(\gamma_{O_2} + \gamma_{N_2} + 20\gamma_{NO} + 20\gamma_O + 20\gamma_N)\gamma_N\gamma_O\alpha^2\frac{R_0}{v},\\
&A_4 = K_4\gamma_{N_2}\gamma_O\alpha\frac{R_0}{v}, \qquad A_{-4} = K_{-4}\gamma_{NO}\gamma_N\alpha\frac{R_0}{v}, \qquad A_5 = K_5\gamma_N\gamma_{O_2}\alpha\frac{R_0}{v},\\
&A_{-5} = K_{-5}\gamma_{NO}\gamma_O\alpha\frac{R_0}{v}, \qquad A_6 = K_6\gamma_{N_2}\gamma_{O_2}\alpha\frac{R_0}{v}, \qquad A_{-6} = K_{-6}\gamma_{NO}^2\alpha\frac{R_0}{v}.
\end{aligned}\right\} \tag{6.12}$$

The quantities γ_O and γ_N may be determined from algebraic equations, one of which expresses the constant ratio of the number of nitrogen atoms to oxygen atoms during the chemical reactions and the other of which relates the initial molecular weight to the molecular weight of the reacting mixture:

$$\frac{\gamma_N + 2\gamma_{N_2} + \gamma_{NO}}{\gamma_O + 2\gamma_{O_2} + \gamma_{NO}} = 3.76, \qquad \sum_i \gamma_i\mu_i = \mu_0.$$

We now write the equations for vibrational relaxation of the molecular components of air, N_2, O_2, and NO:

$$\left.\begin{aligned}
\frac{d\varepsilon_{N_2}}{dy} = \frac{R_0}{v}\Big\{&-(\varepsilon_{N_2} - \varepsilon_{N_2}^0)\alpha\sum_i W_{N_2,i}\gamma_i + W_{N_2,O_2}\gamma_{O_2}\alpha\Big[\varepsilon_{O_2}(\varepsilon_{N_2} +\\
&+ 1)\exp\left(-\frac{E_{N_2} - E_{O_2}}{T}\right) - \varepsilon_{N_2}(\varepsilon_{O_2} + 1)\Big] + W_{N_2,NO}\gamma_{NO}\alpha\Big[\varepsilon_{NO}(\varepsilon_{N_2} +\\
&+ 1)\exp\left(-\frac{E_{N_2} - E_{NO}}{T}\right) - \varepsilon_{N_2}(\varepsilon_{NO} + 1)\Big] + \left(\frac{D_{N_2}}{E_{N_2}} - \varepsilon_{N_2}\right)\frac{1}{\gamma_{N_2}} \times\\
&\times\left[\left(\frac{d\gamma_{N_2}}{dy}\right)_{\text{rec}} + \left(\frac{d\gamma_{N_2}}{dy}\right)_{\text{diss}}\right] + (\varepsilon_{N_2}^{\text{chem}} - \varepsilon_{N_2})\frac{1}{\gamma_{N_2}}\left(\frac{d\gamma_{N_2}}{dy}\right)_{\text{chem}}\Big\},\\
\frac{d\varepsilon_{O_2}}{dy} = \frac{R_0}{v}\Big\{&-(\varepsilon_{O_2} - \varepsilon_{O_2}^0)\alpha\sum_i W_{O_2,i}\gamma_i - W_{N_2,O_2}\gamma_{N_2}\alpha\Big[\varepsilon_{O_2}(\varepsilon_{N_2} + 1)\times\\
&\times\exp\left(-\frac{E_{N_2} - E_{O_2}}{T}\right) - \varepsilon_{N_2}(\varepsilon_{O_2} + 1)\Big] - W_{NO,O_2}\gamma_{NO}\alpha\times\\
&\times\Big[\varepsilon_{O_2}(\varepsilon_{NO} + 1)e^{-\frac{E_{NO} - E_{O_2}}{T}} - \varepsilon_{NO}(\varepsilon_{O_2} + 1)\Big] + \left(\frac{D_{O_2}}{E_{O_2}} - \varepsilon_{O_2}\right)\frac{1}{\gamma_{O_2}}\times\\
&\times\left[\left(\frac{d\gamma_{O_2}}{dy}\right)_{\text{rec}} + \left(\frac{d\gamma_{O_2}}{dy}\right)_{\text{diss}}\right] + (\varepsilon_{O_2}^{\text{chem}} - \varepsilon_{O_2})\frac{1}{\gamma_{O_2}}\left(\frac{d\gamma_{O_2}}{dy}\right)_{\text{chem}}\Big\},\\
\frac{d\varepsilon_{NO}}{dy} = \frac{R_0}{v}\Big\{&-(\varepsilon_{NO} - \varepsilon_{NO}^0)\alpha\sum_i W_{NO,i}\gamma_i - W_{N_2,NO}\gamma_{N_2}\alpha\times\\
&\times\Big[\varepsilon_{NO}(\varepsilon_{N_2} + 1)e^{-\frac{E_{N_2} - E_{NO}}{T}} - \varepsilon_{N_2}(\varepsilon_{NO} + 1)\Big] +\\
&+ W_{NO,O_2}\gamma_{O_2}\alpha\Big[\varepsilon_{O_2}(\varepsilon_{NO} + 1)e^{-\frac{E_{NO} - E_{O_2}}{T}} - \varepsilon_{NO}(\varepsilon_{O_2} + 1)\Big] +\\
&+ \left(\frac{D_{NO}}{E_{NO}} - \varepsilon_{NO}\right)\frac{1}{\gamma_{NO}}\left[\left(\frac{d\gamma_{NO}}{dy}\right)_{\text{rec}} + \left(\frac{d\gamma_{NO}}{dy}\right)_{\text{diss}}\right] +\\
&+ (\varepsilon_{NO}^{\text{chem}} - \varepsilon_{NO})\frac{1}{\gamma_{NO}}\left(\frac{d\gamma_{NO}}{dy}\right)_{\text{chem}} - C_{NO}\varepsilon_{NO}\Big\}.
\end{aligned}\right\} \tag{6.13}$$

TABLE 3. The Probabilities of Vibrational – Translational Relaxation W_{ji} and Vibrational – Vibrational Exchange $W_{jj'}$ Used in the System of Eqs. (6.13) (W in cm^3/sec)

j \ i	N_2	O_2	NO	N	O
N_2	$7.8 \cdot 10^{-12} T e^{-\frac{218}{\sqrt[3]{T}}}$	$1.1 \cdot 10^{-11} T e^{-\frac{225}{\sqrt[3]{T}}}$	$9.2 \cdot 10^{-12} T e^{-\frac{221}{\sqrt[3]{T}}}$	$1.5 \cdot 10^{-12} T e^{-\frac{178}{\sqrt[3]{T}}}$	$2.0 \cdot 10^{-12} T e^{-\frac{186}{\sqrt[3]{T}}}$
O_2	$6.8 \cdot 10^{-13} T e^{-\frac{132}{\sqrt[3]{T}}}$	$1.1 \cdot 10^{-8} e^{-\frac{157}{\sqrt[3]{T}}}$	$7.5 \cdot 10^{-13} T e^{-\frac{134}{\sqrt[3]{T}}}$	$2.3 \cdot 10^{-13} T e^{-\frac{106}{\sqrt[3]{T}}}$	$2.9 \cdot 10^{-13} T e^{-\frac{111}{\sqrt[3]{T}}}$
NO	$1.8 \cdot 10^{-12} T e^{-\frac{165}{\sqrt[3]{T}}}$	$2.2 \cdot 10^{-12} T e^{-\frac{171}{\sqrt[3]{T}}}$	$6.0 e^{\frac{T}{1280}} \cdot 10^{-15}$	$4.8 \cdot 10^{-13} T e^{-\frac{134}{\sqrt[3]{T}}}$	$6.2 \cdot 10^{-13} T e^{-\frac{140}{\sqrt[3]{T}}}$

W_{N_2, O_2}	$W_{N_2, NO}$	W_{NO, O_2}
$6.0 \cdot 10^{-11} e^{-\frac{112}{\sqrt[3]{T}}}$	$1.6 \cdot 10^{-12} \sqrt{T} e^{-\frac{69}{\sqrt[3]{T}}}$	$1.4 \cdot 10^{-12} \sqrt{T} e^{-\frac{48}{\sqrt[3]{T}}}$

Here the subscript i corresponds to the N_2, NO, O_2, N, and O molecules, W_{ji} is the probability of vibrational – translational relaxation of molecule j in a collision with an i molecule, and $W_{jj'}$ ($j \neq j'$) is the probability of exchange of quanta between j and j' (W_{ji} and $W_{jj'}$ were taken from [82] or were calculated using the semiempirical formula given in [97]). A collection of all these probabilities for the system of equations (6.13) is given in Table 3. The quantities E_j and D_j are the energy of a vibrational quantum and the energy of dissociation of a j molecule.

The first terms in each of Eqs. (6.13) describe vibrational – translational relaxation, the next two describe vibrational exchange, the fourth terms describe the change in the stored energy of the molecules due to recombination and dissociation, and the last terms describe the change in ε_j due to exothermic reactions in which part of the chemical energy is released vibrational energy (for example, $N + O_2 \rightarrow NO + O + 1.4$ eV). The last term in the equation for ε_{NO} determines the radiative decay of the levels with probability C_{NO}.

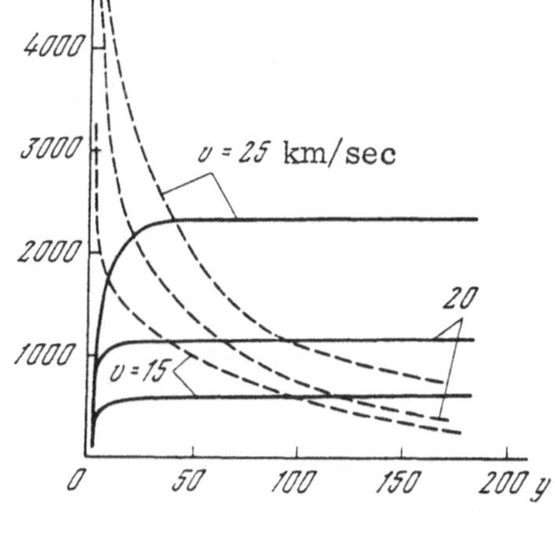

Fig. 43. Dependence of the gas temperature T (dashed curves) and vibrational temperatures T_{NO} (continuous curves) on the dimensionless parameter y for various flight speeds. Flight altitude H = 80 km; object radius R_0 = 100 cm.

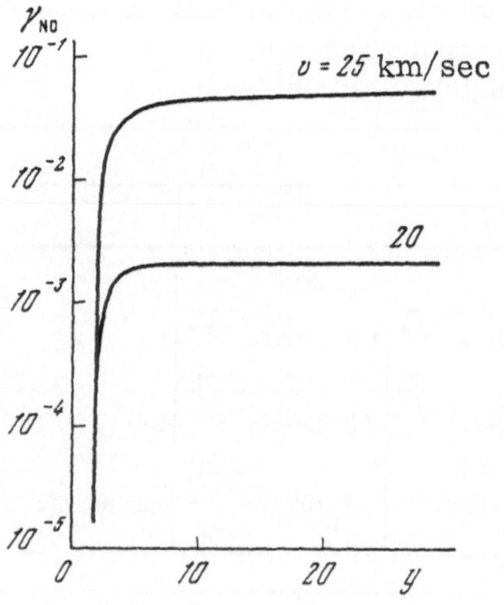

Fig. 44. Dependence of the molar fractions γ_{NO} of nitrous oxide on y for various flight speeds. H = 80 km, R_0 = 100 cm.

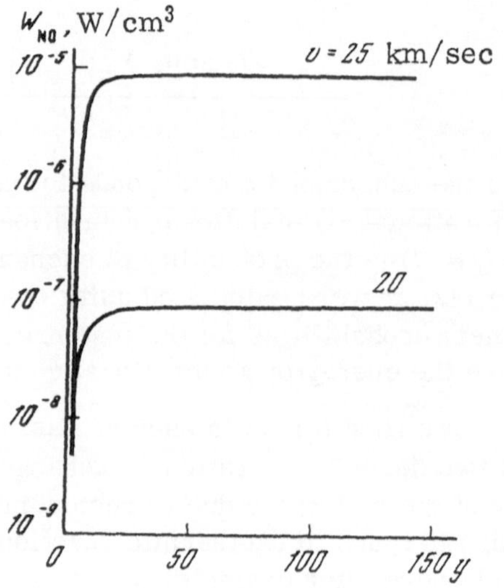

Fig. 45. Dependence of the radiated power W_{NO} per unit volume on y for various flight speeds. H = 80 km; R_0 = 100 cm.

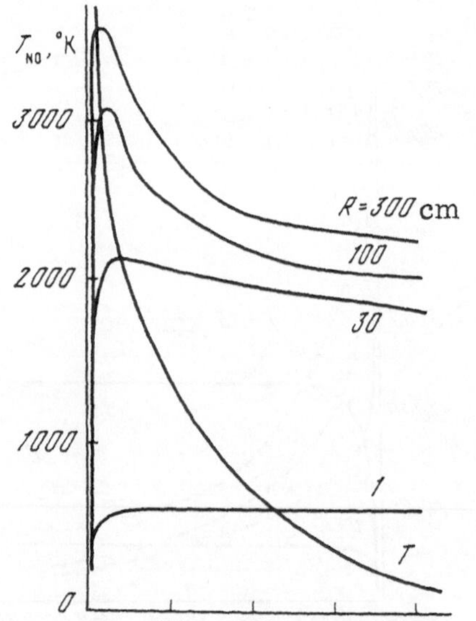

Fig. 46. Dependence of the gas temperature T and vibrational temperatures T_{NO} on y for various object sizes. H = 60 km, v = 20 km · sec^{-1}.

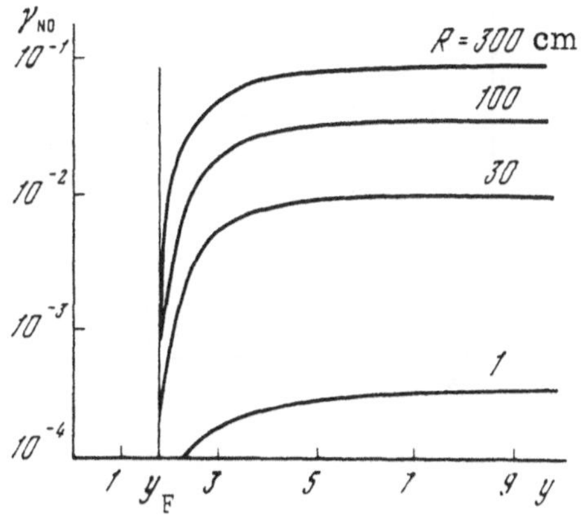

Fig. 47. The function $\gamma_{NO}(y)$ for various object sizes. H = 60 km, v = 20 km/sec.

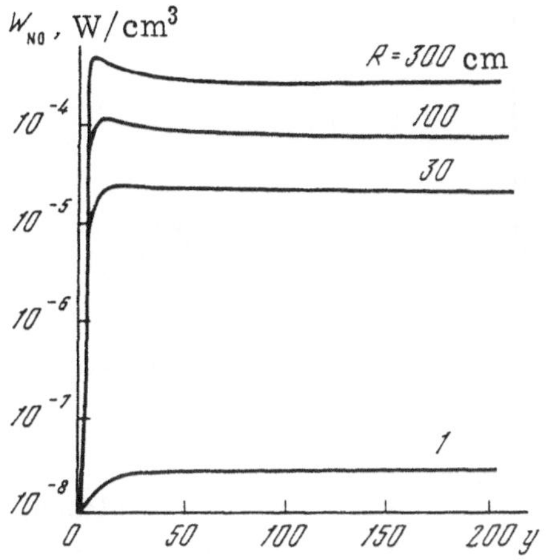

Fig. 48. The radiated power W_{NO} as a function of y for flying objects of various sizes. H = 60 km, v = 20 km/sec.

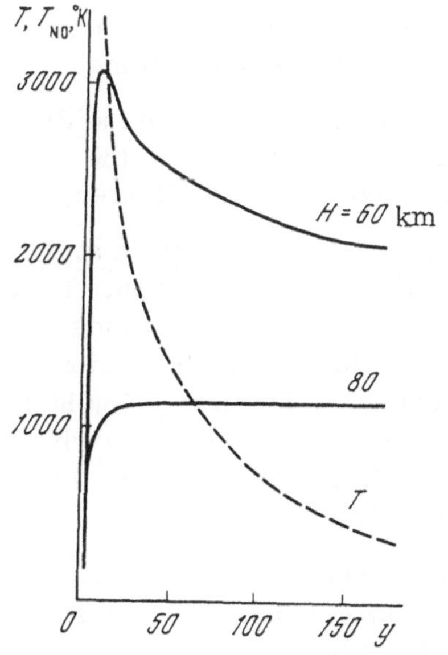

Fig. 49. Dependence of the gas temperature T and the vibrational temperature T_{NO} on y for various flight altitudes H. v = 20 km/sec, R_0 = 100 cm.

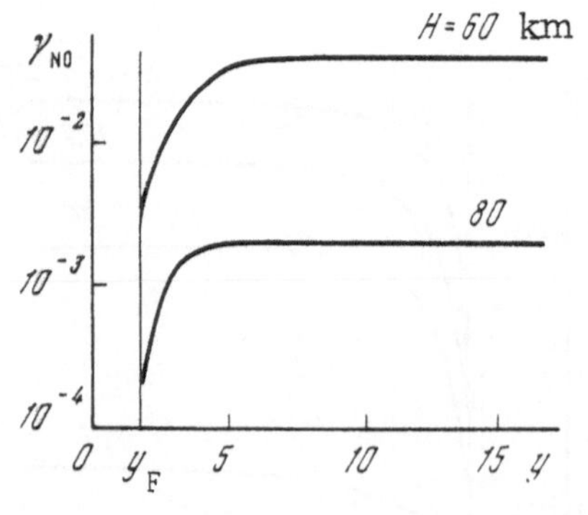

Fig. 50. γ_{NO} as a function of y for various H. v = 20 km/sec, R_0 = 100 cm.

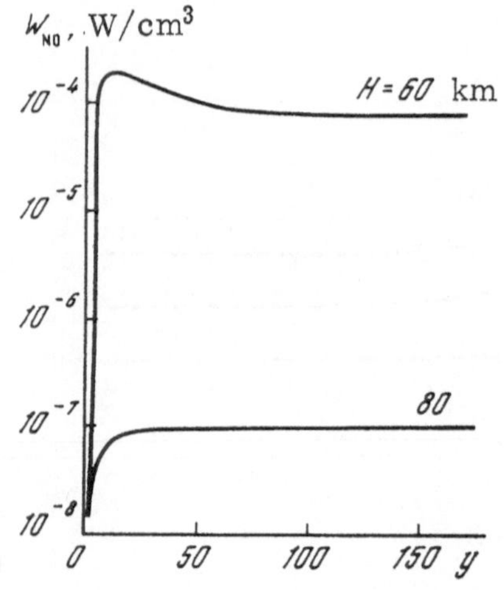

Fig. 51. The radiated power W_{NO} at various altitudes as a function of y. v = 20 km/sec, R_0 = 100 cm.

Equations (6.11) for the chemical kinetics together with Eqs. (6.13) for the vibrational relaxation and Eqs. (6.6), (6.8), and (6.9) for the changes in pressure and gas temperature behind the shock front were solved on a computer. Solutions were obtained for various altitudes H = 20-100 km, flight speeds v = 15-25 km/sec, and object sizes R_0 = 1-300 cm. The basic results obtained for x = 1.5 and γ = 1.4 are shown in Figs. 43-53. From these plots it is clear that equilibrium does not occur in the wake of the body either with respect to the temperatures or with respect to the concentrations. Thus, the IR emission from the NO molecules is also substantially out of equilibrium.

Figures 43-45 illustrate the variation of the gas T and vibrational T_{NO} temperatures of the molar fractions γ_{NO} and of the radiated powers W_{NO} with the dimensionless parameter y for flight speeds v = 15-25 km/sec..

Since the gas temperature at the shock front increases with the speed, vibrational relaxation takes place more rapidly and T_{NO} attains large values. However, due to the rather sharp drop in T and the low gas densities, the vibrational temperatures "freeze out." Subsequently T_{NO} must fall due to slow relaxation and radiation and approach T; however, the relaxation zone may be very large (y ~ 500). For high flight speeds a substantial amount of NO molecules may be formed behind the shock front with a concentration that also does not correspond to equilibrium. The "freezing out" of the vibrational temperatures and the concentrations means that the radiant intensity may substantially exceed (by an order of magnitude) the equilibrium value.

Similar dependences were found as well for different object sizes (Figs. 46-48) and flight altitudes (Figs. 49-51). Here, as in the case of the speed dependence, there is a substantial deviation from thermodynamic equilibrium over a wide range of variation in y with more rapid relaxation the lower the altitude and the larger the radius of the body.

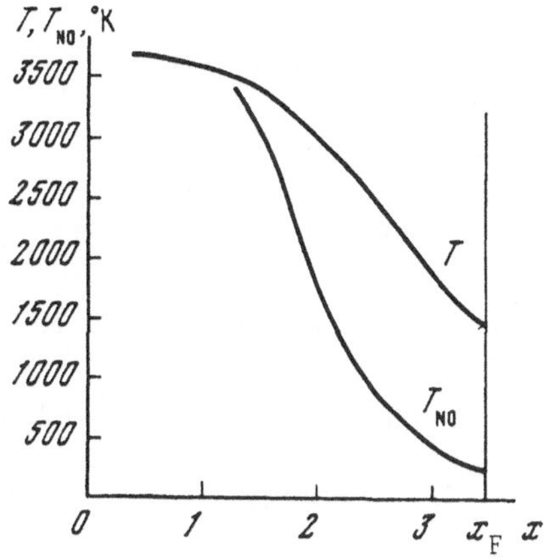

Fig. 52. Profiles of the temperatures T and T_{NO} over x at a distance of 10 radii (y = 10). H = 60 km, v = 20 km/sec, R_0 = 100 cm.

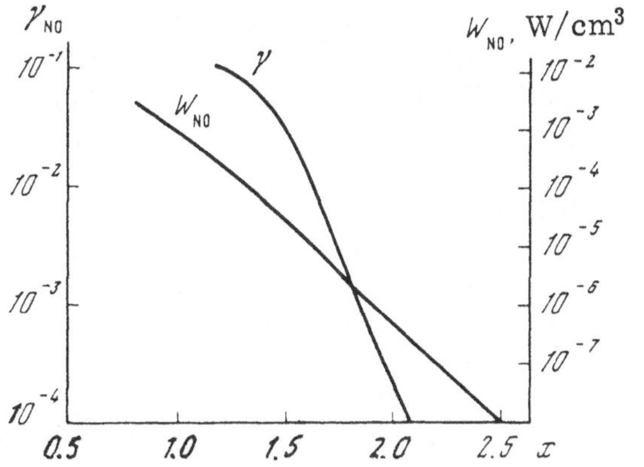

Fig. 53. Profiles of γ_{NO} and W_{NO} over x at a distance of 10 radii (y = 10). H = 60 km, v = 20 km/sec, R_0 = 100 cm.

Profiles of the vibrational temperatures, concentrations, and radiated powers over the cross section of the wake were computed for several parameters. From the results shown in Figs. 52 and 53 it is clear that there may be large deviations from equilibrium over the transverse cross section of the wake.

Therefore, a correct analysis of radiation from the wakes of objects flying in the upper atmosphere requires an examination of vibrational relaxation together with the chemical processes and gas dynamics involved.

In conclusion I wish to express may deep gratitude to my research adviser L. A. Shelepin for his constant interest in and help with this work. I also thank B. F. Gordiets for valuable discussions and help.

LITERATURE CITED

1. N. G. Basov and A. N. Oraevskii, Zh. Éksp. Teor. Fiz., 44:1742 (1963).
2. N. G. Basov, A. N. Oraevskii, and V. A. Shcheglov, Zh. Tekh. Fiz., 37:339 (1967).
3. V. K. Konyukhov and A. M. Prokhorov, Inventor's Certificate No. 223,954, priority November 19, 1966; Pis'ma Zh. Éksp. Teor. Fiz., 3:436 (1966).
4. J. R. Hurle and A. Herzberg, Phys. Fluids, 8:1601 (1965).
5. L. I. Gudzenko, S. S. Filippov, and L. A. Shelepin, Zh. Éksp. Teor. Fiz., 51:1115 (1966).
6. N. G. Basov, V. G. Mikhailov, A. N. Oraevskii, and V. A. Shcheglov, Zh. Tekh. Fiz., 38:2031 (1968).

7. N. I. Yushchenkova and Yu. A. Kalenov, Zh. Prikl. Spektrosk., 11:417 (1969).
8. A. S. Biryukov, B. F. Gordiets, and L. A. Shelepin, Zh. Éksp. Teor. Fiz., 57:585 (1969); Preprint FIAN, No. 41 (1969).
9. J. D. Anderson, Phys. Fluids, 13:1983 (1970).
10. J. Tulip and H. Seguin, J. Appl. Phys., 42:3393 (1971).
11. N. A. Generalov, G. I. Kozlov, and I. K. Selezneva, Zh. Prikl. Mekh. Tekh. Fiz., No. 5, p. 24 (1971).
12. A. P. Dronov, E. V. Kudryavkin, and E. M. Kudryavtsev, Preprint FIAN, No. 103 (1967).
13. D. M. Kuehn and D. J. Monson, Appl. Phys. Lett., 16:48 (1970).
14. E. T. Gerry, IEEE Spectrum, 7:51 (1970); Laser Focus, 6(12):27 (1970); AIAA Paper N 71-23 (1971).
15. A. P. Dronov, A. S. D'yakov, E. M. Kudryavtsev, and N. N. Sobolev, Pis'ma Zh. Éksp. Teor. Fiz., 11:516 (1970).
16. V. K. Konyukhov, I. V. Matrosov, A. M. Prokhorov, D. T. Shalunov, and N. N. Shirokov, Pis'ma Zh. Éksp. Teor. Fiz., 12:461 (1970).
17. A. S. Biryukov and L. A. Shelepin, Author's Certificate No. 1,262,085, priority August 5, 1968; Zh. Tekh. Fiz., 40:2575 (1970).
18. W. H. Christiansen and G. A. Tsongas, Phys. Fluids, 14:2611 (1971).
19. J. D. Anderson, R. L. Humphrey, J. S. Vamos, M. J. Plummer, and R. E. Jensen, Phys. Fluids, 14:2620 (1971).
20. G. Lee and F. E. Gowen, Appl. Phys. Lett., 18:237 (1971).
21. A. S. Biriukov, A. P. Dronov, E. M. Koudriavtsev, and N. N. Sobolev, IEEE J. Quant. Electron., QE-7:388 (1971).
22. S. Yatsiv, E. Greenfield, F. Dothan-Deutsch, D. Chuchem, and E. Bin-Nun, Appl. Phys. Lett., 19:65 (1971).
23. J. Tulip and H. Seguin, Appl. Phys. Lett., 19:263 (1971).
24. V. M. Marchenko and A. M. Prokhorov, Pis'ma Zn. Éksp. Teor. Fiz., 14:116 (1971).
25. M. S. Dzhidzhoev, V. V. Korolev, V. N. Markov, V. G. Platonenko, and R. V. Khokhlov, Pis'ma Zh. Éksp. Teor. Fiz., 14:73 (1971).
26. N. G. Basov, V. V. Gromov, E. L. Koshelev, E. P. Markin, and A. N. Oraevskii, Pis'ma Zn. Éksp. Teor. Fiz., 14:73 (1971).
27. M. S. Dzhidzhoev, M. I. Pimenov, V. G. Platonenko, Yu. V. Filippov, and R. V. Khokhlov, Zh. Éksp. Teor. Fiz., 57:411 (1969).
28. R. Kh. Kurbangalina, E. A. Patskov, L. N. Stesik, and G. S. Yakovlev, Zh. Prikl. Mekh. Tekh. Fiz., No. 4, p. 160 (1970).
29. M. G. Basov, V. I. Igoshin, E. P. Markin, and A. N. Oraevskii, Kvant. Elektron., No. 2, p. 3 (1971).
30. V. I. Igoshin and A. N. Oraevskii, Zh. Éksp. Teor. Fiz., 59:1240 (1970).
31. D. M. Kuehn, Appl. Phys. Lett., 21:112 (1972).
32. J. D. Anderson and F. L. Harris, Laser Focus 8(5):32 (1972); AIAA Paper N 72-143 (1972).
33. N. I. Yushchenkova and Yu. A. Kalenov, Zh. Prikl. Spektrosk., 16:39 (1972).
34. N. A. Generalov, G. I. Kozlov, and I. K. Selezneva, Zh. Prikl. Mekh. Tekh. Fiz., No. 5, p. 33 (1972).
35. G. V. Gembarzhevskii, N. A. Generalov, G. I. Kozlov, and D. I. Roitenburg, Zh. Éksp. Teor. Fiz., 62:844 (1972).
36. V. N. Karnyushin and R. I. Soloukhin, Fiz. Goren. Vzryva, 8:163 (1972).
37. B. F. Gordiets, A. I. Osipov, E. V. Stupochenko, and L. A. Shelepin, Usp. Fiz. Nauk, 108:655 (1972).
38. J. D. Anderson, AIAA J., 8:545 (1970).
39. A. S. Biryukov, B. F. Gordiets, and L. A. Shelepin, Kratk. Soobshch. Fiz., No. 6, p. 13 (1971).

40. B. F. Gordiets, N. N. Sobolev, V. V. Sokovikov, and L. A. Shelepin, Phys. Lett., 26A, 173 (1967).
41. B. F. Gordiets, N. N. Sobolev, and L. A. Shelepin, Zh. Éksp. Teor. Fiz., 53:1822 (1967).
42. B. F. Gordietz, N. N. Sobolev, V. V. Sokovikov, and L. A. Shelepin, IEEE J. Quant. Electron., QE-4:796 (1968).
43. A. S. Biryukov, B. F. Gordiets, and L. A. Shelepin, Zh. Éksp. Teor. Fiz., 55:1456 (1968).
44. P. O. Clark and M. R. Smith, Appl. Phys. Lett., 9:367 (1966).
45. C. K. N. Patel, Phys. Rev., 136A:1187 (1964).
46. C. Grapard, M. Roulot, and X. Ziegler, Phys. Lett., 20:384 (1966).
47. P. K. Cheo, H. Appl. Phys., 38:3563 (1967).
48. N. V. Karlov, G. P. Kuz'min, A. M. Prokhorov, and V. I. Shemyakin, Zh. Éksp. Teor. Fiz., 54, 1318 (1968).
49. A. M. Danishevskii, I. M. Frishman, and I. D. Yaroshetskii, Zn. Éksp. Teor. Fiz., 55:813 (1968).
50. A. E. Hill, Appl. Phys. Lett., 12:324 (1968).
51. A. J. Beaulieu, Appl. Phys. Lett., 16:504 (1970).
52. O. R. Wood, E. G. Burkhardt, M. A. Pollack, and T. J. Bridges, Appl. Phys. Lett., 18:112 (1971).
53. W. B. Tiffany, R. Targ, and J. D. Foster, IEEE J. Quant. Electron., QE-6:5 (1970).
54. R. J. Freiberg and P. O. Clark, IEEE J. Quant. Electron., QE-5, 362 (1969).
55. E. K. Karlova, N. V. Karlov, and G. P. Kuz'min, Kratk. Soobschch. Fiz., No. 11, p. 51 (1970).
56. A. E. Hill, Appl. Phys. Lett., 18:194 (1971).
57. A. S. Biryukov and L. A. Shelepin, Zh. Tekh. Fiz., 43:355 (1973); Preprint FIAN, No. 130 (1972).
58. Yu. A. Kalenov and N. I. Yushchenkova, Dokl. Akad. Nauk SSSR, 189:1041 (1969).
59. J. Sato and S. Tsuchiya, J. Phys. Soc. Jpn., 39:1467 (1971).
60. A. S. Biryukov and B. F. Gordiets, Zh. Prikl. Mekh, Tekh. Fiz., No. 6, p. 29 (1972); Preprint FIAN, No. 32 (1972).
61. A. S. Biryukov, V. K. Konyukhov, A. I. Lukovnikov, and R. I. Serikov, Zh. Éksp. Teor. Fiz., 66:1248 (1974); Preprint FIAN, No. 9 (1973).
62. R. N. Schwartz, Z. J. Slawsky, and K. F. Herzfeld, J. Chem. Phys., 20:1591 (1952).
63. R. N. Schwartz and K. F. Herzfeld, J. Chem. Phys., 22:767 (1954).
64. A. I. Osipov, Zh. Prikl. Mekh. Tekh. Fiz., No. 1, p. 41 (1964).
65. C. E. Treanor, J. W. Rich, and R. G. Rehm, J. Chem. Phys., 48:1798 (1968).
66. N. M. Kuznetsov, Dokl. Akad. Nauk SSSR, 185:866 (1969).
67. B. F. Gordiets, A. I. Osipov, and L. A. Shelepin, Zh. Éksp. Teor. Fiz., 60:102 (1971).
68. B. F. Gordiets, Sh. S. O. Memedov, A. I. Osipov, and L. A. Shelepin, Preprint FIAN, No. 31 (1972).
69. N. M. Kuznetsov, Zh. Éksp. Teor. Fiz., 61:499 (1971).
70. R. Mariott, Proc. Phys. Soc., 83:159 (1964); 84:877 (1964).
71. R. D. Sharma and C. A. Brau, J. Chem. Phys., 50:924 (1969).
72. K. H. Shin, J. Chem. Phys., 47:3302 (1967).
73. E. E. Nikitin, The Theory of Elementary Atom–Molecule Processes in Gases [in Russian], Khimiya, Moscow (1970).
74. C. B. Moore, R. E. Wood, Bei-Lok Hu, and J. T. Yardley, J. Chem. Phys., 46:4222 (1967).
75. W. A. Rosser, A. D. Wood, and E. T. Gerry, IEEE J. Quant. Electron., QE-4:336 (1968).
76. A. S. Biryukov, R. I. Serikov, and E. S. Trekhov, Zh. Éksp. Teor. Fiz., 59:1513 (1970).
77. A. S. Biryukov, V. K. Konyukhov, A. I. Lukovnikov, V. A. Myslin, R. I. Serikov, and E. S. Trekhov, Zh. Prikl. Spektrosk., 16:249 (1972).

78. J. C. Stephenson and C. B. Moore, J. Chem Phys., 52:2333 (1970).
79. J. C. Stephenson, R. E. Wood, and C. B. Moore, J. Chem. Phys., 54:3097 (1971).
80. W. A. Rosser, A. D. Wood, and E. T. Gerry, J. Chem. Phys., 50:4996 (1969).
81. D. J. Seery, J. Chem. Phys., 56:4714 (1972).
82. R. L. Taylor and S. Bitterman, Rev. Mod. Phys., 41:26 (1969).
83. L. Landau and E. Teller, Phys. Z. Sowietunion, 10:34 (1936).
84. K. F. Herzfeld, Disc. Faraday Soc., N 33, p. 22 (1962).
85. K. P. Stanyukovich, Unsteady motion of Continuous Media [in Russian], Gostekhizdat, Moscow (1955).
86. A. S. Biryukov and L. A. Shelepin, Zh. Tekh. Fiz., 44:1232 (1974); Preprint FIAN, No. 59 (1973).
87. E. V. Stupochenko, S. A. Losev, and A. I. Osipov, Relaxation Processes in Shock Waves [in Russian], Nauka, Moscow (1965).
88. C. E. Treanor and P. V. Marrone, Phys. Fluids, 5:1022 (1962).
89. V. N. Kondrat'ev, Rate Constants of Gaseous Phase Reactions, A Handbook [in Russian], Nauka, Moscow (1970).
90. G. N. Abramovich, Applied Gas Dynamics [in Russian], Nauka, Moscow (1969).
91. E. E. Nikitin and S. Ya. Umanskii, Dokl. Akad. Nauk SSSR, 196:145 (1971).
92. J. Wieder, Phys. Lett., 24A:759 (1967).
93. B. F. Gordiets, A. I. Osipov, and L. A. Shelepin, Zh. Éksp. Teor. Fiz., 59:615 (1970).
94. B. F. Gordiets, A. I. Osipov, and L. A. Shelepin, Zh. Éksp. Teor. Fiz., 61:562 (1971).
95. A. S. Biryukov and L. A. Shelepin, Zh. Prikl. Mekh. Tekh. Fiz., No. 4, p. 25 (1973).
96. P. F. Zittel and C. B. Moore, Appl. Phys. Lett., 21:81 (1972).
97. R. C. Millikan and D. R. White, J. Chem. Phys., 39:3209 (1963).
98. J. D. Teare and R. L. Taylor, Nature, 225:240 (1970).
99. R. L. McKenzie, Appl. Phys. Lett., 17:462 (1970).
100. W. D. Breshears and P. F. Bird, J. Chem. Phys., 50:333 (1969).
101. Hao-Lin Chen, J. C. Stephenson, and C. B. Moore, Chem. Phys. Lett., 2:593 (1968).
102. E. B. Gordon, A. F. Dodonov, G. L. Lavrovskaya, I. I. Morozov, A. N. Ponomarev, and V. L. Tal'roze, International Symposium on Chemical Lasers, Collected Papers, Moscow (1969).
103. G. J. Schulz, Phys. Rev., 135A:988 (1964); M. J. Boness and G. J. Schulz, Phys. Rev. Lett., 21:1031 (1968).
104. A. S. Biryukov, B. F. Gordiets, and L. A. Shelepin, IX Intern. Conference on Phenomena in Ionized Gases, Collected Papers, Bucharest (1969).
105. H. Seguin and J. Tulip, Appl. Phys. Lett., 21:414 (1972).
106. N. G. Basov, É. M. Belenov, V. A. Danilychev, O. M. Kerimov, I. V. Kovsh, and A. F. Suchkov, Pis'ma Zh. Éksp. Teor. Fiz., 14:421 (1971).
107. B. F. Gordiets, I. A. Dymova, and L. A. Shelepin, Zh. Prikl. Spektrosk., 14:205 (1971).
108. B. Lewis and G. Von Elbe, Combustion, Flames, and Explosions of Gases, Academic Press (1961).
109. S. B. Brown, Elementary Processes in Gaseous Discharge Plasmas [Russian translation], Atomizdat, Moscow (1961); G. Hasted, Physics of Atomic Collisions, American Elsevier (1972).
110. A. S. Biryukov, V. M. Marchenko, and L. A. Shelepin, Tr. FIAN, 83:87 (1975) (next article in this volume).
111. B. F. Gordiets, M. N. Markov, and L. A. Shelepin, Kosmich. Issled. 8:437 (1970).
112. S. Feldman, J. Aerospace Sci., 28:433 (1961).
113. S. Feldman, ARS J., 30:463 (1960).
114. S. B. Koleshko, Yu. P. Lun'kin and F. D. Popov, Aerophysical Studies of Supersonic Flows [in Russian], Nauka, Moscow (1966).
115. O. M. Belotserkovskii and V. K. Dushin, Zh. Vych. Mat. Mat. Fiz., 4:61 (1964).

116. Yu. P. Lun'kin and F. D. Popov, Zh. Vych. Mat. Mat. Fiz., 4:896 (1964).
117. A. G. Hall, A. Q. Eschenroeder, and P. V. Marrone, J. Aerospace Sci., 29:1038 (1962); L. A. Young, J. Quant. Spectrosc. Radiat. Transfer, 8:105 (1968).
118. M. A. Tsikulin, Shock Waves due to Motion of Large Meteoritic Objects in the Atmosphere [in Russian], Nauka, Moscow (1969).
119. L. I. Sedov, Similarity and Dimensional Analysis in Mechanics [in Russian], Gostekhizdat (1957).

ELECTRICAL GAS-DYNAMIC LASERS WITH THERMAL IONIZATION

A. S. Biryukov, V. M. Marchenko, and L. A. Shelepin

The possibility of a gas-dynamic laser with electrical excitation of a thermally ionized gas is studied in detail. The conditions for initiating a spatially uniform discharge in a supersonic flow without the use of external preionizers are discussed. It is shown that energy transfer from the electrons to the vibrational degrees of freedom of molecules is highly efficient. The efficiency of lasers using this effect is three times that of ordinary gas-dynamic lasers.

1. In this article we examine the physical processes which take place as an electric current passes through supersonic flows of a weakly ionized molecular gas and the mechanisms for producing a population inversion in the vibrational levels of the molceules. The discussion is limited to mixtures which are thermally ionized in shock tubes, in explosions, and in high-temperature combustion. The supersonic flows are produced as heated gases escape through a slit or nozzle and by free expansion. An electric current is created in the gas flow by means of an external electric field.

Our purpose is to find the conditions for existence of a spatially uniform, unconstricted discharge in a supersonic flow of a relaxing, thermally ionized plasma and to determine the efficiency of energy transfer to the vibrational degrees of freedom of molecular gases (using the example of CO_2 molecules in a mixture with other gases).

2. Increasing the electric field strength E leads to a rise in the energy transfer from the discharge to the internal degrees of freedom of the gas molecules; however, this increase is limited by the requirement that static breakdown not occur, i.e., that an electron avalanche not be formed [1], or

$$\xi(E/p)\,d < 20, \tag{1}$$

where $\xi(E/p)$ is the first Townsend coefficient (which describes the multiplication of electrons in a field E at pressure p) and d is the distance between the electrodes. Since an avalanche develops over 10^{-7}-10^{-8} sec [2], the effect of charge removal from the field region by the gas-dynamic motion (with a characteristic speed of about 10^5 cm/sec) may be neglected and the breakdown may be regarded as static. Condition (1) for the absence of static breakdown is fulfilled in many cases of practical importance for the gas-dynamic laser arrangement being discussed here. Another requirement that limits the magnitude of the field is that a constricted arc discharge must not develop. This type of thermal equilibrium discharge is caused by a temperature instability of the plasma and is due to its finite thermal conductivity. A current fluctuation causes local heating of the plasma and, therefore, increases its electrical conductivity by increasing the electron density n_e. In order to avoid the ionization-overheating instability [3], the power Q absorbed per unit volume of the discharge gap and going into excitation of molecules must satisfy the condition

$$Q = en_e v_d E' < \frac{\gamma}{\gamma-1}\frac{p}{\tau_E}. \tag{2}$$

Here e and $v_d = v_d(E/p)$ are the charge and drift velocity of the electrons, γ is the adiabatic index, E' is the field inside the discharge gap outside the cathode fall region, and τ_E is the time the plasma spends in the field. In view of the limitation on E', the energy input can be increased only by increasing n_e.

We shall examine the possibility of obtaining high values of n_e in an adiabatically expanding, thermally ionized plasma flow due to the finite rate of electronic recombination. The initial equilibrium electron density or degree of ionization i is determined from the Saha formula with a characteristic temperature T_0 depending on the specific means of heating the gas and on the ionization potential. In particular, for a CO_2-N_2-He mixture consisting half of molecules we find $i_0 = 10^{-8}$-10^{-9} when $T_0 = 3600°K$ with $p_0 = 50$ atm at the critical cross section of the slit or nozzle and including dissociation, assuming that the most easily ionized components have an ionization potential I = 10 eV.

It was noted in [4] that the region where the field is applied in a gas-dynamic laser with supplementary electronic excitation must lie at the place where the gas temperature has fallen to $T \leqslant 800°K$. It is inexpedient to apply the field earlier because of the very intense vibrational relaxation and, therefore, the low efficiency of energy transfer to the vibrational degrees of freedom. At lower temperatures the density of working particles is low so the optimum location for the discharge gap depends on the specific initial conditions (parameters) of the system.

3. The problem is to determine the conditions under which the degree of ionization remaining in the discharge gap during nonequilibrium recombination of a thermally ionized plasma is sufficient for triggering a spatially homogeneous discharge without use of an external preionizer.

We shall consider flow from a plane nozzle with a small aperture angle as modeled by adiabatic plane parallel escape of gas into a vacuum through a long slit with half-width r_0. For the density and temperature we have [5]

$$N = N_0 \frac{r_0}{r_0 + v_0 t}, \qquad T = T_0 \left(\frac{r_0}{r_0 + v_0 t}\right)^{\gamma-1}, \tag{3}$$

where T_0 and N_0 are the temperature and density at the critical cross section and v_0 is the propagation speed of the leading layers of the gas [5]. Assuming that the chemical composition of the gas mixture does not change beyond the critical cross section ("freezes out") [6], the equation for the degree of ionization may be written in the form

$$di/dt = -(\beta_1 + \beta_2)\, i - \beta_3 N i^2, \tag{4}$$

where β_1, β_2, and β_3 are the diffusion, attachment, and recombination coefficients, respectively. Using the experimental temperature and pressure dependences of these coefficients [7, 8] we can show that dissociative recombination is the principal electron loss mechanism for $i_0 \sim 10^{-8}$ up to distances of 10 cm from the critical cross section.

It follows from the solution of Eq. (4) that for $r_0 = 0.1$ cm i varies weakly along the expansion axis and is about 10^{-10} as early as ~3-5 cm from the critical cross section where $N \approx 1\text{-}5 \cdot 10^{18}$ cm^{-3}. At these distances, as a rule, a divergence develops between the vibrational temperatures of the antisymmetric and deformed vibrational modes of CO_2 which is sufficient to produce an inversion, and at the same time there is a substantial (10^8-10^9 cm^{-3}) density of free electrons. With them it is possible to produce a spatially uniform discharge which ensures additional energy transfer from the external field to the vibrational degrees of freedom of the molecules. The electrons in such a discharge may be "warmed up" to energies of 1-2 eV which corresponds to the maximum cross sections for electron impact excitation of molecular vibrations.

The degree of ionization in the discharge gap may be increased by three or four orders of magnitude by adding impurities with low ionization potentials such as $[CH_3(CH_2)_2]_3N$ [9] (I = 7.23 eV), CH_3NH-NH, CH_3N-NH_2, $(CH_3)_2N$, $(CH_3)_2N-NH$ (all with I ~ 5 eV [10]), or small amounts of alkali metals at the heater. In these cases the initial temperatures T_0 may be below 3000°K.

Depending on the initial degree of ionization in the discharge gap, the situation is analogous to gas-dynamic lasers with external ionization sources [9, 11, 18]; however, in our case the gas does not have to be preionized to produce sufficient conductivity, and a quasicontinuous regime may be achieved with a substantially greater population inversion than in a gas-dynamic laser without a discharge.

4. The electrons can be "warmed up" with either a dc or rf electrical discharge. In the constant field case the discharge may have a longitudinal [19] or transverse configuration (a scheme is shown in Fig. 1). In the longitudinal configuration (Fig. 1a) the electron emitter is a hot gaseous cathode and the anode is either a grid mounted perpendicular to the flow some distance from the critical cross section or plates oriented along the flow. In the transverse configuration, a discharge is initiated between metal electrodes (Fig. 1b) positioned along the flow and insulated from the nozzle or slit. Their length l must ensure closure of the current (that is, the ions must not be carried away by the flow and removed from the discharge zone). This requirement may be written approximately in the form $l \geq (v/v_i)d$, where v is the flow speed, v_i is the speed of the ions along the field, and d is the distance between the electrodes.

We note that the most effective energy transfer takes place when the bulk of the applied voltage is in the discharge gap and only a small part of it goes into the cathode region. Estimates of the cathode potential fall φ using the equations given in [18] for E/p = 20-60 V · cm^{-1} · torr^{-1}

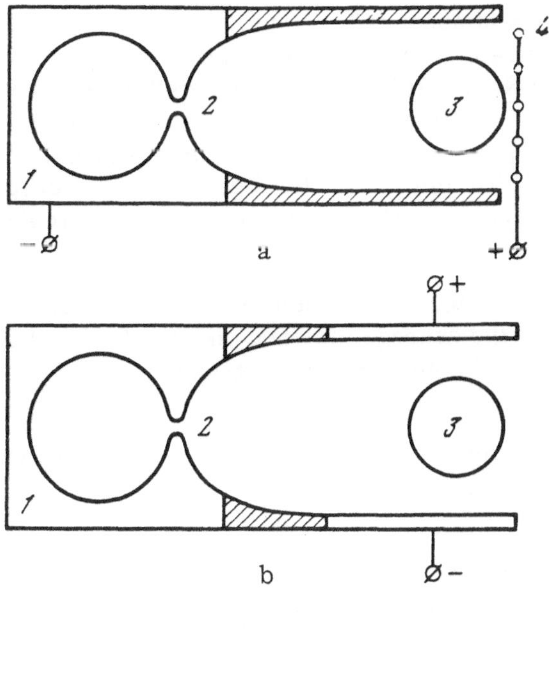

Fig. 1. A schematic diagram of the nozzle system. 1) Heater; 2) nozzle (slit); 3) optical cavity; 4) grid.

and $n_e = 10^{10}$ cm^{-3} yield values of φ of 20-25% of the applied voltage (here p = 100 torr and d = 3 cm). Increasing n_e, E/p, or d (for fixed n_e and E/p) leads to a reduction in the relative share of the cathode fall.

An rf field must evidently be introduced into the system (array) of nozzles along the individual cells in a traveling wave regime while the optical axis is situated perpendicular to the array. This makes it possible to avoid spatial modulation of the electron density. An electron gains energy in a rf field during collisions with heavy particles due to transformations of ordered vibrational motion into chaotic motion. This process lasts until the energy of an electron is sufficient to excite internal degrees of freedom of the gas particles and the rate of energy gain drops. We shall estimate the field amplitude E and frequency ω required to heat the electrons to energies of 1-2 eV. When an rf field acts on a plasma the electron temperature T_e may be found from the equation [20, 21]

$$\frac{dT_e}{dt} + \delta\nu (T_e - T) = \frac{e^2E^2}{3km_e f_0} \frac{\nu}{\nu^2 + \omega^2 (h_0/f_0)^2}, \tag{5}$$

where δ is the fraction of energy lost by an electron in a single collision with a heavy particle, $\nu = \nu(T_e)$ is the frequency of collisions between electrons and heavy particles, T is the gas temperature, f_0 and h_0 are functions (close to unity) determined by the relations among ω, ν, ν_{ei}, and ν_{em} and by the dependence of ν on the electron velocity [21] (ν_{ei} and ν_{em} are the collision frequencies of electrons with ions and neutral molecules), and m_e is the mass of an electron.

For low degrees of ionization $\nu = \nu_{ei} + \nu_{em} \approx \nu_{em}$. The dependence of ν on the speed of the electrons is usually represented in the form $\nu \propto v_e^r$, where r is a constant. It was observed in [22] that in the case of a molecular plasma $r \approx 1$ and the expression

$$\nu = \frac{4\pi}{3} a^2 N v \approx 5.5 \cdot 10^{-10} N \sqrt{T} \left(\frac{T_e}{T}\right)^{1/2} \tag{6}$$

may be used for ν, where a is the radius of a molecule.

The principal error in the estimates of the electron temperature, as in the case of a constant field, is due to the uncertainty in δ. It is known that in a molecular gas with rotational and vibrational degrees of freedom δ is much larger than the relative energy loss in elastic collisions ($\delta_{elas} = 2m_e/M \approx 10^{-5}$) and attains values of about 10^{-2}. Thus, using Eq. (5) we shall estimate the lower bound on the stationary value of T_e^0 for $\delta = 10^{-2}$ (setting $dT_e/dt = 0$). For field frequencies $\omega \approx \nu (f_0/h_0)$, T = 500°K, and N = $2 \cdot 10^{18}$ cm^{-3}, we obtain $T_e^0/T \approx 5 \cdot 10^{-2}$ E from Eq. (5) using Eq. (6). A field amplitude E ≈ 1 kV/cm ensures that $T_e^0 \sim 2$ eV. (This value of E is much less than the breakdown rf field.) Here, if ω is higher than the collision frequency between electrons and molecules and $n_e \approx 10^9$ cm^{-3}, then the field is not screened and easily penetrates the plasma since $\omega \gg \omega_p$, where ω_p is the plasma frequency. Increasing n_e further would require use of microwaves.

Analogous estimates of the stationary value of T_e^0 for a dc electric field showed that a field strength E ~ 1 kV/cm is sufficient to maintain the average electron energy at a level of 2 eV in this case as well. The time to reach T_e^0 from the moment an electron enters the discharge gap is only 1-5 · 10^{-9} sec.

These estimates of n_e and T_e demonstrate that conditions are fully realizable under which it is possible to transfer additional energy from an electric field to the internal degrees of freedom of the molecules without breakdown.

5. An experiment was set up to study the possibility of obtaining an unconstricted electrical discharge in a supersonic flow. A heated mixture of molecular gases was prepared by

exploding [23] a stoichiometric mixture of acetylene and oxygen half diluted with nitrogen. A high degree of ionization of the detonation products is characteristic of hydrocarbon mixtures and it exceeds the thermal equilibrium value [24] by several orders of magnitude. In the chosen mixture values of $n_e \approx 10^{12}$ cm^{-3} and $T_0 \approx 2.8 \cdot 10^{3\circ}$K may be achieved. The detonation products were carried along a tube from the high-pressure chamber and were spread out in the perpendicular direction in front of the entrance to a wedge-shaped nozzle with a width of 37 cm and a full aperture angle of 30°. The critical cross section of the nozzle was 2.5 mm and the length along the flow, 0.8 cm. The nozzle formed an extended stream of relaxing ionized gas. Cylindrical copper electrodes were mounted transverse to this stream at a distance of 7 cm from the critical cross section. In this region the pressure was about 100 torr and the gas temperature was about 1000°K. The electrodes were connected to a 1200-μF condenser. The spatial configuration of the discharge was studied by means of its luminosity using a high-speed framing camera. The pictures were taken upstream of the flow (Fig. 2). The discharge between the electrodes took place as soon as the stream of ionized gas reached the interelectrode zone. Depending on how much the gas has spread out in front of the slit, the discharge takes up more space and is cut off due to a reduction in the voltage on the electrodes. No signs of constriction of the discharge were observed when the field was increased to 750 V/cm.

6. We now go to a more detailed examination of the physical kinetics and the mechanisms which determine additional energy transfer to a gas-dynamic laser with a supersonic nozzle. We shall determine the inverted populations in the CO_2 molecule due to "warming" of the thermal electrons in the discharge.

First of all we shall write the equation for the variation in the electron density n_e along the axis of the nozzle when there is a discharge gap in it (Fig. 1), without specifying the nozzle configuration. This equation is nothing other than the continuity equation with sources:

$$\partial n_e/\partial t+\nabla(\mathbf{v}_d+\mathbf{v})\,n_e=A+\nabla(D\nabla n_e). \tag{7}$$

Here A is a symbolic representation of the electron sources and sinks, $\mathbf{v}$ is the directed velocity of the gas, $\mathbf{v}_d$ is the drift velocity of the electrons in the presence of the field, and D is the diffusion coefficient. We now introduce the notation $i = n_e/N$, where N is the density of heavy particles, and transform the left-hand side of Eq. (7) assuming there are no chemical reactions. Then, in place of Eq. (7) we have

$$\frac{di}{dt}=\frac{A}{N}-\nabla i\mathbf{v}_d-i\mathbf{v}_d\frac{\nabla N}{N}+\frac{\nabla(D\nabla iN)}{N}. \tag{8}$$

We shall consider two variants of applying the field (See Fig. 1a and b) and assume that the flow is quasi-one-dimensional along the nozzle.

Case a. The drift velocity and the flow velocity are parallel and $|\mathbf{v}_d| = v_d(x)$ or $v_d = v_d[(E/p)(x)]$. Usually in the discharge zone E/p = const and $v_d \approx$ const. Then from Eq. (8), we have

$$\frac{di}{dx}=\frac{1}{v_d+v}\left[\frac{A}{N}-iv_d\frac{1}{N}\frac{dN}{dx}+\frac{\nabla(D\nabla iN)}{N}\right]. \tag{9}$$

The equation of mass conservation has the form Nvf = const, where $f = S(x)/S_0$, with S(x) being the area of the perpendicular cross section of the nozzle and S_0 the area of the critical cross section. Then

$$\frac{1}{N}\frac{dN}{dx}=-\left(\frac{1}{v}\frac{dv}{dx}+\frac{1}{f}\frac{df}{dx}\right)$$

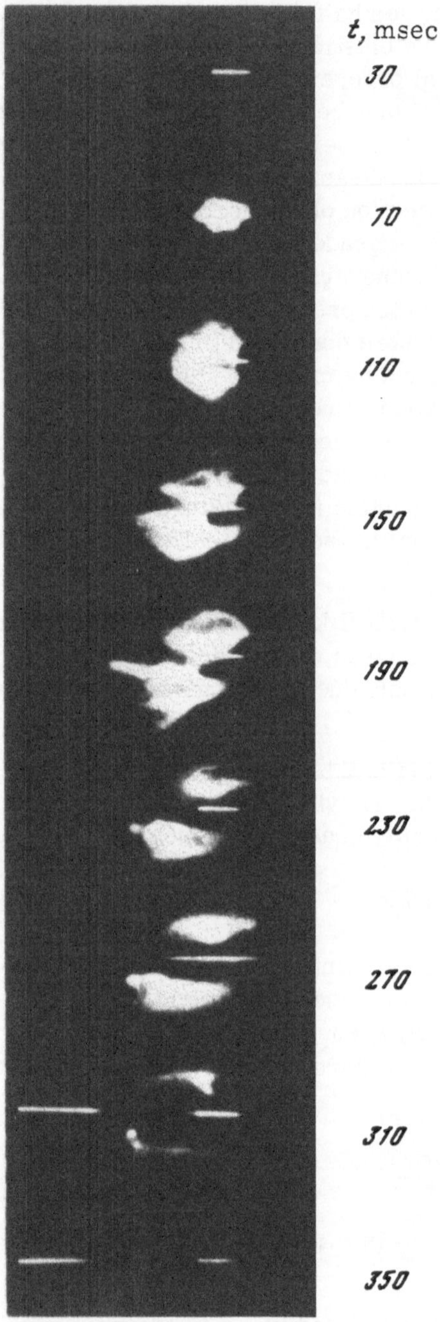

Fig. 2. High-speed framing camera pictures of the luminosity of an electrical discharge in a supersonic flow of explosion products from a $C_2H_2 + 2.5O_2 + 3.5N_2$ mixture between cylindrical electrodes with an initial field strength of 280 V/cm. The anode is located above. The luminosity of the gas through the nozzle slit is visible in the center portion of the pictures.

and the equation for i takes the form

$$\frac{di}{dx}=\frac{1}{v}\left[\frac{A_1}{N}+\frac{\nabla(D\nabla iN)}{N}\right] \tag{10a}$$

outside the discharge zone, and

$$\frac{di}{dx}=\frac{1}{v_d+v}\left[\frac{A_2}{N}+iv_d\left(\frac{1}{v}\frac{dv}{dx}+\frac{1}{f}\frac{df}{dx}\right)+\frac{\nabla(D\nabla iN)}{N}\right] \tag{10b}$$

in the discharge zone. Expressions for A_1 and A_2 will be given below.

Case b. $\mathbf{v}_d \perp \mathbf{v}$. We shall assume that N = const (constant nozzle cross section) and E/p = const in the discharge zone. In the steady-state regime there is a slight gradient in n_e along the field; however, the distribution is such that in the main region $n_e = j/v_d$ = const [25] (j is the current density). Thus, v_d = const and has only a component perpendicular to the nozzle axis. In case b, Eq. (8) takes the form

$$\frac{di}{dx}=\frac{1}{v}\left[\frac{A_1}{N}+\frac{\nabla(D\nabla iN)}{N}\right] \tag{11a}$$

outside the discharge zone and

$$\frac{di}{dx}=\frac{1}{v}\left[\frac{A_2}{N}+\frac{\nabla(D\nabla iN)}{N}\right] \tag{11b}$$

within the discharge zone.

Furthermore, assuming approximately that the nozzle can be represented by a plane-parallel channel along an arbitrarily small section Δx, and that D = const in this section, we obtain

$$\nabla(D\nabla n_e)=D\nabla^2 n_e=-D\frac{n_e}{(D\tau)}\approx-\frac{Dn_e}{\Lambda^2},$$

where τ is the damping time, determined by the intensity of diffusion over the characteristic length, and Λ is the characteristic diffusion length for the lowest harmonic (see, for example, [7, 8]). In the case of a plane-parallel geometry, $\Lambda^{-1} = \pi/L$, where L is the distance between the plates. The diffusion term in Eqs. (10) and (11) may be approximately written in the form

$$\frac{\nabla(D\nabla n_e)}{N}=-D_e i\left(\frac{\pi}{L}\right)^2,$$

where D and L are already taken to be functions of x, i.e., $D_e = D_e(T_e)$ and $L = y(x) = r_0 f$, where r_0 is the half-width of the critical cross section. Here D_e is the coefficient of free (and not ambipolar) diffusion since space charge cannot develop because of the small degree of ionization ($i \ll 1$). We can find an expression for D_e in [7]: $D_e = (v^2/3\nu)_{avg}$, where $v(T_e)$ is the average electron speed and $\nu(T_e)$ can be determined from Eq. (6).

7. Next we shall find expressions for the electron source powers, A_1 and A_2. To do this we write the processes which determine the density of the electron component in a mixture containing both atomic and molecular gases. The following processes are included:

$$\begin{array}{ll}
1) & A^+ + e + e \rightleftarrows A + e,\\
2) & AB^+ + e + M \rightleftarrows AB + M,\\
3) & AB^+ + e \rightleftarrows A + B,\\
4) & A + e + M \rightleftarrows A^- + M,\\
5) & AB + e \rightleftarrows A^- + B.
\end{array} \tag{12}$$

From the plasma quasineutrality condition we have $n_e + n_- = n_+$ (however, for very small degrees of ionization this relation may not hold).

We now estimate the recombination rates of the positive ions in reactions (1)-(3) and compare them:

$$\begin{array}{l}
1)\; k_1 n_+ n_e^2 = (n_+ n_e) k_1 n_e = (n_+ n_e) L (n_e/N)(N/L) k_1 = k_1 (n_+ n_e) i \delta_1 L,\\
2)\; k_2 n_+ N n_e = (n_+ n_e) \delta_1 L k_2,\\
3)\; k_3 (n_+ n_e) = (n_+ n_e) k_3,
\end{array}$$

where L is the Loschmidt number (2.687 · 10^{19} cm^{-3}). The values of the rate constants k_1 and k_2 and their temperature dependences are taken from [26], and k_3 is taken from [27]:

$$k_1 = 10^{-26} T_e^{-9/2},\ cm^6/sec; \quad k_2 = 10^{-30} T_e^{-5/2}, cm^6/sec; \quad k_3 = 3 \cdot 10^{-9} T_e^{-3/2},\ cm^3/sec.$$

It is clear that reaction (3) predominates over reaction (1) if

$$i\delta_1 < 10^{-2} T_e^3 \quad (T_e, eV).$$

Even for relatively high values of i (up to 10^{-6}) and large densities ($\delta_1 \sim 10$, $N \sim 3 \cdot 10^{20}$ cm^{-3}) this relation is satisfied over a wide range of temperatures ($T_e \gtrsim 0.13$ eV). An increase in i or δ_1 requires an increase in T_e.

Reaction (3) predominates over (2) as well. For this to be so the condition $\delta_1 < 10^3 T_e$ must hold, as it does over a very wide range of δ_1 and T_e. At the same time reactions (1) and (2) often have comparable rates. Thus, for simplicity we shall consider the case in which the only easily ionized component in the mixture is molecular, so there are no positive atomic ions (or very few of them). In this case it can be assumed that recombination of the positive ions takes place mainly through reaction (3).

The mechanism by which negative ions are formed may have a substantial effect on ionization relaxation and on the electron density [26]. Thus, Eq. (4) is fundamental. The largest value of k_4 given in the currently available literature is the rate constant in molecular oxygen, which for $T_e \sim 0.1$ eV is satisfactorily written in the form [26]

$$k_4 \approx 5 \cdot 10^{-\ 1} T_e^{-1/2},\ cm^6/sec.$$

Because of a lack of experimental data on the attachment cross sections, in calculating the lower limit of i (or n_e) we have assumed that attachment to CO and O takes place with the same efficiency as to O_2, that the N_2 and He components of the mixture do not participate in electron capture [27] (they take part only as a third particle), and that the rate constant for attachment to CO_2 and to an easily ionized impurity is equal to $5 \cdot 10^{-32} T_e^{-1/2}$. (The maximum capture cross section in CO_2 is roughly an order of magnitude lower than for O_2.).

As for reaction (5), it has a large energy threshold determined by the lowest electron kinetic energy at which this reaction is possible. The binding energy of the electron in a negative ion is equal to $\varepsilon_{AB^-} = D_{AB^-} - \varepsilon_{min}$, where D_{AB} is the dissociation energy of the molecule. From this $\varepsilon_{min} = D_{AB} - \varepsilon_{AB^-}$ [27]. As a rule, for molecules $D \sim 5$ eV and $\varepsilon_{AB^-} \sim 1$ eV (electron affinity); hence, $\varepsilon_{min} \gtrsim 4$ eV (an exception is the halogens, for which $\varepsilon_{AB^-} \gtrsim D$). Thus, reaction (5) may be neglected in the kinetics of a negative charge. The reverse reaction is also weak [26].

According to [26] (for O_2 or air at high initial temperatures) we may conclude that breakdown of the "equilibrium" ratio of the densities of positive ions and electrons and breakdown of the ratio $i_-/i = x$ take place almost simultaneously. For the times over which this ratio is destroyed we obtain values of $\sim 10^{-3}$ sec. For gas speeds of about 10^5 cm/sec, this corresponds to distances from the nozzle throat of roughly 10^2 cm; that is, in our case the relation $i_- = i\varkappa$ must clearly be regarded as valid over the entire length of the nozzle.

Therefore, the expression for A_1/N may be written in the form

$$\frac{A_1}{N} = k_3 N (1 + \varkappa)(i^{*2} - i^2) + N^2 (i^* - i) \sum_M \gamma_M k_4^M,$$

where i* is the equilibrium ($T_e = T$) degree of ionization. This expression is valid both for very small and for very large deviations of i from i*. In the intermediate case it is an approximation. If we introduce the notation $k_4N^2 = \nu_a$ (sec^{-1}) (attachment, or capture, rate) and $k_3N = \beta$ (sec^{-1}) (recombination coefficient [7]), we obtain

$$\frac{A_1}{N} = \beta(1+\varkappa)(i^{*2}-i^2)+(i^*-i)\sum_M \gamma_M \nu_a^M, \tag{13}$$

where γ_M is the molar fraction of the M-th component of the mixture.

8. It should be further noted that in a discharge gap where it is assumed that there are electrons with an average energy of about 2 eV, the gas can be ionized by electron impact. If we denote the ionization frequency in the field by ν_i, then the effect of this process is included in the equation for n_e by adding terms of the type $\nu_i^M iN\gamma_M = \xi^M v_d iN\gamma_M$, where ν_i^M is the ionization frequency of the M-th component and ξ is the first Townsend ionization coefficient for the M-th component. Thus,

$$\frac{A_2}{N} = \frac{A_1}{N} + iv_d \sum_M \xi^M \gamma_M. \tag{14}$$

In computing A_2 we have used the experimental dependences $\xi^M(E/p)$ given in [7], which we have represented in analytic form by

$$\xi^M(E/p) = pa\exp[-b(E/p)^{-1}],$$

where *a* and *b* are constants chosen for each component to make the best fit to the experimental curve.

9. The initial conditions for Eqs. (10)-(11) correspond to complete thermodynamic equilibrium at the critical cross section of the nozzle for p_0 and T_0, that is, $i(0) = i^*(T_0, p_0)$ and $T_e(0) = T_0$, while the equilibrium chemical composition of the gas, consisting of CO_2 and N_2 molecules and He, was found from the system of equations

$$K_1 = \frac{\gamma_{CO}\gamma_O}{\gamma_{CO_2}}N, \quad K_2 = \frac{\gamma_O^2 N}{\gamma_{O_2}}, \tag{16}$$

where

$$\gamma_O + \gamma_{CO} + \gamma_{CO_2} + \gamma_{O_2} = 1 - \gamma_{N_2} - \gamma_{He} = \text{const},$$

$$K_1 = 3.57 \cdot 10^{28} T^{-0.7} \exp(-63\,260/T);$$

$$K_2 = 6.21 \cdot 10^{26} T^{-0.5} \exp(-59\,380/T).$$

K_1 and K_2 are the equilibrium constants (in cm^{-3}) for the respective reactions [28]

$$CO_2 + M \rightleftarrows CO + O + M,$$

$$O_2 + M \rightleftarrows 2O + M.$$

The equilibrium ionization i* in the presence of only one easily ionized component B with molar fraction γ_B was calculated using the formula [26]

$$(i^*)^2 \approx \frac{225\gamma_B T^{3/2}}{\delta_1}\exp\left(-\frac{I}{T}\right)(1+\varkappa)^{-1}, \tag{17}$$

where $\varkappa = i_-^*/i^* \approx (\delta_1/225T^{3/2}) \sum_M \gamma_M \exp(-\varepsilon_M/T)$, ε_M is the electron affinity of the M-th component [10], T is given in electron volts, and J = 5 eV.

10. We again consider the vibrational relaxation equations. The basic equations without pumping sources for the vibrational levels have been obtained in [29]. Here we shall use the system of equations (2.28) in Biryukov's article [30] assuming there are no chemical reactions, but with a source for electronic pumping of the vibrational levels. The energy flux from the electrons into the j-th vibrational degree of freedom, as in [30], has the form

$$W_{e,j}\omega_j \frac{1}{F} iN\left(1 - e^{\frac{\theta_j}{T_e}} x_j\right), \tag{18}$$

where F is the vibrational statistical sum of the molecule, $x_j = \exp(-\theta_j/T_j)$, T_j is the vibrational temperature of the j-th vibrational mode, $\theta_j = h\nu_j/k$, ω_j is the average number of quanta $h\nu_j$, excited during a single electronic impact [5, 27], $W_{e,j} = \langle \sigma_{e,j} v \rangle$ is the probability of electronic excitation of vibrations, and averaging is done over a Maxwellian distribution because the mean energy of the electrons corresponds to the maximum cross section for excitation of vibrations, $\sigma_{e,j}$, and the form of the velocity distribution function does not have a strong effect on $W_{e,j}$.

The dependences of $\sigma_{e,j}$ on the energy of the electrons $\mathcal{E}$ were taken from [31-33] and represented analytically as follows:

$$\begin{aligned} \sigma_{e,N_2} &= 2.85 \cdot 10^{-13} (\mathcal{E} - 1.2)^7 e^{-6,9(\mathcal{E}-1,2)}, \\ \sigma_{e,CO} &= 5.5 \cdot 10^{-13} e^{-8.53\mathcal{E}} \mathcal{E}^{15}, \\ \sigma_{e,CO_2}(00^0v) &= 4.6 \cdot 10^{-15} \mathcal{E}^{2.352} e^{-2.69\mathcal{E}} \qquad (\mathcal{E}, \text{eV}). \end{aligned} \tag{19}$$

Using Eq. (19) for $W_{e,j}$ we find

$$\begin{aligned} W_{e,N_2} &= 0.1156 T_e^{-3/2} e^{-1,2/T_e} \left(6.9 + \frac{1}{T_e}\right)^{-9} \left(13.57 + \frac{1}{T_e}\right), \\ W_{e,CO} &= 7.7 \cdot 10^8 T_e^{-3/2} \left(8.53 + \frac{1}{T_e}\right)^{-17}, \\ W_{e,CO_2}(00^0v) &= 2.923 \cdot 10^{-6} T_e^{-3/2} \left(2.69 + \frac{1}{T_e}\right)^{-4.352} (W_{e,j}, \ \text{cm}^3/\text{sec}). \end{aligned} \tag{20}$$

The temperature variation of $W_{e,12}$ for excitation of the lower levels of CO_2 was chosen in accordance with [34]. The gas-dynamic equations were written as in [6, 30]; however, the terms describing the change in the molecular weight and chemical composition as the gas expands were left out. At the same time, the expression for the flux of energy into translational motion of the particles from the vibrational degrees of freedom has the same form as before since the energy is transferred from the electrons to translational degrees of freedom through excitation of vibrations and direct transfer may be neglected.

By analogy with [35] (see also Eq. (5)] the equation for T_e in a constant field is written

$$\frac{dT_e}{dx} = \frac{1}{v}\left[-\delta\nu(T_e - T) + \frac{e^2E^2}{3km_e\nu}\right], \tag{21}$$

where $\nu = \nu(T_e)$ is found from Eq. (6) and δ is taken to be $2.6 \cdot 10^{-2}$. As noted already, the stationary value of T_e in the discharge gap is reached after a very short time (about 10^{-9} sec); thus, T = const over practically all the discharge region. It is clear from Eq. (20) that the maxima of the probabilities for electron impact excitation of molecular vibrations lie within the limits T_e = 0.7-1.6 eV. Hence in our calculations we assumed T_e = 1.2 eV, and the value of the field strength E required for this was determined by the equation

$$E = \frac{\nu}{e} \sqrt{3km_e\delta (T_e - T)}. \tag{22}$$

The calculation showed that outside the discharge gap T_e is close to T.

11. The system of equations (11) and (21), together with the equations of vibrational relaxation and gas dynamics [6, 30] including Eqs. (6), (13)-(20), and (22), was solved on an electronic computer. The calculations were not intended to optimize this system with supplementary pumping in the discharge. Our purpose was to demonstrate the feasibility of efficient energy transfer from the applied field to vibrational degrees of freedom and to obtain the characteristic dependences of the population inversion and vibrational temperatures on the system parameters. Thus, we have only analyzed a transverse discharge configuration in a nozzle with a logarithmic expansion section [6, 30]:

$$f(x) = \frac{m}{r_0} \ln (px + q).$$

Beginning some distance x = x* from the throat, the nozzle became a constant-cross-section channel and the gas passed into a discharge gap of length $l_2 - l_1$ (l_1 = x*). The results of the calculation are shown in Figs. 3-5. It was assumed that there is 5% of an easily ionized impurity with an ionization potential of 5 eV in the CO_2-N_2-He mixture so the initial degree of ionization i* determined by Eq. (17) reaches values $\gtrsim 10^{-7}$ even at T_0 = 2000°K and including the possible dissociation of this impurity.

Figure 3 shows typical dependences of the inverted population ΔN, the vibrational and gas temperatures, and the degree of ionization on the distance along the nozzle axis. It is clear that during expansion the gas temperature falls rapidly while T_3 and T_4, beginning at x ~ 1 cm, reach their "frozen" values (Fig. 3a). The degree of ionization decreases by roughly three orders of magnitude and then changes little, remaining considerably higher than the equilibrium value.

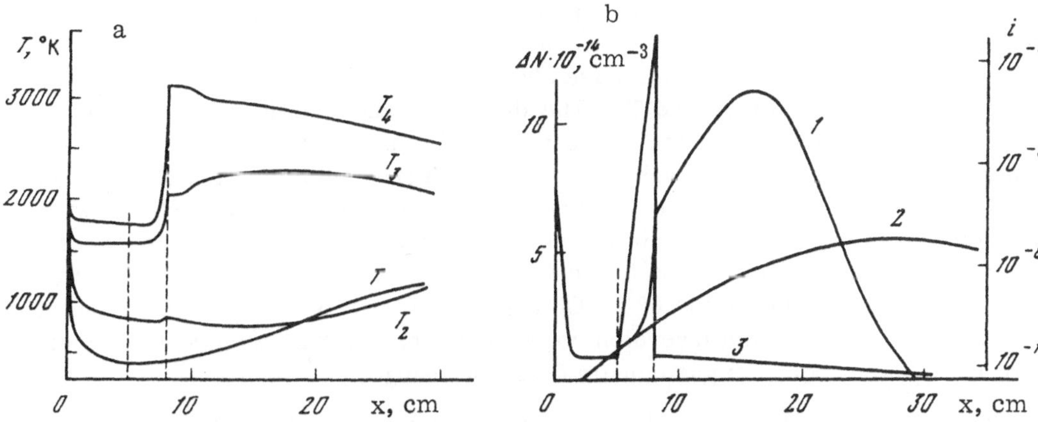

Fig. 3. Typical variations of the vibrational and gas temperatures, the inversion ΔN (curve 1), and the degree of ionization (curve 3) with distance along the nozzle. The initial composition of the mixture in the heater is $CO_2:N_2:He$ = 0.15:0.3:0.5 (5% impurity). T_0 = 2000°K; p_0 = 10 atm; tan θ/m = 0.2175; r_0/m = 8.75 · 10^{-3}; m = 4. The dashed lines denote the discharge region (l_1 = x* = 5 cm, l_2 = 8 cm). Curve 2 gives the results of calculations with E = 0.

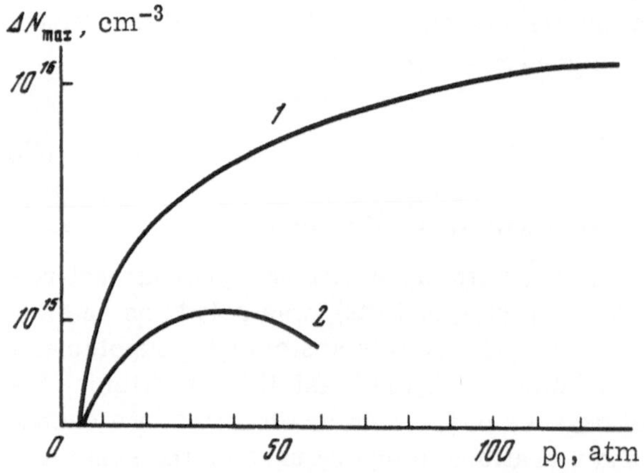

Fig. 4. Dependence of the maximum inversion (curve 1) on the total pressure p_0 at the critical cross section for the conditions of Fig. 3. Curve 2 gives the results of calculations with E = 0.

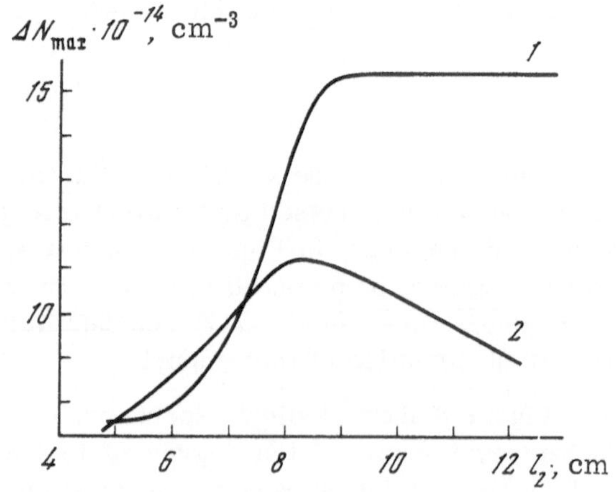

Fig. 5. Dependences of ΔN_{max} on the length $(l_2 - l_1)$ and location of the discharge zone on the nozzle axis for the conditions of Fig. 3. 1) $l_1 = x^* = 2$ cm; 2) $l_1 = x^*$, $l_2 - l_1 = 3$ cm.

From the moment the gas reaches the discharge gap, a rapid multiplication of the electrons in the field sets in. The multiplication rate is substantially limited by recombination and attachment (capture). It is shown in [4, 30] that electron densities $\gtrsim 10^{11}$ cm^{-3} are necessary for efficient excitation of vibrational states in an expanding relaxing CO_2-N_2 mixture. It is clear from Fig. 3 that until the degree of ionization in the discharge gap reaches a certain value ($\sim 10^{-6}$) the vibrational temperatures T_3 and T_4 are independent of the presence of the electron component. Then T_3 and T_4 increase sharply with i, and toward the end of the discharge gap they may noticeably exceed T_0. The deformed and symmetric vibrational modes are also excited by electrons; however, because of the high relaxation rate, the increase in T_2 is insignificant. After the discharge gap, the degree of ionization falls rapidly (over a distance of less than 0.1 cm) to about 10^{-10} and then decreases slowly with increasing x. The increase in the inverted population in this section of the nozzle is due, on the one hand, to the decrease in T_2, and on the other, to the fact that because of the difference between T_3 and T_4 nonresonant exchange between the 00^01 CO_2 and v = 1 N_2 levels becomes important and leads to a growth in T_3. The maximum inversion corresponds to reaching the greatest divergence between T_3 and T_2 since here the gas density does not change.

Despite the fact that the energy defect between the upper levels of N_2 and CO_2 is insignificant (~30°K), intense nonresonant exchange is the principal reason for the growth in the gas temperature up to distances of about 20 cm. An increase in T leads to an increase in the rate of relaxation of all vibrational modes and the system approaches equilibrium at an ever increasing rate.

For clarity we show the inversion distribution for the same gas-dynamic parameters without a discharge in Fig. 3b. It is clear that in this case the maximum inversion in the field increases by a factor of 2 compared to an ordinary gas-dynamic laser.

Figure 4 illustrates the dependence of the maximum inversion on the pressure at the critical cross section of the nozzle. In the presence of a field, as the pressure increases ΔN_{max} may exceed the inversion in a gas-dynamic laser by more than an order of magnitude. Heating the electrons and maintaining T_e at a level of 1.2 eV requires (depending on the pressure) a field strength of 50-300 V/cm. Under these conditions more than 30% of the total number of CO_2 and N_2 molecules may be in excited levels (00^0v CO_2 and $v \geqslant 1$ N_2) as opposed to $\leqslant 10\%$ of the molecules when the disequilibrium is produced by purely gas-dynamic means. This is evidence of the possibility of producing highly efficient energy transfer to the vibrational degrees of freedom of molecules by including supplementary electrical pumping in a gas-dynamic laser.

Figure 5 shows a typical dependence of the maximum inversion on the length of the discharge gap and its location along the nozzle (for constant length). The longer the discharge region, the greater the energy delivered to the vibrational levels; however, in this case the flux of energy from vibrational to translational motion is important and in the end leads to the existence of an optimum gap length (curve 1).

The optimum location of the discharge region (curve 2) is determined by two factors: on the one hand, the fact that for small l_1 the contribution of relaxation is large (the gas density and temperature are high), and, on the other, that for large l_1 the number of working particles per unit volume decreases. Therefore, in each specific case the efficiency of energy transfer depends on the size and location of the discharge zone, and the excitation conditions must be chosen taking the relaxation processes into account.

12. The efficiency of such a system can be estimated using the formula

$$\eta_{max} = \frac{\frac{h\nu_3 - h\nu_1}{k}(x_3 - x_2^2)}{(2.5 + 2.5k_{N_2} + 1.5k_{He})\Delta T + \sum_j \frac{h\nu_j}{k}(\varepsilon_j^0 - \varepsilon_j')k_j + \frac{\nu e i v_d E' \cdot 10^7}{k\gamma_{CO_2}(1 + k_{N_2})v}},$$

where $\Delta T = T_0 - T_{init}$; $T_{init} = 300°K$ is the gas temperature in the heater before the initial state with p_0 and T_0 is created in the system, $k_j = N_j/N_{CO_2}$, $\varepsilon_j^0 = \varepsilon_j(T_0)$ is the average number of vibrational quanta $h\nu_j$ per molecule, $\varepsilon_j' = \varepsilon_j'(T_{init})$, and k is the Boltzmann constant. Estimates of the energy transfer due to thermal heating of the gas and discharge show that they are roughly the same for an applied electrical power of $Q = 10^4$ (W/cm^3). If we assume that the efficiency of an ordinary gas-dynamic laser without a discharge is of order about 1% and the inversion in an electrical gas-dynamic laser averages about an order of magnitude more, then the efficiency of an electrical gas-dynamic laser is three to five times higher than that of a gas-dynamic laser.

LITERATURE CITED

1. I. S. Marshak, Usp. Fiz. Nauk, 71:631 (1960).
2. H. Raether, Electron Avalanches and Breakdown in Gases [Russian translation], Mir, Moscow (1968).
3. E. P. Velikhov, S. A. Golubev, Yu. K. Zemtsov, A. F. Pal', I. G. Persiantsev, V. D. Pis'mennyi, and A. T. Rakhimov, Zh. Éksp. Teor. Fiz., 65:543 (1973).
4. A. A. Biryukov and L. A. Shelepin, Zh. Tekh. Fiz, 43:355 (1973); Preprint FIAN, No. 130 (1972).
5. A. S. Biryukov and L. A. Shelepin, Zh. Prikl. Mekh. Tekh. Fiz., No. 4, p. 25 (1973).
6. A. S. Biryukov and L. A. Shelepin, Zh. Tekh. Fiz., 44:1232 (1974); Preprint FIAN, No. 59 (1973).

7. S. C. Brown, Elementary Processes in Gaseous Discharge Plasmas [Russian translation] Atomizdat, Moscow (1961).
8. J. B. Hasted, Physics of Atomic Collisions, American Elsevier (1972).
9. J. S. Levine and A. Javan, Appl. Phys. Lett., 22:55 (1973).
10. V. I. Vedeneev, L. V. Gurvich, V. N. Kondrat'ev, V. A. Medvedev, and E. L. Frankevich, Dissociation Energies of Chemical Bonds; Ionization and Electron Attachment Potentials [in Russian], Izd. AN SSSR, Moscow (1962).
11. V. M. Adriyakhin, E. P. Velikhov, S. A. Golubev, S. S. Krasil'nikov, A. M. Prokhorov, V. D. Pis'mennyi, and A. T. Rakhimov, Pis'ma Zh. Éksp. Teor. Fiz., 8:346 (1968).
12. H. Seguin and J. Tulip, Appl. Phys. Lett., 21:414 (1972).
13. O. P. Judd, Appl. Phys. Lett., 22:95 (1973).
14. R. K. Gransworthy, L. E. S. Mathias, and C. H. H. Carmichael, Appl. Phys. Lett., 19:506 (1971).
15. C. A. Fenstermacher, M. J. Nutter, J. P. Rink, and K. Boyer, Bull Am. Phys. Soc., 16: 42 (1971).
16. C. A. Fenstermacher, M. J. Nutter, W. T. Leland, and K. Boyer. Appl. Phys. Lett., 20:56 (1972).
17. A. Crocker, H. Foster, H. M. Lamberton, and J. H. Holiday, Electron. Lett., 8:460 (1972).
18. N. G. Basov, E. M. Belenov, V. A. Danilychev, O. M. Kerimov, I. B. Kovsh, A. S. Podsosonnyi, and A. F. Suchkov, Zh. Éksp. Teor. Fiz., 64:108 (1973).
19. V. P. Chebotaev, Dokl. Akad. Nauk SSSR, 206:334 (1972).
20. A. V. Gurevich, Zh. Éksp. Teor. Fiz., 35:392 (1958).
21. J. Shkarovsky, T. Johnson, and M. Bachynsky, Particle Kinetics of Plasmas, Addison-Wesley (1966).
22. Ya. L. Al'pert, V. L. Ginzburg, and E. L. Feinberg, The Propagation of Radio Waves [in Russian], Nauka, Moscow (1963).
23. Yu. A. Bokhon, I. I. Davletchin, V. M. Marchenko, A. M. Prokhorov, A. I. Serbinov, and Ya. K. Troshin, Kratk. Soobschch. Fiz., No. 11, 52 (1972).
24. S. A. Miller, Acetylene: Its Properties, Manufacture, and Uses, Vol. 1, Academic Press (1964).
25. B. M. Smirnov, The Physics of Weakly Ionized Gases [in Russian], Nauka, Moscow (1972).
26. N. M. Kuznetsov, Zh. Prikl. Mekh. Tekh. Fiz., No. 5, p. 42 (1966).
27. B. M. Smirnov, Atomic Collisions and Elementary Processes in Plasmas [in Russian], Atomizdat, Moscow (1968).
28. L. V. Gurvich et al., Thermodynamic Properties of Pure Materials; A Handbook [in Russian], Izd. AN SSSR, Moscow (1962).
29. A. S. Biryukov and B. F. Gordiets, Zh. Prikl. Mekh. Tekh. Fiz., No. 6, 29 (1972); FIAN Preprint FIAN, No. 32 (1972).
30. A. S. Biryukov, Tr. FIAN, 83:13 (1975) (this volume, preceding article).
31. G. J. Schulz, Phys. Rev., 135A, 988 (1964).
32. N. N. Sobol'ev and V. V. Sokovikov, Usp. Fiz. Nauk, 91:425 (1967).
33. M. J. Boness and G. J. Schulz, Phys. Rev. Lett., 21:1031 (1968).
34. A. S. Biryukov, B. F. Gordiets, and L. A. Shelepin, IX Intern. Conf. on Phenomena in Ionized Gases, Collected Articles, Bucharest (1969).
35. A. V. Gurevich, Zh. Éksp. Teor. Fiz., 38:116 (1960).

THE THEORY OF PLASMA LASERS

L. I. Gudzenko, L. A. Shelepin, and S. I. Yakovlenko

A review of electronic-transition lasers in a plasma medium is given. Plasma lasers using a recombination flux are analyzed in detail, and their advantages are pointed out. The basic types of plasma lasers are discussed. They use electrical-discharge (including those utilizing electron beams) plasmas, moving plasmas, expanding (plasma-dynamic lasers) and compressed (pinch discharge lasers) plasmas, and plasma chemical reactions (including lasers employing molecular electronic transitions with dissociating lower terms). Short-wavelength lasers, in particular the possible construction of an x-ray laser, are discussed. Experimental research is also analyzed.

INTRODUCTION

Plasmas are of interest as an amplifying medium primarily for the following two reasons.

First, as opposed to solids, liquids, or gases, a plasma does not change its state of aggregation when a high-density energy is delivered to it; thus, it may be hoped that plasma lasers will have much higher energies than lasers using other media.†

Second, in a plasma it is possible to efficiently populate excited electronic states of atoms, ions, and molecules, transitions from which lie in the short-wavelength range. This opens up the prospect of lasing, not only in the visible or ultraviolet, but even in the x-ray range.

A highly ionized plasma was first considered as a lasing medium in 1963 [2]. Then, and in a number of subsequent articles mentioned below, it was shown that substantial gain coefficients could be obtained in a dense, recombining plasma. Lasers using a recombining plasma were called plasma lasers.

At the same time, the development of lasers using various kinds of gaseous discharges also led to the appearance of lasers using plasmas. Such lasers, in which amplification takes place when the gas is in an ionization regime, are known as gaseous discharge lasers [3]. Here we shall retain this terminology, despite the fact that in recent years, as the power of gaseous lasers has been increased, the degree of ionization of the amplifying medium has remained large.

Therefore, two types of plasma lasers must be distinguished: a) gaseous, in which the active medium amplifies the radiation under ionization conditions, and b) plasma, in which the amplifying medium is a recombining plasma. This complex terminology also reflects the fact that in the recombination regime the plasma properties show up more strongly than in the ioni-

† We note that not only with solid-state (ruby, neodymium, semiconductor, etc.) and liquid lasers, but even with CO_2 lasers, experiments are approaching a limit beyond which the working material is changed chemically (destroyed) [1] if the power is increased.

zation regime. One can say that the qualitative difference between plasma and gaseous lasers is that the deviation of their active media from thermodynamic equilibrium is in opposite directions. If in a gas laser the temperature of the free electrons T_e in the medium is greater than the equilibrium temperature T_u at which the degree of ionization coincides with the actual value, then for a plasma laser $T_e < T_u$. In each specific case this qualitative difference determines the means of producing the amplifying medium. Thus, if a pulsed gas laser operates in the front of the current pulse, then a plasma laser operates in the afterglow (Section 9); if an electron beam is sent into a rarefied medium to make a gas laser, then to make a plasma laser, the beam should be sent into a dense medium (Section 10), and so on.

Why, with the promising properties of plasmas, are the energies and powers now achieved in electronic-transition lasers so meager compared with molecular lasers using vibrational transitions (in particular, with gas-dynamic CO_2 and hydrogen halide chemical lasers)?

This is first of all due to technical difficulties. The electron relaxation times in a dense plasma are smaller than the vibrational–rotational relaxation times in a gas; thus, there are sharply increased demands on the intensity of pumping to the upper working level and of depopulating the lower level in order to produce an efficient amplifying plasma. The very methods of producing a strongly recombining plasma in large volumes, in particular the corresponding means of energy input, have not yet been extensively developed. Choosing the chemical composition and parameters of the plasma requires a rather detailed theoretical analysis of the relaxation processes.

Here it is impossible to underestimate the subjective factors as well. The naturalness and comparative ease of obtaining a population inversion and lasing in traditional gaseous discharge devices have led to a one-sided orientation in experiments on ionization (gas) lasers, in which, as will be explained below (Section 5), the advantages of the plasma medium mentioned here are hardly used. This is reflected in the fact that quite a few original papers, review articles, and monographs deal with gas lasers. The problem of recombining plasma lasers has only received attention in a comparatively small number of theoretical papers until recently. This review (the first devoted to plasma lasers) is intended to fill that gap to some extent, to draw the attention of experimenters to this timely topic, and to outline some already planned trends in experimental research.

In addition, we present the data from the theory of relaxation processes needed to analyze the operation of plasma lasers and briefly discuss the latest achievements in ionization lasers so as to more clearly understand the role and place of plasma lasers and the advantages of the recombination regime.

CHAPTER I

RELAXATION PROCESSES IN A DENSE, LOW-TEMPERATURE PLASMA

1. Plasma Lasers and Problems in Kinetics

Any analysis of the prospects for recombining plasma lasers is closely associated with a discussion of relaxation processes involving a large number of discrete electronic levels. Relaxation in plasmas is the topic of a number of reviews (see, for example, [4-7]) dealing mainly with methods for calculating the recombination and ionization coefficients. The foundations of the theory were laid here during the 1960's for solving problems in research on gaseous discharges, for design of MHD devices and plasma engines, and for studies of flight in the atmosphere. One of the basic approaches was developed by Bates, Kingston, and McWhirter

[8-10], who found the unknown coefficients by numerically solving a system of balance equations for the atomic level populations. Another approach, which has its origin in the work of Belyaev and Budker [11] and Gurevich and Pitaevskii [12-13], is based on representing the motion of the electrons among the excited states as diffusion in energy space. The discreteness of the spectrum has been taken into account by Biberman, Vorobev, and Yakubov (for details, see [7]).

In a study of the prospects for amplification it is necessary to have available much more detailed information than when calculating the recombination and ionization coefficients. For example, it is important to know the populations of the low-lying levels, which usually have no effect on the recombination rate. At the same time there is interest in only a certain range of temperatures and free electron densities for a given plasma chemical composition (determined by the existence of an inversion and the technical possibilities). The details of any approximation scheme used to analyze the gain properties are associated with these features.

An investigation of plasma behavior requires that the variations of a large number of interrelated characteristics be taken into account. It is made much easier by the fact that in most cases of practical interest the characteristic relaxation times of a dense plasma differ greatly from one another. One of the fastest processes is usually establishment of a distribution over the translational degrees of freedom. The populations of the lower discrete levels of the atoms and ions reach their quasistationary values much more slowly since, as opposed to the interaction of free electrons, the principal role in energy exchange between them and tightly bound electrons is played by short-range collisions. For highly excited states lying at a depth less than kT_e below the continuum, the contribution of long-range collisions is important and these states relax almost simultaneously with the establishment of the free electron energy distribution. Thus, they can be joined in a single distribution with the continuous spectrum (which amounts to a reduction in the ionization potential) and referred to as "states in the quasiequilibrium spectrum."

The balance equations (see Section 3, paragraph a), including various elementary processes, are usually used to examine relaxation of the lower levels. In the following paragraph we shall discuss the probabilities of these processes.

2. The Probabilities of Elementary Processes

(a) Radiative Transitions. The rates of spontaneous radiative transitions are determined by the Einstein coefficients $A_{mm'}$. The methods of calculating them are discussed in some detail in several books [4, 14, 15]. The most complete data are given in [16, 17]. For transitions $m \to m'$, about which such data are lacking, it is possible to make an approximate calculation using the Bates–Damgaard tables [15, 18]. For high levels (with effective principal quantum numbers $n > 6$) which are not included in these tables [15, 18], the method of quantum defects and the tables of [19], which make it possible to do the calculations up to $n = 12$, may be used.† Finally, we introduce a simple formula [20] which is valid for highly excited states of hydrogen, but qualitatively describes single electron transitions in various atoms as well:

$$A_{nn'} = 1.5 \cdot 10^{10} n'^{-3} n^{-5} (n^{-2} - n'^{-2})^{-1}, \text{ sec}^{-1}. \quad (2.1)$$

Usually a plasma is optically thin for transitions between excited levels since they are lightly populated, but transitions to the ground states are, as a rule, strongly reabsorbed in a large-volume, dense plasma. Reabsorption by a medium of its own radiation may be convenient-

† In a dense plasma radiative transitions have almost no effect on the populations of these levels, but in certain approximations the cross section for inelastic collisions with electrons is associated with the oscillator strength (and, therefore, with the Einstein coefficient).

ly taken into account by the Biberman–Holstein approximation [21, 22]. This method consists of replacing the actual value $A_{mm'}$ by an effective value $\tilde{A}_{mm'} = A_{mm'}\theta_{mm'}$, where $\theta_{mm'}$ $(0 \leq \theta \leq 1)$ characterizes the probability of escape of a photon beyond the bounds of the plasma. For a Doppler line shape and a long tube of radius R,

$$\theta_{mm'} = \{4\sqrt{\pi}k_{mm'}R[\ln(k_{mm'}R)]\}^{-1/2}, \tag{2.2}$$

where

$$k_{mm'} = \frac{\lambda^3}{8\pi^{3/2}}\frac{g_m}{g_{m'}}\frac{A_{mm'}}{v_T}N \tag{2.3}$$

is the absorption coefficient in the center of the line. Here λ is the wavelength of the transition, g_m and $g_{m'}$ are the statistical weights of the levels, $v_T = (2T/M)^{1/2}$ is the thermal speed of the atom, M is its mass, and N_m is the population of the state m. It is often assumed that $\theta_{1m} = 0$.

(b) Collisions with Electrons. The rates of collisional atomic transitions,

$$X(m') + e \rightarrow X(m) + e \tag{2.4}$$

are characterized by averaging the product of the cross section and the velocity over the free electron distribution:

$$V_{mm'} = \langle\sigma_{mm'}v\rangle \equiv \int_0^\infty \sigma_{mm'}(v)f(v)\,dv, \tag{2.5}$$

where $f(v)$ is the velocity distribution function of the electrons normalized to unity. From the principle of detailed balance it follows for a Maxwellian distribution that

$$V_{mm'} = \frac{g_m}{g_{m'}}V_{mm'}\exp\left(\frac{E_{mm'}}{T_e}\right). \tag{2.6}$$

Here $E_{mm'} = E_m - E_{m'}$ and E_m is energy of the level relative to the continuum. We shall denote the rate of ionization of the m-th level and the rate of the inverse process of three-body recombination

$$X(m) + e \rightleftharpoons X^+ + e + e \tag{2.7}$$

by V_{em} and V_{me}, respectively. For a Maxwellian electron distribution

$$V_{me} = V_{em}\frac{g_m}{g_e g_+}\left(\frac{2\pi\hbar^2}{m_e T_e}\right)^{3/2}\exp\left(\frac{|E_m|}{T_e}\right). \tag{2.8}$$

Here g_e and g_+ are the statistical weights of an electron and an X^+ ion and m_e is the mass of an electron.

In order to write the matrix $V_{mm'}$ for the collisional transitions we must know the excitation and ionization cross sections for the discrete levels being considered. There are a number of thorough reviews devoted to various aspects of the problem of inelastic collisions of electrons with atoms and ions: ionization [23, 24], excitation [25], collisions of electrons with ions [26, 27], strong coupling methods [28], resonances in electron–atom collisions [29], practical computation of cross sections [30], and experimental evaluation of cross sections [31, 32]. Unfortunately, the existing computational methods do not give reliable results in the low-temperature energy range $m_e v_e^2 \leqslant E_{mm'} \leqslant 10$ eV of greatest interest for our purposes. There are insufficient experimental data on transitions between excited levels. Until now there is not even a reliable general method of measuring these cross sections. Thus, so-called semiempirical formulas are widely used which satisfy the available experimental data and yield the correct (in agreement with the Bethe–Born theory) asymptotic behavior of the cross section.

A number of formulas are approximations for the low-energy region [33, 34] while many (Bethe–Born type) give good results for high energies [35-37]. Drawin's approximation, which combines the merits of the Thomson (low energies) and Bethe (high energies) formulas, is convenient [38]. Attempts to improve the accuracy of Drawin's formulas [39, 40] lead to cumbersome and less universal expressions. Based on systematic calculations with the Born and Coulomb–Born approximations, formulas have been proposed for atomic and ionic excitation in which the parameters were determined from computational data obtained by the method of least squares (see the tables in [41]). A major shortcoming is the absence of tabulated values for many transitions of practical interest.

Ordinarily the choice of formulas is determined by the relation between T_e and $E_{mm'}$. In the case of low temperatures ($T_e < E_{mm'}$) it is important to know the slope of the curve $\sigma(v)$ at the threshold and, for high temperatures ($T_e > E_{mm'}$), to know the location of the maximum and the asymptotic behavior. We now give the result of averaging the equations of [15] for excitation and ionization of hydrogenlike states over a Maxwellian distribution:

$$V_{nn'} = 1.7 \cdot 10^{-7} n'^{-5} n^{-3} \left(\frac{Ry}{T_e}\right)^{9/2} \left(\frac{E_{nn'}}{T_e}\right)^{-4} \exp\left(-\frac{|E_{nn'}|}{T_e}\right) W\left(\frac{|E_{nn'}|}{T_e}\right), \text{ cm}^3/\text{sec},$$
$$V_{en} = 8.7 n^{-5} \left(\frac{Ry}{T_e}\right) \exp\left(-\frac{|E_{n,\,n+1}|}{T_e}\right) I\left(\frac{|E_{n,\,n+1}|}{T_e}\right), \text{ cm}^3/\text{sec}. \tag{2.9}$$

Here $W(x) = 1 + xe^x \mathrm{Ei}(-x)$ and $I(x) = [(6 - x)W(x) + x - 2]/12x^3$. The oscillator strengths were computed using a formula analogous to Eq. (2.1).

(c) Dissociative Recombination. At relatively low degrees of ionization in a dense medium dissociative recombination, including a reaction involving negative ions, may play an important role. These questions have been investigated by Tunitskii and Cherkasov (see [42]) and Smirnov [6]. In the present article, where we mainly deal with high degrees of ionization, dissociative recombination will be discussed only in connection with lasers using dissociating (disintegrating) molecules.

(d) Collisions with Heavy Particles. At high degrees of ionization relaxation is usually completely determined by electronic collisions and radiative transitions. According to [12, 13] the quenching efficiencies due to electrons and atoms are comparable when $N_e/N \sim 10^{-6}$ for highly excited states. Thus, when $N_e/N \sim 10^{-5}$ the electronic collisions play a controlling role and heavy particles can affect the populations only by resonance processes [6] (charge exchange, Penning effect, resonance transfer of excitation, etc; for details see Section 8). An estimate of the contribution of nonadiabatic transitions [43] indicates that the efficiency of collisions with heavy particles and with electrons may be comparable when $N_e/N \sim 10^{-4}$ in special cases (crossing of terms). The role of collisions with heavy particles depends on the chemical composition of the plasma and will be discussed below as specific mixtures are analyzed.

3. Population and Temperature Relaxation Equations

We shall write down the relaxation equations for a uniform atomic plasma of simple chemical composition assuming that Maxwellian distributions† with temperatures T_e and T

† A more detailed statement of the problem including spatial inhomogeneities and transport processes is given in [44].

have already been established among the translational degrees of freedom of the electrons and heavy particles. The system includes the balance equations for the populations N_m with an additional condition at the boundary between the continuous and discrete spectra of the electrons (for levels with $m > m_1$ a Boltzmann distribution is valid), an equation for conservation of the number of particles,

$$N = N_+ + \sum_{m<m_1} N_m = \text{const}, \tag{3.1}$$

including the quasineutrality of the plasma ($N_+ = N_e$), and equations for the temperatures of the electrons and heavy particles.

(a) The Equations for the Populations. These equations may be written in the form

$$\frac{dN_m}{dt} = \sum_{m \leqslant m_1} K_{mm'} N_{m'} + D_m \equiv \Gamma_e \quad (m < m_1). \tag{3.2}$$

We shall call the matrix $K_{mm'}$ the relaxation matrix in the following. Its elements give the average number of transitions of an atom per unit time interval from state m' to state m. The diagonal element K_{mm} determines the total loss of particles (per unit time) from state m. The quantity D_m characterizes arrival of particles from the continuum. Including only radiative transitions and collisions with electrons (see Section 2), we write

$$\begin{gathered} K_{mm'} = V_{mm'} N_e + A_{mm'} \quad (m' \neq m), \\ -K_{mm} = K_m = \sum_{m' \leqslant m_1} K_{mm'} + V_{em} N_e = V_m N_e + A_m. \end{gathered} \tag{3.3}$$

In a dense, low-temperature plasma it is possible to neglect radiative recombination compared to three-particle recombination [5] when

$$N_e\,(\text{cm}^{-3}) < 3 \cdot 10^{13}\,[T_e\,(\text{eV})]^{3.75} \tag{3.4}$$

and assume $D_m = V_{em} N_e^2 N_+$.

The effective boundary of the continuous spectrum m_1 is assumed to be at that place where its choice has practically no effect on the results of a calculation of the populations. To find it control calculations were done for various values of m_1 when solving the system on a computer. It is clear beforehand that the "radius" of the electron orbit in the m_1 state is less than the Debye radius; thus, for hydrogenlike states ($m = n$, $m_1 = n_1$),

$$n_1 < n_D \equiv \left(\frac{2r_D}{a_0}\right)^{1/2}, \quad r_D = \left(\frac{T_e}{8\pi N_e^2}\right)^{1/2}, \quad a_0 = \frac{\hbar^2}{me^2}. \tag{3.5}$$

The energy E_{m_1} may be found more accurately knowing that its value lies somewhat above the "bottleneck" E_{m^*} which determines the recombination rate. (In the region $m \approx m^*$ transitions "downward" predominate over transitions "upward.") The recombination coefficient (meaning the flux from the continuum Γ_e as well) may be found by using the diffusion approximation or, directly, from tabulated values. Then the location of E_{m^*} is determined from the equality

$$g_{m^*} \exp\left(\frac{E_{m^*}}{T_e}\right) \sum_{m<m^*} V_{mm^*} = \Gamma_e. \tag{3.6}$$

Therefore, the boundary of the hydrogenlike states ($m = n$) to be considered here lies in the region

$$n^* \leqslant n_1 < n_D, \quad n^* = \sqrt{\frac{Ry}{E_{n^*}}}, \quad Ry = 13.6 \text{ eV}. \tag{3.7}$$

(b) The Thermal Balance Equations. For T_e and T in a homogeneous medium these equations have the form

$$\frac{3}{2} N_e \frac{dT_e}{dt} = \frac{3}{2} T_e \sum_{m \leqslant m_1} \Gamma_m + Q_{\text{inel}} - Q_{\Delta T}; \tag{3.8}$$

$$\frac{3}{2} N \frac{dT}{dt} = Q_{\Delta T}. \tag{3.9}$$

Here the heat transfer to the electron component of the plasma due to inelastic collisions is given by

$$Q_{\text{inel}} = N_e \sum_{m,\, m' < m_1} E_{m'm} V_{m'm} N_m = \sum_{m < m_1} |E_m| \Gamma_m - Q_{\text{rad}}, \tag{3.10}$$

where

$$Q_{\text{rad}} = \sum_{m' < m_1} E_{m'm} A_{mm'} N_{m'} \tag{3.11}$$

is the energy lost as radiation. The quantity

$$Q_{\Delta T} = \frac{2m_e}{M} \nu_{\text{el}} N_e (T_e - T) \tag{3.12}$$

characterizes the heat exchange between electrons and heavy particles (of mass M) in elastic collisions and ν_{el} is the frequency of these elastic collisions. The system of equations (3.1), (3.2), (3.8), and (3.9) complete the initial conditions, which correspond to the moment the sources of disequilibrium cease to act on the plasma.

4. Population Relaxation. Basic Approximations

The system of relaxation equations may be greatly simplified by taking into account the large differences among the relaxation times characteristic of this problem. We shall dwell on the case in which the electron temperature changes slowly.

(a) The Constant (or Stationary) Sink Approximation. This approximation consists of setting the time derivatives of the left-hand side of Eq. (3.2) equal to zero for all states except the ground (m = 1) state. The populations of the excited states as functions of N_e and T_e are found by solving the system of linear algebraic equations

$$\Gamma_m = \sum_{1 < m' < m_1} K_{mm'} N_{m'} + D_m = 0 \quad (m \neq 1). \tag{4.1}$$

The differential equation

$$\frac{dN_1}{dt} = \beta N_e^3 - S N_1 N_e = -\frac{dN_e}{dt} \tag{4.2}$$

is retained for the population of the ground state, where

$$\beta = \frac{\Gamma_e}{N_e^3} = \sum_{m > 1} \left(V_{me} - \frac{1}{N_e^2} N_m V_{em} \right) = \frac{\Gamma_1}{N_e^3} = \frac{1}{N_e^3} \sum_{m > 1} K_{1m} N_m \tag{4.3}$$

is the recombination coefficient and

$$S = V_1 = \sum_{m>1} V_{m1} + V_{e1} \tag{4.4}$$

is the ionization coefficient.

The constant-sink approximation is valid if the characteristic time for changes in the free electron density $\tau_{N_e} = \left|\frac{1}{N_e}\frac{dN_e}{dt}\right|^{-1}$ is much greater than the characteristic relaxation times $\tau_m \sim K_m^{-1}$ of the excited levels. During intense recombination the second term in Eq. (4.2) may be neglected; then $\tau_{Ne} = (\Gamma_e/N_e)^{-1} = (\beta N_e^2)^{-1}$ and the condition for applicability of the constant-sink approximation takes the form

$$\beta N_e^2 \ll \sum_m K_m. \tag{4.5}$$

Since in the stationary state $N_m/\tau_m \leqslant N_e/\tau_e$ (equality corresponds to the case where the entire flux of recombining electrons passes through the m-th level), condition (4.5) leads to the relation

$$\sum_m N_m \ll N_e. \tag{4.6}$$

Beginning with [8] the coefficients β and S have been tabulated numerically using the stationary-sink approximation and have been widely circulated. Computations have been done for helium [45, 46], lithium [47, 48], sodium [49], cesium [50, 51], and argon [52] plasmas. The time-dependent equation (3.2) was solved numerically in [53]. The problem of establishing a stationary sink for the populations was studied in [54]. The solution of the nonstationary equation

$$\frac{d}{dt}\delta N_m(t) = \sum_{m'} K_{mm'}\delta N_{m'} \tag{4.7}$$

for $\delta N_m(t) = N_m(t) - N_m$, the deviations of the populations from the constant-sink values, is written in the form

$$\delta N_m(t) = \sum_{m',m''} \eta_{mm'} e^{-\lambda_{m'} t} \eta^{-1}_{m'm''} \delta N_{m''}(0). \tag{4.8}$$

Here $\lambda_{m'}$ are the characteristic values of the relaxation matrix $\|K_{mm'}\|$, and $\|\eta_{mm'}\|$ is the

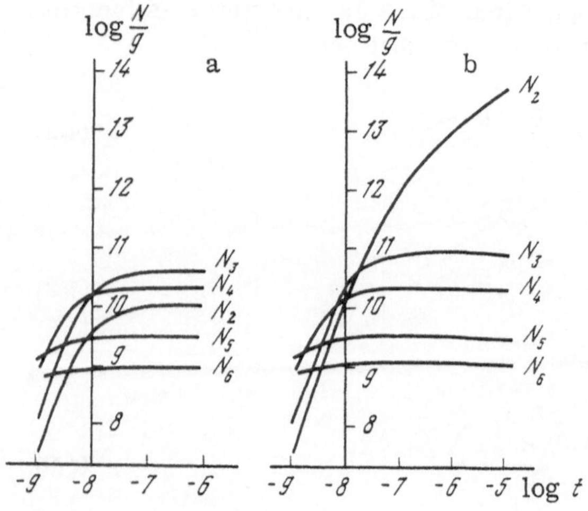

Fig. 1. Buildup of a stationary sink. a) Optically thin hydrogen plasma: $T_e = 0.1$ eV, $N_e = 10^{14}$ cm^{-3}; b) including reabsorption of radiation: $T_e = 0.1$ eV, $N_e = 10^{14}$ cm^{-3}, $A_{1n} = 0$.

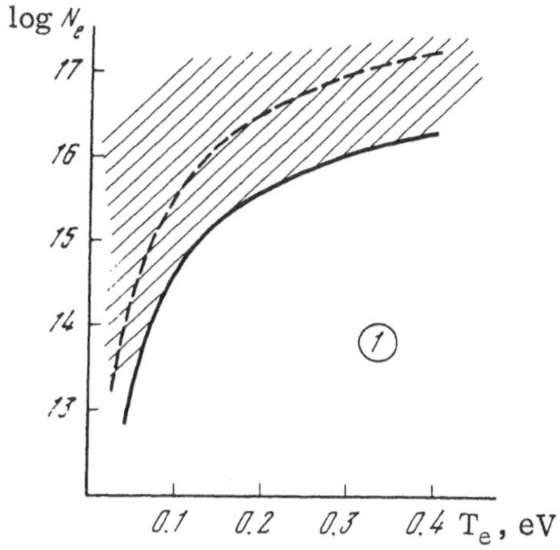

Fig. 2. Values of the plasma parameters for which the "stationary-sink" approximation is valid. The dashed curve corresponds to failure of the inequality $\tau_m < \tau_{Ne}$.

matrix whose columns are the characteristic vectors of $\|K_{mm'}\|$ [55]. The problem reduces to finding these quantities. Figure 1 illustrates the approach of the hydrogen-level populations to a constant sink. The region of applicability of the constant-sink approximation for an optically thin hydrogen plasma is shown in Fig. 2.

A number of approximations are directly connected with the structure of the relaxation matrix. The fact that collisional transitions to nearby levels are, as a rule, more probable than transitions with a large energy defect is widely used.

(b) The Single-Quantum Approximation. This approximation consists of taking into account only those transitions between nearest levels. In the corresponding relaxation matrix only those elements on the principal and the two neighboring diagonals are nonzero. In the time-independent case such a system may be solved by the "screw die" method of [55]; the result may be found in [7, 56].

An analytic solution of the nonstationary equations (3.2) was found in [57] using the single-quantum approximation by expanding the population distribution in a series of time derivatives of the ground state,

$$N_m = \sum_{i=0}^{m-1} \alpha_m^{(i)} \frac{d^i}{d\tau^i} N_1, \tag{4.9}$$

where $\tau = \int_0^t N_e(t')\, dt'$. For $\alpha_n^{(i)}$ in the single-quantum approximation, neglecting radiative transitions the recurrence relation

$$\alpha_m^{(i)} = \sum_{m'=i+1} \frac{g_m \exp\left(-\frac{E_{mm'}}{T_e}\right)}{g_{m'} V_{m-1,m}} \sum_{m''=i}^{m'-1} \alpha_{m''}^{(i-1)} \tag{4.10}$$

was obtained, where $\alpha_m^{(0)} = g_m \exp(-E_{m_1}/T_e)$. In the first approximation, including a stationary sink, the expression for the temperatures θ_n which specify the ratio of the populations of neighboring levels may be written in the form

$$\theta_n = T_e\left[1 + \frac{T_e}{E_{n+1,n}} \ln\left(\frac{1+\chi\beta_n}{1+\chi\beta_{n+1}}\right)\right]^{-1}, \tag{4.11}$$

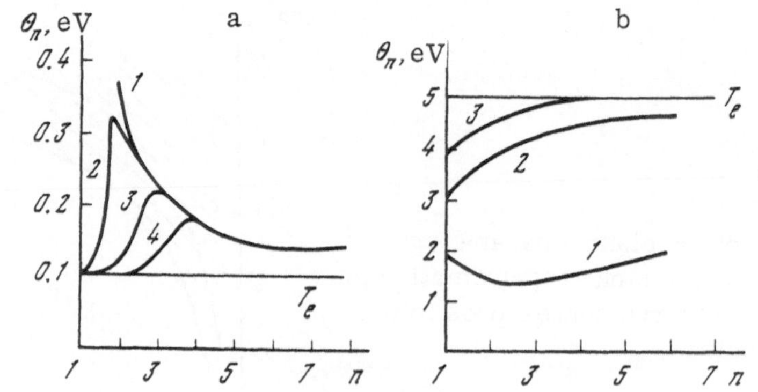

Fig. 3. Quasistationary distributions of the temperatures θ_n over the levels.

where

$$\chi(\tau) = \frac{dN_1}{N_1 d\tau} = 4V_{12}\left[\exp\left(-\frac{E_{21}}{\theta_1}\right) - \exp\left(-\frac{E_{21}}{T_e}\right)\right],$$
$$\beta_n = \sum_{m=2}^{n} \frac{\exp(E_m/kT_e)}{g_m V(m, m-1)}. \qquad (4.12)$$

The parameter χ depends on the nature of the disequilibrium; during recombination it is positive, and during ionization $\chi < 0$. Figure 3 shows the quasistationary distributions of the temperatures over the levels. The sequence of curves 1, 2, and 3 corresponds to the approach to the equilibrium state. Recombination corresponds to Fig. 3a and ionization to Fig. 3b. In [57] Eq. (4.12) is also generalized to relaxation of multielectron atoms and radiative transitions are also taken into account. The resulting stationary values of the populations (assuming a Boltzmann distribution, for each n, over the orbital quantum number l) are in accordance with numerical calculations for Ar [52], He [45], and Li [47]. These distributions are similar to the Treanor distribution which is widely used in analyzing molecular vibrational kinetics [58]. Due to inclusion of the time dependences of the populations they may be useful for analyzing the nonstationary regimes of plasma lasers.

(c) The Diffusion Approximation. As already noted in Section 1, this approximation makes it possible to analyze the population kinetics of highly excited states. Regarding the principal quantum number (and the energy of the state) as a continuously variable quantity and assuming that transitions take place mainly between neighboring levels, it is possible to reduce the system of equations (3.2) to a second-order differential equation of the Fokker–Planck type. Its stationary solution yields [5, 12]

$$\beta_z(T_e) = 8.8 \cdot 10^{-27} \Lambda z^3 [T_e(\text{eV})]^{-9/2} = 5.2 \cdot 10^{-23} \Lambda z^3 \, [T_e \text{ (thous. deg)}]^{-9/2}, \qquad (4.13)$$

where z is the charge of the recombining ion and $\Lambda \sim 1$ is (a form of) the Coulomb logarithm. This expression is valid for low temperatures T_e when the "bottleneck" ($n = n^*$) in the population distribution is sufficiently high.

(d) The Preferred-Sink Approximation.† This approximation involves neglecting upward transitions from low-lying states. In accordance with this (for small T_e) the relaxation matrix is triangular thereby greatly easing the calculations. Thus, for $m = m_0$ we find, in conjunction with the single-quantum approximation, that

$$N_{m_0} = \beta N_e^3 / K_{m_0}. \qquad (4.14)$$

†This name was proposed by Zemtsov [46].

The following scheme is convenient for plasma laser applications. In the region of the "bottle-neck" the diffusion approximation is valid and the recombination coefficient is given by Eq. (4.13). Subsequently, the single-quantum approximation is valid up to some level m_0, and beginning with m_0 the preferred-sink approximation is valid. Then for $m = m_0$ Eq. (4.14) is correct and the populations of the low-lying levels are found by means of the simple recurrence relation

$$N_m = \frac{1}{K_m} \sum_{m'>m} K_{mm'} N_{m'} \qquad (1 < m < m_0). \tag{4.15}$$

Usually the levels m are split into a number of sublevels μ. If this splitting is comparable with T_e and at the same time is much less than the distance between the levels, then for a sufficiently large density N_e the population distribution over the sublevels is close to a Boltzmann distribution

$$N_{m\mu} = N_m g_\mu \exp\left(\frac{E_{m\mu}}{T_e}\right) \Big/ \sum_{\mu'} g_{\mu'} \exp\left(\frac{E_{m\mu'}}{T_e}\right), \tag{4.16}$$

where g_μ is the statistical weight of sublevel μ. The previous equations and approximations, in which

$$K_{mm'} = \sum_{\mu\mu'} K_{m\mu,\, m'\mu'} g_{\mu'} \exp\left(\frac{E_{m'\mu}}{T_e}\right) \Big/ \sum_{\mu''} g_{\mu''} \exp\left(\frac{E_{m'\mu''}}{T_e}\right), \tag{4.17}$$

are true for the total level populations N_m.

The approximations discussed here may be used to examine specific plasma lasers even when the system of equations (3.2) is generalized to processes in mixtures and in the case of self-consistent problems where the interrelated relaxation equations for the plasma and the external disequilibrium source are simultaneously taken into account. In a number of cases application of a system of equations of the type of Eqs. (3.2) to ionization lasers will require additional consideration of the non-Maxwellian velocity distribution of the free electrons.

CHAPTER II

PLASMA LASERS

5. Ionization (Gaseous Discharge) Lasers

Quite different methods of producing disequilibrium may be used to prepare a plasma as a lasing medium: gaseous discharges (in a longitudinal or transverse electric field relative to the amplification direction; with dc, pulsed, or microwave fields), electron or fast ion beams, modulated magnetic fields or pinches, chemical reactions, or optical pumping. The kinetics will mainly be determined by the chemical composition, density, and temperature of the gas, so light amplification may take place on atomic, molecular, or multiply charged ion transitions. However, as a basis for classifying plasma lasers we here choose the direction of the electron flux over the discrete levels. As was noted in the Introduction, it is just in this respect that ionization (or gas) lasers are distinguished from recombination (or plasma) lasers.

As the degree of ionization is increased with constant electron temperature and free electron density, the gain begins to fall after passing through an optimum since the conditions for ionization will disappear. (The medium begins to approach thermodynamic equilibrium.) The degree of nonequilibrium and the gain performance of a recombining plasma, on the other hand, will increase with N_e; i.e., for amplification on electronic transitions there are power limitations connected with the very principle of operation of the ionization laser. In other words, in gas lasers, as opposed to plasma lasers, the very important advantages of plasmas as amplifying media noted in the Introduction can hardly be used. These general assertions require analysis, which will be done for gas lasers in this section and for plasma lasers in the course of examining specific setups in the following sections of this review. Since gas lasers using electronic transitions have been the topic of a number of reviews and monographs (see, for example, [3, 59-62]), we shall mostly limit ourselves here to evaluating the overall situation, only mentioning important results from the latest research.

The maximum average powers are obtained from ion lasers (using electronic transitions), many of which operate in a continuous regime. A detailed review of the features of lasers using the 4p → 4s transitions of Ar II ions has been written by Kitaeva et al. [62]. At present argon lasers with cw powers of up to 175 W have been built [63], and it is considered possible to raise that up to 1 kW with amplifiers [64]. In recent years improvements in the operation of such lasers [65] have mainly been due to development of a new type of cathode [66], use of ceramic tubes [67], and improved heat removal. The operating mechanism of argon lasers has been analyzed in several papers (see, for example, [52, 62, 67]). An investigation of the cw regime of this kind of laser is the topic of [52]. There the population balance equations for the lower nine levels (n, l) were solved together with the equations for N_e, T_e, N, and T. Along with collisional mechanisms and ion losses due to recombination at the walls, radiative processes complicated by reabsorption were taken into account. The controlling role in filling the upper working level of Ar II (4p) is played [52] by transitions from the ground state of Ar II. For the customary parameters of the discharge tube and gas density, an increase in the current density from 50 to 500 A/cm^2 still leads to a substantial (three to four orders of magnitude) increase in the inverted population as a result of increases in both N_e and T_e (Fig. 4). For a 3.5-mm-diameter tube reabsorption of radiation begins to have an effect at a gas pressure of 0.6 torr and a current density j = 250 A/cm^2. When the diameter is increased slightly lasing is stopped due to radiative capture in the 4s → 3p transition. Figure 5 shows the dependence of the population inversion on the gas pressure (p_0). The existence of an optimum pressure is due to competition between the growth in N_e and the fall in T_e as the pressure p

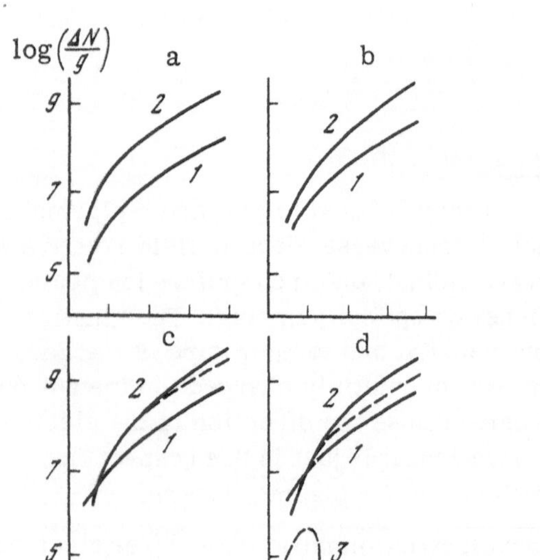

Fig. 4. Dependence of the inversion in Ar II on the density of the discharge current for tubes of various diameters. a) D = 1 mm; b) D = 2.8 mm; c) D = 3.5 mm; d) D = 4 mm. 1) p = 0.1; 2) p = 0.6; 3) p = 0.7 torr. The dashed parts of the curves show the effect of reabsorption of radiation on the inversion.

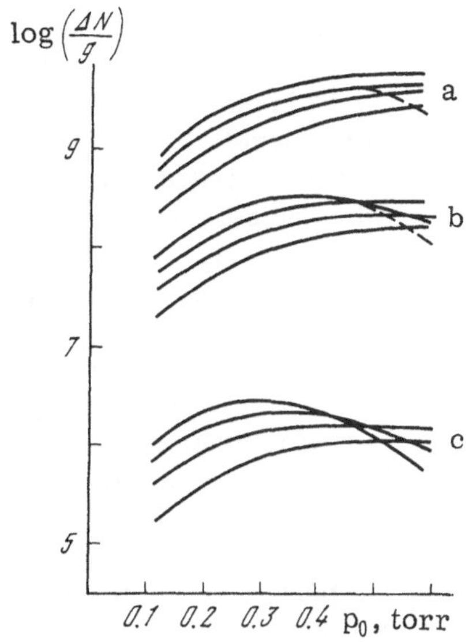

Fig. 5. Dependence of the inversion in Ar II on the filling pressure of the gas at various current densities. a) j = 500 A/cm^2; b) j = 200 A/cm^2; c) j = 50 A/cm^2. 1) D = 1 mm; 2) D = 1.9 mm; 3) D = 2.8 mm; 4) D = 4 mm. The dashed parts of the curves show the effect of reabsorption on the inversion.

is increased. It is also clear from Fig. 5 that at small pressures ($p_0 \lessdot 0.4$ torr) the inverted population increases with the diameter of the discharge tube (see also [69]). At large current densities j and tube diameters reabsorption begins to have an effect, thus limiting the inversion.

Besides Ar II transitions, up to now lasing has been obtained on about 500 ionic transitions of more than 30 elements. The characteristics of cw lasers on a number of lines are shown in Table 1 [70].

Marked success has been achieved in recent years in research on pulsed ion lasers (including doubly ionized). Thus, in a pulsed Xe IV laser (λ = 0.5395; 0.5353; 0.5260 μ) using a discharge current density j = 1500 A/cm^2 at a gas pressure of p = 0.01-0.02 torr, a power P_{pulse} = 20 kW was obtained [71]. Detailed studies have been made of other inert gas ion lasers [73-75]. Besides pure substances such mixtures as Cd–He [76-78] (lasing on Cd II), Se–He [79] (lasing on Se at a number of lines between 4605 and 6444 Å), and He–Hg [80] (lasing on Hg II at the 6149.9 and 7944 Å transitions) were used.

In each of these lasers the free electrons must be heated when their density is not high enough to maintain the upward ionization flux over the discrete electron energy levels. Production of a flux of a certain intensity (for filling the upper working level) is related to the existence of substantial inhomogeneities due to which the volume ionization is compensated by wall recombination. Then the inversion is maintained by depopulation of the lower working level through radiative decay. Both of these mechanisms lose effectiveness when the gas density, electron density, and diameter of the gaseous discharge tube are increased.

TABLE 1

Working substance	Wavelength, Å	Output power P, W	Efficiency, %
Ar II	4880; 5145	120	0.16
Kr II	6471	10	0.01
Cd II	4416; 3250	0.2; 0.02	0.09; 0.01
Ar III	3638; 3511	5	0.01
Xe IV	4954; 5395	0,5	0.01

Production of an inverted population in the ionization flux in the pulsed regime does not have to be directly due to wall recombination of the plasma and radiative depopulation of the lower working level. In pulsed lasers using self-limiting transitions in metal vapors [60, 83-85] the upper working level [8] is filled only while the external applied field is rising, that is, in the front of the current pulse. Since this is excitation from the ground state, to obtain a population inversion the collisional probabilities must satisfy the inequality $V_{b1} > V_{a1}$, where a is the lower working level, most often a metastable state. Depending on how developed the relaxation processes are, the level a is filled. Even in this arrangement for gas lasers, increasing the gas and electron densities and the transverse dimension of the plasma causes a drop, and then a cessation, of amplification. Here the advantages of a plasma are not used; moreover, the manifestation of plasma properties cuts lasing off. A typical pressure is $p \sim 1$ torr and the degree of ionization is very small. The risetime of the current pulse must not exceed the time to fill the lower working level and is usually (for example, in the case of Pb and Tl vapors) about 10^{-8} sec and in the case of copper vapor it is about 10^{-7}-10^{-6} sec. To increase the efficiency the pulse repetition rate is increased. In a copper vapor laser an average power of about 15 W has been obtained in this way [83]. We note that the maximum efficiency obtained for pulsed lasers using self-limiting transitions is greater than for the gas ion lasers discussed above and is about 1%.

6. Recombination (Plasma) Lasers

After what has been said in Section 5, it is not surprising, despite the thermal stability of a dense plasma and the weak dependence on the temperature T_e (over a wide range of its values), the nature of the quasistationary distributions of the electron level populations, and the practical possibilities for various intense interactions with a dense plasma, that the maximum average powers obtained from CO_2 lasers using vibrational transitions [58] are almost three orders of magnitude (!) greater than from gas lasers using electronic transitions. We now turn to research on plasma (recombination) lasers.

The first work was devoted to explaining the fundamental possibility of constructing plasma lasers. Initially the simplest (theoretical) situation was discussed – pulsed recombination of an atomic hydrogen plasma following rapid cooling of its free electrons [2, 53, 84, 85]. Then an analysis was made of several electron cooling mechanisms [86, 87]: in the course of ambipolar diffusion when wall effects are important; by means of adiabatic expansion of a plasma burst into a vacuum; and, due to collisions of electrons with cold heavy particles. After this a discussion was begun of the relaxation of the electronic populations in a decaying plasma of more complex chemical composition [88]: atomic lithium [47], argon [52], helium [46], sodium [49], helium with an impurity [89], atomic hydrogen with impurities [54], ionic beryllium [90, 91] and helium [92], multiply charged ions [91], disintegrating molecules [93, 94], etc. The possibility of making a dense steady-state, recombining plasma and, using it, a cw plasma laser [94-96], was pointed out.

The results of these and a number of other papers on the problems associated with plasma lasers will be discussed in the following sections of this review. In this section it is reasonable to dwell on the simplest relaxation scheme in order to evaluate the gain properties of a decaying, dense plasma. We shall discuss an "open two-level model" [97] of a strongly recombining plasma that amplifies radiation in the $b \to a$ transition, for which it is assumed that the condition for a stationary preferred sink (Section 4, paragraphs a and d) is satisfied. Assuming that transitions from states lying above the upper working level b to the lower working level a and into states lying below a can be neglected, we write

$$\frac{dN_b}{dt} = N_+\tau_+^{-1} - K_bN_b = 0, \qquad \frac{dN_a}{dt} = K_{ab}N_b - K_aN_a = 0, \tag{6.1}$$

where $\tau_+ \equiv \left|\frac{1}{N_+}\frac{dN_+}{dt}\right|^{-1} = (\beta N_e^2)^{-1}$ is the characteristic decay time.† Here it is not necessary that $K_b = K_{ab}$. An inversion of the populations of the working levels is obtained when

$$\delta_{ab} \equiv \frac{N_a}{g_a}\Big/\frac{N_b}{g_b} = \frac{K_{ab}g_b}{K_a g_a} < 1. \tag{6.2}$$

In the case of Eq. (6.1) the expression for the unsaturated gain coefficient [3]

$$\varkappa = \sigma_{ab}^{\mathrm{ph}} N_b (1 - \delta_{ab}) \tag{6.3}$$

(where $\sigma_{ab}^{\mathrm{ph}} = \lambda_{ab}^2 A_{ab}/4\Delta\omega_{ab}$ is the photoabsorption cross section at the center of the line, λ_{ab} is the wavelength of the amplified radiation, and $\Delta\omega_{ab}$ is the linewidth) takes the form

$$\varkappa = \sigma_{ab}^{\mathrm{ph}} \frac{N_+}{K_b \tau_+}(1 - \delta_{ab}). \tag{6.4}$$

The limiting output power,

$$P_{\max} = \hbar\omega_{ab} \frac{N_+}{\tau_+}(1 - \delta_{ab}), \tag{6.5}$$

per unit volume of active medium is also related to the characteristic decay time of the plasma.

From Eqs. (6.1), (6.4), and (6.5) it is clear that the problem of filling the upper working level under the conditions discussed here in fact reduces to obtaining a sufficiently dense supercooled plasma. Actually, the gain coefficient and specific power are proportional to the ion density and inversely proportional to the characteristic ion recombination time. The time τ_+ falls rapidly with a reduction in the free electron temperature [Eq. (4.13)]. It will be shown below that in a dense gas the value of T_e, and therefore of τ_+, is fairly small due to collisions between free electrons and cold heavy particles. This means that methods for uniformly introducing large amounts of energy into substantial volumes of dense plasma must be used, such that the energy delivered to the plasma goes mainly into ionizing the gas without significantly heating the heavy particles. We now consider schemes to ensure a population inversion between selected pairs of electronic levels as a plasma relaxes.

CHAPTER III

RECOMBINATION MECHANISMS FOR POPULATION INVERSION IN ATOMS AND IONS

7. Population Inversion in a Plasma of Simple Chemical Composition

Production of an efficient amplifying medium is the result of the simultaneous action of two seemingly contradictory mechanisms, one of which fills the upper working level, the other of which empties the lower level. Since under dense, intensely recombining plasma conditions

† The ion density N_+ (recombination of which leads to filling of the upper working level) must be distinguished from N_e in general since in the following sections we shall examine recombination of multiply charged ions and of plasmas consisting of a mixture of several chemical elements.

there is no special need to worry about filling the upper level (Section 6), it remains to discuss the possible mechanisms for emptying the lower working level in such a plasma. For simplicity we shall assume here that the temperature and density of the free electrons vary slowly. This statement of the problem corresponds either to the constant-sink (see Section 4, paragraph a) stage (for example, pulsed ionization of the gas followed by cooling of the free electrons) or to recombination disequilibrium being held stationary by an external source (Section 9). At this time it is clear that an inverted population can be achieved for any atom (ion, molecule) over a substantial number of transitions in the recombination regime. But from the standpoint of constructing an efficient laser the greatest interest is in situations where, first, the ratio E_{ba}/I (the energies of the transition and of ionization of the atom) is large and, second, it is possible to empty the a (lower) level by a mechanism that has little effect on the population N_b of the upper level and operates reliably in large volumes of dense plasma. Our analysis of the possible mechanisms for this consists of examining the results of a numerical solution of the system of equations (3.2) for specific cases and of using the simple scheme described in Section 6 in order to understand the basic qualitative features.

(a) Radiative Depopulation of the Lower Working Level. This aspect of the problem was analyzed first [2, 53, 84]. It follows from Eq. (6.2) that for $A_a > (g_b/g_a) \times K_{ab}$ a population inversion develops in the $b \rightarrow a$ transition. However, the right-hand side of this inequality depends on the free electron density, and in going to high densities, $N_e > N_{e0}$, where $N_{e0} \equiv A_a g_a / V_{ab} g_b$, radiative decay can no longer ensure a population inversion. Obviously, the radiative mechanism is extremely sensitive to reabsorption of radiation. For increasing density of the medium and minimal transverse plasma dimension spontaneous decay becomes ineffective. Reabsorption has an especially strong effect on the prospects for obtaining an inverted population among atoms (or atomic ions) when the lower working level is a resonance level. The effective decay probability, $\theta_{1a}\ A_{1a}$, of such a level decreases with increased population of the ground state [Eqs. (6.2) and (6.3)].

Calculations for a highly ionized atomic hydrogen plasma [53] confirm these qualitative considerations. During recombination at $T_e = 0.05\text{-}0.2$ eV and $N_e = 10^{12}\text{-}10^{14}$ cm^{-3} an inverted population is realized on several transitions due to radiative decay of the lower levels. Reabsorption of resonance radiation changes the level population picture considerably, and already in a container with the smallest linear dimension, about 1 cm, no inversion takes place in transitions into n = 2 when the population of the ground state $N_1 \gtrsim 10^{15}$ cm^{-3}. The population inversion among higher-lying levels disappears for $N_e \gtrsim 5 \cdot 10^{13}$ cm^{-3} Observation of a population inversion in the $5 \rightarrow 4$ transition in experiments on the expansion of an argon–hydrogen plasma stream [98] and of lasing during pulsed breakdown of a hydrogen–helium mixture [99] is apparently due to the mechanism being considered here. The impossibility of using high densities (of the gas and electrons) and volumes is an important shortcoming of radiative depopulation of the lower working level of atoms.

(b) Deexcitation by Electrons. These limitations are removed if depopulation of the lower working state is controlled by its deexcitation by cold plasma electrons. According to Eq. (6.2) this mechanism may ensure an inverted population. For this, it is necessary that there be a sufficient density of free electrons and that the condition

$$V_{a}g_{a}/V_{a.}g_{b} > 1 \tag{7.1}$$

which determines the choice of levels and is not directly dependent on the plasma parameters N, N_e, and T_e be satisfied. Such levels a and b cannot be found for all atoms. Calculations [53, 46] show that electron impact deexcitation does not produce an inversion in H atoms, for example. The situation improves substantially on going to atoms with large splitting in the sublevels $\{\alpha\}$ and $\{\beta\}$. Thus, if the sublevel populations have a Boltzmann distribution, then

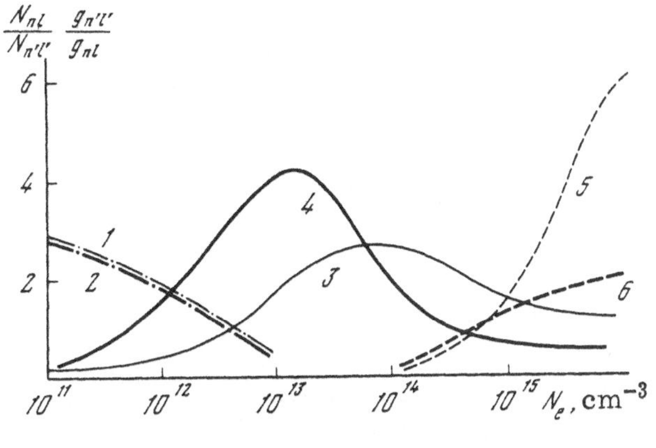

Fig. 6. Dependence of the relative populations of the levels on the free electron density for $T_e = 0.2$ eV. Curves 1 and 2 correspond to the 3p−3s levels; curves 3 and 4, to the 4s−3p levels; curves 5 and 6, to the 3s−2p levels. Curves 1, 3, and 5 were obtained using collisional probabilities calculated according to [33], and curves 2, 4, and 6 according to [41].

Eq. (4.17) implies that in place of the condition (7.1) we have

$$\exp\left(\frac{\Delta E_{b\beta} - \Delta E_{a\alpha}}{T_e}\right) \frac{V_{a}g_{a}}{V_{ab}g_{b}} > 1. \tag{7.2}$$

It is clear that for the "inner" sublevels of states *a* and *b* this is a weaker condition than Eq. (7.1).

The possibility of efficiently emptying the lower working level by deexciting it by means of the free electrons was first demonstrated by Gordiets et al. [47] for the example of a lithium plasma. The system of relaxation equations (3.2) was solved for the nine lower excited levels including reabsorption. The results are shown in Fig. 6. It is clear that as opposed to the 3p → 3s transitions, the inversions in the 4s → 3p and 3s → 2p transitions are due to collisional decay of the lower working level. This is already clear from the fact that an inverted population in these levels develops only after a certain density N_e (for 4s → 3p, when $N_e > 10^{12}$ cm^{-3}; for 3s → 2p, when $N_e > 10^{15}$ cm^{-3}) is attained and does not cease even for complete reabsorption of the radiation. Certain properties of the collisional mechanism are immediately clear from Eqs. (7.1) and (7.2). For example, Eq. (7.2) and the structure of the alkali metal atom terms show why an inversion is formed over the (n + 1) s → np transitions in a dense alkali metal plasma. Moreover, in the calculation for an atomic lithium plasma, this has been confirmed for the sodium atom [49], where an inversion was obtained in the 5s → 4p transition, and for the alkalilike beryllium ion [90, 91], for which the possibility of lasing in the vacuum ultraviolet (λ = 1776 Å; for details, see Section 14) on the 3s → 2p transition has been demonstrated. Calculations show [53] that collisions with cold electrons lead to an inverted population in argon atoms over the 5s → 4p transition for $T_e = 0.2$ eV and $N_e = 10^{14}$ cm^{-3}; here, for $N_e = 10^{15}$ cm^{-3}, the gain coefficient reaches ≃10 cm^{-1}. Therefore, emptying the lower working level by means of deexciting collisions with free electrons is a very general and effective mechanism for transitions with strongly ($\Delta E > T_e$) split levels. Quite recently this conclusion has received confirmation in a series of experiments [100-102] which we shall discuss in more detail (Section 9). The principal advantage of the collisional mechanism is the fact that it makes it possible to use large volumes of a dense medium. This means that a plasma of simple chemical composition is already a possible medium for an efficient recombination laser.

8. Plasmas with Two Chemical Constituents

A plasma of complex composition is a more "flexible" material for use as an active medium since in it populations of individual discrete levels of an atom (ion, molecule) X may be created by using inelastic interactions with heavy particles from an impurity Y chosen for this purpose [88]. Here large cross sections correspond to collisions in which energy is exchanged

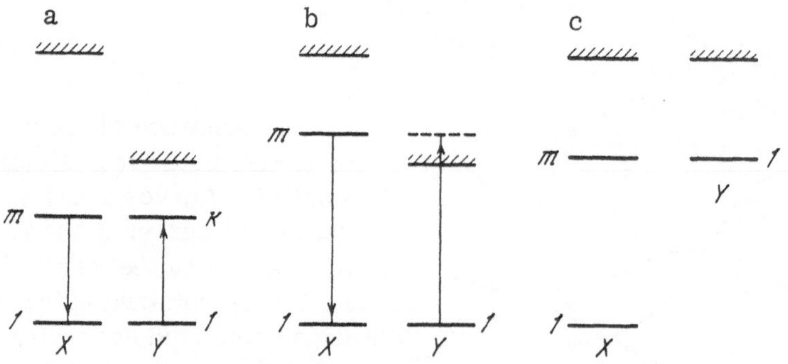

Fig. 7. Excitation transfer reactions: a) Resonance transfer; b) Penning effect; c) resonance charge exchange.

between electron shells [6]. These are usually resonance cross sections,† and including their effects is of primary importance in an analysis of relaxation in the plasma.

(a) Effect of an Impurity on the Character of the Nonequilibrium. The following reactions may have the most efficient influence on the population of the levels X(m):

resonance excitation transfer (Fig. 7a),

$$X(m)+Y(1)\rightleftarrows X(1)+Y(k), \quad (8.1)$$

ionization during excitation transfer – the Penning effect (Fig. 7b),

$$X(m)+Y(1)\rightleftarrows X(1)+Y^{+}+e, \quad (8.2)$$

resonance charge exchange (Fig. 7c),

$$X^{+}+Y(1)\rightleftarrows X(m)+Y^{+}. \quad (8.3)$$

The possibility of three-body chemical reactions will be discussed briefly in Section 12. We shall not consider the charge exchange reaction at this time either (see Section 15), noting only that in the recombination regime it fills the state X(m) [88]. Quasiequilibrium states (Section 1) are efficiently exchanged with the continuum; thus, if reaction (8.1) takes a Y atom into one of these states it is kinetically equivalent to reaction (8.2) for ionization of a Y atom. The effective cross section for such a process (following [88], we shall call it resonance ionization) is, as a rule, much larger than that for "ordinary" ionization [reaction (8.2)]. We shall trace how both reactions of type (8.2) affect the population distribution of an X atom. We shall consider in detail the case in which the constant preferred-sink approximation (see Section 4, paragraphs a and d) is valid for the low-lying excited levels. Including reactions (8.2), we rewrite Eq. (4.1) in the form

$$D_m+\sum_{m'=m+1}^{m_1} K_{mm'}N_{m'}\equiv I_{m,\,m+1}=K_m(1+\eta_m)N_m. \quad (8.4)$$

Here the quantity $I_{mm'}$ characterizes the recombination flux to the m-th level from all levels $k \geq m'$; $\eta_m = (q_m/K_m)N_Y$ characterizes the efficiency of reactions (8.2) for the m-th level; $q_m \equiv \langle\sigma_m v\rangle$ is the rate of these reactions‡; and N_Y is the impurity concentration.

Reactions (8.2) have a substantial effect on levels for which $\eta \gg 1$ and (because of changes in the recombination flux) on lower levels. Usually for the first excited levels $K_m \sim 10^7$-10^8

† That is, they have a sharp maximum when the energy lost by the shell of one heavy particle is equal to the energy gained by the electrons of the other particle.

‡ In Eqs. (8.4) the fact that the reverse reactions of (8.2) may be neglected under these conditions is taken into account (for details, see [88]).

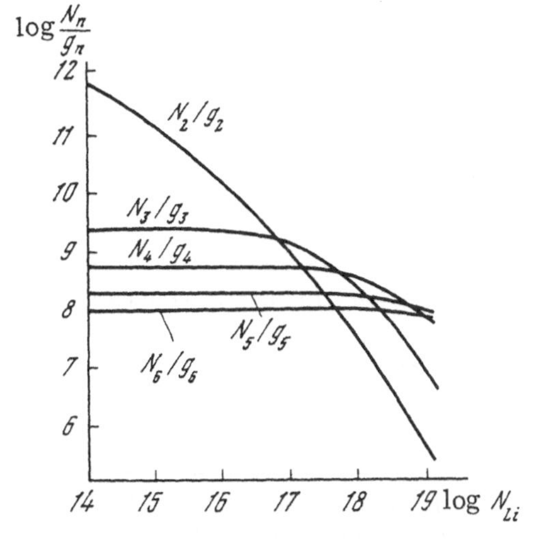

Fig. 8. Dependence of the level populations N_n on the impurity density N_{Li}: $T_e = 0.1$ eV, $N_e = 10^{14}$ cm^{-3}.

sec^{-1}, $q_m \sim 10^{-9}$–10^{-10} cm^3/sec, and reactions (8.2) are important even for $N_Y \sim 10^{16}$–10^{18} cm^{-3}. It is necessary to include the fact that the collision rates V_m as opposed to q_m increase sharply on going to the upper levels while the values of η_m fall sharply. For levels $m \approx m^*$ in the region of the "bottleneck" (see Section 3, paragraph c), reactions (8.2) are usually already unimportant and the presence of the impurity has almost no effect on the value of the recombination coefficient for X. As the impurity concentration N_Y is increased, the populations N_m do not change uniformly as when N_e is increased, but, "in turn," upward from below (see Fig. 8). There are ranges of N_Y in which reaction (8.2) is important for N_m but has no noticeable effect on populations $N_{m'}$, where $m' > m$. In this case $N_m \approx \tilde{N}_m(1 + \eta_m)^{-1}$, where $\tilde{N}_m$ is the population in the absence of an impurity. The presence of an appropriately chosen impurity thus makes it possible to control the populations of individual lower levels.

(b) The Population Inversion. In the recombination regime reactions (8.2) may produce an inverted population. Actually, using Eq. (8.4) with $a = m$ and $b = m + 1$, we write

$$\frac{N_a}{g_a} : \frac{N_e}{g_e} = \frac{g_b}{g_a}\left[\frac{I_{a,a+2}}{I_{b,b+1}}\frac{K_b}{K_a}\frac{1+\eta_b}{1+\eta_a} + \frac{K_{ab}}{K_a}\frac{1}{1+\eta_a}\right]. \tag{8.5}$$

For sufficiently large N_Y we have $\eta_a \gg 1$ and $\eta_b \gg 1$, and, therefore,

$$\frac{N_a}{g_a} : \frac{N_b}{g_b} \simeq \frac{I_{a,a+2}}{g_a q_a} : \frac{I_{b,b+1}}{g_b q_b}. \tag{8.6}$$

As a rule, $g_a I_{b,b+1} \gg g_b I_{a,a+2}$ (we recall that in the single-quantum approximation $I_{m,m+2} = 0$). At the same, usually the q_m are larger for the lower levels; at least it can be assumed that $q_a \not< q_b$. Taking a sufficiently large impurity concentration, we can almost always achieve a population inversion over the transition $X(b) \to X(a)$. Here the question soon becomes one of whether the impurity densities required to obtain an inverted population are reasonable. Of course, the condition $\eta_b \gg 1$ is not necessary for making these estimates. Furthermore, excessively large values of N_Y result in a drop in the gain.

We shall estimate the optimum value with the following assumptions: the single-quantum preferred-sink approximation is valid for the levels a and b (see Section 4, paragraphs b and d); reactions (8.2) are unimportant for levels above level b; and the lower level is emptied only by reaction (8.2). We shall refer to this scheme as the open two-level model. For it the gain coefficient has the form [89]

$$\varkappa = \tilde{\varkappa}\left(1 - \frac{g_b}{g_a}\frac{K_b}{q_a N_Y}\right)\Big/\left(1 + \frac{q_b N_Y}{K_b}\right)^{-1}, \tag{8.7}$$

where $\tilde{\varkappa} \equiv \sigma^{ph} N_{+} \beta N_e^2 / K_b$ is the gain coefficient obtained formally for an empty a level if, in addition, reaction (8.2) does not affect the population N_b. From Eq. (8.7) we immediately find the impurity density N_Y^0 required for a population inversion, $N_Y^0 = g_a q_a (g_b K_b)^{-1}$, and the density N_Y^{opt} corresponding to a maximum in $\varkappa$.

$$N_Y^{opt} = \left(1 + \sqrt{1 + \frac{g_a q_a}{g_b q_b}}\right) \frac{g_b K_b}{g_a q_a}.$$

Relaxation in decaying plasmas made up of the mixtures H + Xe and H + Li have been analyzed numerically by Gudzenko et al. [54]. Depletion of the H(n) levels by reaction (8.2),

$$H(n) + Li \rightarrow H(1) + Li^+ + e \quad (n \geqslant 2)$$
$$H(n) + Xe \rightarrow H(1) + Xe^+ + e \quad (n \geqslant 3)$$

was taken into account by adding the terms $q_n^Y N^Y$ to the diagonal elements of the relaxation matrix. The expression for the cross sections obtained in [103] and Eq. (3.9) for the relaxation matrix were used to compute q_m^Y. We have examined the dependence of the populations on the impurity density. As can be seen from Fig. 8, the populations of the levels do not change together. The intersection points of the curves correspond to the appearance of a population

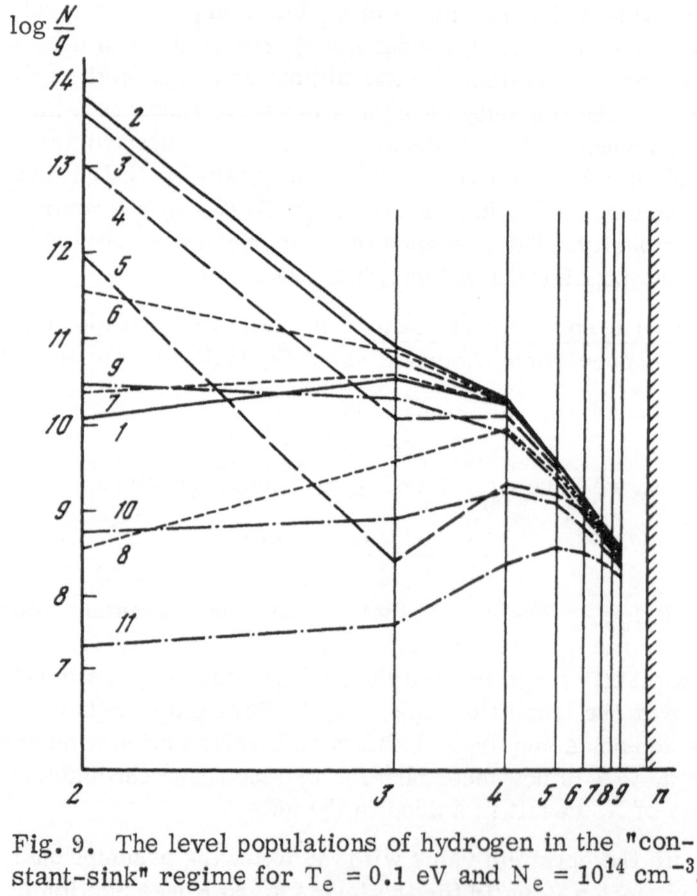

Fig. 9. The level populations of hydrogen in the "constant-sink" regime for $T_e = 0.1$ eV and $N_e = 10^{14}$ cm^{-3}. 1) Neglecting reabsorption of radiation; 2) including reabsorption; 3, 4, 5) with a xenon impurity $N_{Xe} = 10^{17}$, 10^{18}, 10^{19} cm^{-3}; 6, 7, 8) with a lithium impurity N_{Li} = 10^{17}, 10^{18}, 10^{19} cm^{-3}; 9, 10, 11) including the effect of a chemical reaction $\alpha = 10^8$, $5 \cdot 10^8$, and $1.5 \cdot 10^9$ sec^{-1}.

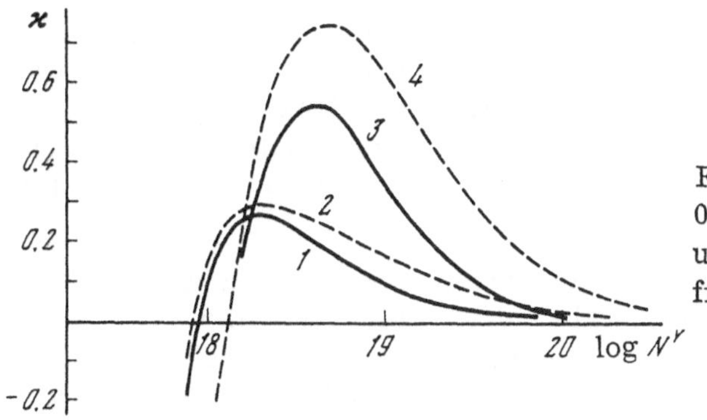

Fig. 10. Dependence of $\varkappa_{n,\,n+1}$ on N_Y for T_e = 0.1 eV and $N_e = 10^{14}$ cm^{-3}. 1, 2) $\varkappa_{23}$ with a lithium impurity; 3, 4) $\varkappa_{34}$ with a xenon impurity; 1, 3) from a computer calculation; 2, 4) from estimates.

inversion in the respective transition and the impurity has little effect on the more highly excited levels. A calculation of the recombination coefficient showed that the existence of impurity has practically no effect on it. The values of N_Y^0 and N_Y^{opt} and the variation of $\varkappa(N_Y)$ are close to the estimates made using Eq. (8.7). An example of a population distribution is given in Fig. 9. Figure 10 shows the variation in the gain coefficient with the impurity density.

The possibility of lasing in a decaying helium plasma with an impurity was examined in a note [89] on the basis of the open two-level model with the working transitions He(3) → He (2). Practically any gas can serve as an impurity: Ne, H_2, N_2, O_2, Kr, Xe, and so on. The most interesting is neon† since for each He state with n = 2 there are several resonant highly excited Ne states, so reactions (8.2) should be highly efficient. An estimate for N_Y^{opt} yields

$$\varkappa \sim 0.02\ \text{cm}^{-1} \text{ for } N_e = 10^{14}\ \text{cm}^{-3},\ T_e = 0.3\ \text{eV};$$
$$\varkappa \sim 0.2\ \text{cm}^{-1} \text{ for } N_e = 10^{15}\ \text{cm}^{-3},\ T_e = 0.1\ \text{eV};$$
$$\varkappa \sim 2\ \text{cm}^{-1} \text{ for } N_e = 10^{16}\ \text{cm}^{-3},\ T_e = 0.3\ \text{eV}.$$

When H_2 is used for Y an inversion sets in for $N_Y^0 \sim 10^3 N_e$. The splitting of the n = 3 level, which would improve the gain characteristics of the plasma, was neglected in [89].

From these results it is clear that the lower working level may be depopulated by ionizing an impurity in the case of dense media. This method is more convenient than electronic deexcitation in many cases. First, obtaining a population inversion in this way depends weakly on the ratio of the cross sections, and, second, the inverted population depends less strongly on the free electron temperature T_e.

CHAPTER IV

ATOMIC AND IONIC PLASMA LASERS

9. Lasing in the Afterglow of an Electric Field Pulse

We emphasize once again that preparing a dense supercooled plasma (along with rapidly depopulating the lower working level in it by means of the technically simple prescriptions given above) is now actually the only problem which, if solved, would permit construction of efficient plasma lasers using electronic transitions of various atoms, molecules, and ions. We

† It is well known that gas lasers operate in a He + Ne mixture with Ne, rather than He, as the active substance. In them the upper working levels of neon are populated by transfer of excitation via the helium metastables (2^1S, 2^3S).

therefore proceed to examine specific methods of obtaining this kind of plasma which differ primarily in the methods of depositing energy in the medium and of cooling the free electrons. We shall first consider relaxation in a plasma ionized by a sufficiently short risetime pulse. After fast cutoff of the ionizing pulse, electron cooling and volume recombination take place in the plasma. To obtain a population inversion it is necessary that cooling take place over a time much less than the relaxation time of the electron density N_e.

In a motionless macroscopic plasma two channels for cooling the electrons are realized: 1) due to wall recombination and the ambipolar flux, and 2) determined by collisions with heavy (cold) particles. The first cooling channel leads to intense recombination of the plasma on the "vessel" walls. Fast electrons hit the wall first, so the plasma takes on a potential φ relative to the container. In the resulting field the ions are accelerated (toward the walls), while the electrons are slowed down until they are so slow that they cannot reach the wall and do not take part in wall recombination. In a steady-state regime with sufficient plasma densities the fluxes of positive and negative particles are balanced and ambipolar diffusion takes place. Since it removes the fastest electrons from the plasma, this mechanism results in wall cooling of the electron component. Biondi has been able to obtain an electron temperature smaller than the ion temperature in a capillary tube [104]. In the steady-state ambipolar diffusion regime the potential φ for a Maxwellian charged particle velocity distribution is given by $\varphi = (T_e/2e)\ln(T_e M_i/T_i m_e)$, where e and m_e are the electronic charge and mass, and M_i is the (singly charged) ion mass. According to [86] the electron cooling rate is then given by

$$\frac{dT_e}{dt} = -\frac{\mu_i}{e\Lambda^2}\ln\left(\frac{T_e M_i}{T_i m_e}\right) T_e (T_e + T_i), \tag{9.1}$$

where μ_i is the ion mobility and Λ is the diffusion length ($\Lambda = R/2.4$ for a tube of radius R). The wall-cooling channel has two important shortcomings: 1) useless (from the standpoint of light amplification) loss of charged particles during wall recombination, and 2) a loss in efficiency as the density of the medium and the transverse dimensions of the container are increased. The second cooling channel is a volume effect and is due to collisions (elastic and inelastic) of free electrons with heavy particles whose temperature is much less than T_e.

The shortcoming of electron cooling by collisions with heavy particles is the requirement that the heavy particles have a low temperature and the plasma have a relatively low degree of ionization. Thus, to produce recombination conditions in a multiply charged ion plasma there remains only the prospect of cooling the medium by expansion and, possibly to some extent, by wall effects. In many cases it is appropriate to make the active medium of a plasma laser a mixture in which, besides atoms or ions X for directly amplifying the radiation [and, possibly, impurity atoms Y for depopulating the lower working level of X by means of reaction (8.2)], there is a buffer gas at a substantial concentration in order to enhance cooling of the electrons on heavy particles without changing the remaining features of the population kinetics of X. Usally helium [47] is most suitable for this purpose since it is chemically inert, has a high ionization potential, and has light atoms that aid energy transfer in elastic collisions with electrons.

Latush and Sem have done some convincing experiments on the construction of plasma lasers in the afterglow of a field pulse [100-102]. In their work a high-voltage longitudinal field pulse was used to produce double ionization in alkaline earth metal vapors. Helium was usually used as a buffer gas, and neon was used (with the least effect) in some experiments for comparison. Lasing was obtained on transitions of the singly charged ions Mg II, Ca II, Sr II, and Ba II, and was observed on infrared, visible, and ultraviolet lines. The results of [103] are of greatest interest: Lasing was obtained in a superradiant regime with a single mirror at λ = 4350 Å in Sr II with a gain coefficient of $\varkappa$ = 28-30 dB/m for a pulsed output power of P_{pulse} = 50 W at helium pressures of above 17 torr. The Ca II λ = 3737 Å transition yielded a gain coefficient of $\varkappa$ = 10-12 dB/m with P_{pulse} = 20 W.

Estimates by Latush and Sem [102] have shown that the lower working states are depopulated by inelastic collisions of the second type with cooled plasma electrons. Experiments were set up to verify directly the recombination character of the process whereby the upper working level is filled by applying a short field pulse to the supercooled plasma to heat the free electrons. This small-amplitude pulse not only cut off laser action while it acted but also caused a dip in (the intensity of) all the luminescence lines from highly ionized states of both the metal ion and the helium atom observed in the afterglow. The powers obtained from plasma lasers are low. This is due to their design and, principally, to the fact that applying a longitudinal field pulse over a 1-m-long discharge tube is not a reasonable method of delivering energy to a dense plasma. Thus, the results of [103] are far from optimal for the chosen chemical composition. However, these papers are important, mainly as publications about successful experiments intended specifically for the construction of plasma lasers.

Reports on observations of lasing in the afterglow of a longitudinal field pulse appeared even earlier. However, in none of the published works known to us were the plasma parameters during lasing estimated or the type of plasma nonequilibrium analyzed. Thus, the role of various processes in populating and emptying the working levels was not explained. For example, a note [105] reported lasing on the $3^3S \rightarrow 2^3P_{1/2}$ (λ = 7065 Å) transition of the helium atom in the afterglow of a high-voltage (U = 50-120 kV) pulsed ($\tau_{pulse} \sim 10^{-8}$ sec) discharge in a He + H_2 ($N_{He} = 5 \cdot 10^{17}$ cm^{-3}, $N_{H_2} \simeq 10^{17}$ cm^{-3}) mixture in a 5-mm-diameter, 80-cm-long tube. Lasing lasted $\tau_{las} \simeq 10^{-6}$ sec and the gain was ~30-40% with a peak power of $P_{peak} \sim 10$ W. In a diametral cross section the glow region had the shape of a ring, bright at the edge and dark near the tube axis. Since without a hydrogen impurity lasing was not observed, Pirton and Fowles [105] explained the inversion by an ion–ion recombination process involving the negative (atomic) hydrogen ion,

$$H^-(1^1S_0) + He^+(1^2S_{1/2}) \rightarrow H(1^2S_{1/2}) + He(3^3S_1), \tag{9.2}$$

which populates the upper working level. However, under these experimental conditions the rate of formation of the H^- ion is too small [106] and cannot compete with three-body recombination.

The inadequacy of the data given in [105] does not allow us to uniquely interpret the details; however, the data are fully consistent with population of the upper level by means of three-body electron–ion recombination and emptying of the lower level by reactions of the type (8.2) between helium and H_2 molecules. The annular luminosity is apparently due to the wall cooling of electrons discussed at the beginning of this section. We also note [107, 108], where lasing was observed in argon ions, not only at the beginning of the current pulse, but at its end, and in the afterglow.

There are a number of papers (see [42]) where the mechanisms for populating the upper working levels are controlled by dissociative and ion–ion recombination. Then lasing was observed in the afterglow as well as in the ionization regime. The role of negative ions in producing a population inversion in the ionization regime was discussed in [61].

Experiments with plasmas of a number of chemical elements confirm the theoretical conclusions presented above about the role of various mechanisms for producing inverted atomic level populations in the afterglow of an ionizing pulse. Thus, two recent articles [109, 110] present detailed measurements of the time variation of the populations of several levels, along with that of the density and temperature of the free electrons. Population inversions due to radiative depopulation of lower levels were observed for several transitions at low pressures. In particular, at a pressure of about $5 \cdot 10^{-3}$ torr the inversion in the $9D_{5/2} \rightarrow P_{3/2}$ transition of cesium varied within the limits $3 \cdot 10^6$-$5 \cdot 10^5$ cm^{-3}, falling with increasing gas density. When the free electron density was increased along with the gas pressure, inversions

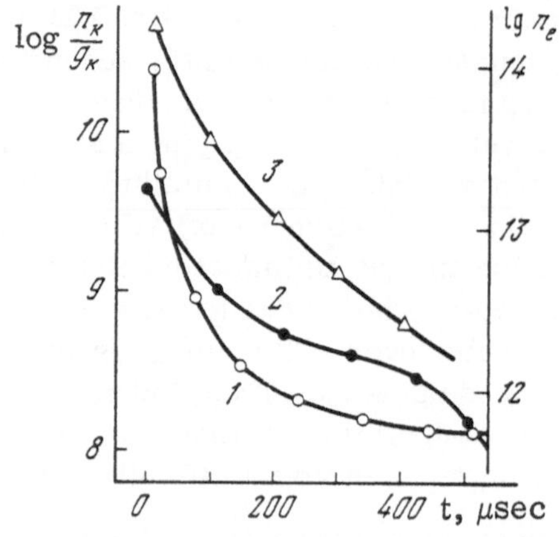

Fig. 11. Time dependences of the reduced populations N_n/g_n of the $5P_{3/2}$ and $6S_{1/2}$ levels of potassium and of the electron density. $p = 3 \cdot 10^{-1}$ torr, $T_e(0) = 2900°K$.

due to emptying of the lower levels by the plasma electrons were observed in several levels. Figure 11 shows the time variation of the reduced populations of the $5P_{3/2}$ (curve 1) and $6S_{1/2}$ (curve 2) levels of potassium, as well as $N_e(t)$ (curve 3) for a pressure of $p = 0.3$ torr and $T_e(0) = 0.24$ eV.

Plasma lasers operating in the afterglow of a longitudinal electric field pulse have a number of attractive features, among which we note the relative technical ease of realizing them and the wide range of choice in the chemical composition of the active medium and in the output frequency. Among the shortcomings we must include the low energy yields of such systems. They are limited, first of all, by the interrelated field and discharge geometry, which make it impossible to pump a high energy density in a large volume. Second, because of the inertia of large volume energy sources special unwieldly devices are required to form intense, short, field pulses. Besides, it is already possible at this time to substantially increase the energy of this type of plasma laser by causing pulsed breakdown with a transverse electric field or a longitudinal high-current electron beam.

10. Continuously Pumped Plasma Lasers

In the last ten years a whole series of theoretical and experimental papers have been devoted to electron level population kinetics and the prospects for pulsed amplification in a freely decaying plasma (after the ionizing pulse is shut off and the free electrons have undergone rapid cooling). Until quite recently, no analysis has been made of the methods for creating a stationary supercooled plasma or of the conditions for continuous maintenance of an inverted population in it. As already noted, in traditional gaseous discharges [11] the free electrons are heated by the electric field in such a way that their temperature T_e exceeds the thermodynamic equilibrium value T_u corresponding to their density and the degree of ionization; that is, the plasma in a gaseous discharge is superheated. In order to steadily maintain a supercooled intensely recombining plasma, in which the mean electron energy obeys the inequality $T_e < T_u$, the electron distribution must deviate from Maxwellian in such a way that there is a sufficient concentration of fast electrons (to produce the given degree of ionization). Thus, we must point out the conditions for forming such a distribution, in particular, the means of delivering energy to a dense plasma. Since ionization is usually accompanied by stationary excitation of levels, including the lower working level, we must make the conditions for a population inversion more precise. The basic problems in computing the electron relaxation and in making a theoretical analysis of the possibility of amplification by a supercooled plasma are practically independent of whether the auxiliary ionization source is a charged particle (electron or ion) beam injected into the medium [94, 95], external x-radiation, or showers of nuclear fragments formed in the medium during nuclear reactions [96].

We shall discuss qualitatively the features of forming an energetic distribution of free electrons in a dense atomic gas that is ionized and excited by a hard external source. The free electrons formed in this manner are divided into fast (with energies $\gtrsim I$, where I is the ionization energy) and plasma (cooled in collisions with heavy particles) groups. Collisions of fast electrons with one another and with plasma electrons may be neglected compared with their inelastic interaction with atoms. The cross sections for Coulomb collisions between plasma electrons are large. This allows us to introduce their temperature T_e, which is close to the comparatively low temperature T of the dense gas; thus, the plasma electrons recombine strongly. Even without special cooling, at the low degrees of ionization $\alpha \lesssim 10^{-3}$ of interest in this problem the temperature T changes much more slowly than the temperature T_e and density N_e of the free electrons. This makes it possible to speak of a quasistationary regime for producing a supercooled plasma. If sufficiently effective cooling of the heavy particles (at the vessel walls by pumping the gas through) is ensured, then the process is stationary. Then we may set $dN_e/dt = 0$ and $dT_e/dt = 0$ in the equations [cf. Eqs. (3.8) and (4.2)]

$$\frac{dN_e}{dt} = \nu_u N - \beta N_e^3, \quad \frac{3}{2} N_e \frac{dT_e}{dt} = E_{\text{pair}} \nu_u N - Q_{\Delta T}. \tag{10.1}$$

Here ν_u is the ionization rate due to the auxiliary source and the fast plasma electrons and E_{pair} is the energy expended by the auxiliary source to produce an electron–ion pair. Proceeding from Eq. (10.1) we can write the following for the times to establish the electron density and temperature:

$$\tau_{N_e} = \frac{\alpha}{\nu_u}, \quad \tau_{T_e} \simeq \tau_{N_e} \frac{T_e}{E_{\text{pair}}}. \tag{10.2}$$

Taking into account the relations $\alpha \lesssim 10^{-3}$, $T_e/E_{\text{pair}} \lesssim 10^{-2}$ we find from this that

$$\tau_{T_e} \ll \tau_{N_e} \ll 1/\nu_u. \tag{10.3}$$

The steady-state values of the plasma parameters [according to Eq. (1.1)] are

$$N_e = \left(\frac{\nu_u N}{\beta}\right)^{1/3}, \quad T_e = T = E_{\text{pair}} \frac{\nu_u}{\alpha \frac{2m_e}{M} \nu_{\text{elas}}}. \tag{10.4}$$

We shall examine the conditions for producing a population inversion on transitions of an X atom in an X + Y mixture, where the Y atoms deplete the lower $X(a)$ and upper $X(b)$ working levels in the course of reaction (8.2). Under the conditions of applicability of the open two-level model (see Section 6), including only the recombination channel for filling the $X(b)$ level and assuming the $X(a)$ level to be filled by excitation from the ground state X(1) (by the external source and the fast electrons), as well as by transitions from $X(b)$, and neglecting all mechanisms for emptying $X(a)$ except reaction (8.2), we may write

$$N_b(q_b N_Y + K_{ab}) = \nu_u N_X, \quad N_a q_a N_Y = \nu_a N_X + K_{ab} N_b.$$

Here N_X and N_Y are the densities of the atoms in the ground state, $q_a = \langle \sigma_a V \rangle$ and $q_b = \langle \sigma_b v \rangle$ are the rates of the reactions (8.2), ν_a is the excitation rate of the $X(a)$ state, and K_{ab} is the rate of the transition $X(b) \to X(a)$ due to radiation and inelastic collisions with plasma electrons. According to Eq. (7.3) we find

$$\varkappa_{ab} = \sigma_{ab}^{\text{ph}} I_{ab}, \quad I_{ab} \equiv N_b - \frac{g_b}{g_a} N_a \tag{10.5}$$

for the gain coefficient.

Using Eq. (10.4) the density of the inversion I_{ab} may be written in the form

$$I_{ab}=N_X\frac{\nu_u}{K_{ab}}\left[\frac{1-\frac{\nu_a}{\nu_u}\frac{q_b}{q_a}\frac{g_b}{g_a}}{1+\frac{q_bN_Y}{K_{ab}}}-\frac{1+\frac{\nu_a}{\nu_u}}{\frac{q_aN_Y}{K_{ab}}\left(1+\frac{q_bN_Y}{K_{ab}}\right)}\right]; \qquad (10.6)$$

I_{ab} may be positive for sufficiently large densities N_Y provided the following inequality holds:

$$R\equiv\frac{\nu_u}{\nu_a}\frac{q_a}{q_b}\frac{g_a}{g_b}>1. \qquad (10.7)$$

This condition limits the choice of working transitions for steady-state amplification. If it is satisfied, then there is a threshold value $N_Y^{(0)}$ of the impurity concentration beyond which I_{ab} and $\varkappa_{ab}$ are positive. The increase in the inversion with N_Y weakens monotonically, and for $N_Y^{(m)} = N_Y^{(0)}[1+(1+K_{ba}/q_bN_Y)^{1/2}]$ the amplification reaches a maximum. Further increasing the impurity density leads to an unbounded drop in the gain coefficient due to the increasing rate of emptying the upper working level with N_Y.

(a) Electron Beam Plasma Lasers. A conceptual scheme for a laser in which the active medium is continuously produced by ionization of a cold gas with an electron beam is discussed in [94, 95]. Estimates for helium with an impurity [94] indicate the feasibility of obtaining a lasing medium by this means which would efficiently amplify radiation in He (3) → He (2) transitions. For temperatures $T_e \leqslant 0.3$ eV the value of the threshold impurity density (which in this case may be almost any gas: H_2, Ne, Xe) is $N_Y \sim 10^{18}$ cm^{-3} and the optimum density is $N_Y \sim 2\text{-}3 \cdot 10^{18}$ cm^{-3}. For a relatively low beam current density j = 10 A/cm^2, particle energies W ~ 10-100 keV, and a helium density $N_x \sim 10^{19}$ cm^{-3}, an estimate of the gain coefficient yielded $\varkappa \sim 2 \cdot 10^{-2}$ cm^{-1}.

The results of a computer simulation of the populations of the hydrogen atom in stationary beam-ionized mixtures of H + Li and H + Xe are given in [95]. In a dense, low-temperature, hydrogen plasma it is difficult in practice to maintain the atomic composition. Furthermore, it clearly is not an optimum medium for amplification because (as opposed to He, for example) its frequency of excitation by hard (fast) electrons to the resonance state is equal to the ionization frequency and the n = 2 and n = 3 levels are not split. It is all the more significant that these calculations show that with a lithium impurity even in atomic hydrogen it is possible to achieve an inverted population on the H (3) → H (2) transition sufficient for construction of a laser.

Most laboratory studies involving the use of electron beams are devoted to lasers using vibrational–rotational transitions of molecules (excitation of N_2 [112], H_2 [113], stimulation of the $H_2 + F_2$ reaction [114], lasing with CO_2 [115-117]). High-current, fairly hard (W ~ 1 MeV) beams are now used for this and the gas pressure is taken up to p ~ 50 atm. Published experiments on amplification on electronic transitions of atoms and ions are limited as yet to ionization laser schemes [118-119]. In them the beam energy goes mainly not into ionization but into strong heating of the plasma electrons due to collective interactions. In the work of Fainberg and colleagues [118] quasistationary lasing has been observed on lines of the ions Ar^+, Kr^+, Xe^+, and N^+ in a low-density plasma. When N was increased to above $2 \cdot 10^{13}$ cm^{-3} the electron temperature fell due to cutoff of the instability and lasing ceased. Emission spectra have been obtained from the afterglow of a recombining helium plasma excited by an electron beam at high pressure [120].

(b) Atomic Reactor Laser. At present various types of laser designs with nuclear pumping are being considered. These include covering a tube containing the working gas with a layer that emits heavy particles due to neutron bombardment or radioactive decay, and mixing the working gas with another gas which serves as a source of heavy particles (see the review [121]). For efficient nuclear pumping it is necessary that a substantial fraction of the energy of the nuclear particles be absorbed in the gas (plasma).

As opposed to the papers reviewed in [121], the idea was proposed in [96] of directly using a nuclear reactor so that the energy released from nuclear decay can be removed in the form coherent light. It is proposed to use a reactor fueled with a gaseous uranium compound, such as UF_6. The bulk of the energy of the fission products ($W \sim 160$ MeV) goes into ionizing the filler gas which might be helium. An energy $E_{pair} = 46$ eV is used on the average per act of ionization; that is, each uranium fission is accompained by an ionization number $z = 3.5 \cdot 10^6$. Let there be G disintegrations per unit time per cubic centimeter of the active medium; then there will be zG ionizations per cubic centimeter per unit time. Calculations show that for ensured population inversion conditions, for example, on the He (3) → He (2) transition, the linear gain coefficient in dense helium reaches $\varkappa\ (cm^{-1}) = 3 \cdot 10^{-20}\ zG\ (cm^3 \cdot sec)^{-1}$. Thus, to reach $\varkappa \sim 10^{-3}\ cm^{-1}$ requires $G \sim 10^{10}$ fissions/$cm^3 \cdot$ sec. This requirement can be greatly eased if lasing is in a pulsed regime with steady-state energy pumping by the reactor. An inverted population may be assured by emptying the lower working level during the ionization reaction by means of an impurity which, in particular, may be UF_6.

11. Lasing in a Moving Plasma

We now turn to the features of a recombining plasma as an active medium when it undergoes several simple types of macroscopic motion: expansion, channel flow, and compression. In a number of cases the motion of the plasma sharply changes the character of the population kinetics of the electronic levels.

(a) Plasma-Dynamic Lasers.

The idea of such a laser was formulated in [87], where a scheme to create an inverted population among the discrete electronic states by adiabatically expanding a dense, in particular, a magnetized, plasma was considered. The significance of using the expansion of a plasma burst or the rapid expansion into a vacuum of a plasma stream is primarily to cool the medium and thereby create the recombination conditions. In addition, since the decay of different parts of an expanding plasma takes place at different times, the recombination process unfolds in time and the radiating medium is carried out of the active zone. Thus, it is possible to obtain cw lasing in a plasma stream flowing constantly out of a slit or nozzle. The condition for amplification on the electronic transitions of a z-fold ion may be given the form $\tau_{cooling} \lesssim \tau_{N_{z+1}}/10$, where τ_{N_z} is the recombination time of a z-fold ion. The characteristic time for cooling of the plasma during expansion, for example, of a cylindrical burst of diameter d is given by $\tau_{cooling} \sim d/2v$, where v is the expansion velocity.

To estimate the temperature variation of a highly ionized plasma before it begins to recombine we can use the equation [5]

$$\frac{T_e(t)}{T_0} \sim \left[1 - \frac{4v_0 t}{d(\gamma - 1)}\right]^{\mu(\mu-1)}.$$

Here $v_0 = (\gamma T_0/M)^{1/2}$ is the speed of sound in the initial (up to time t = 0) motionless plasma, T_0 is the initial temperature, $\gamma = 5/3$, t is the time since the beginning of the expansion, μ is the number of dimensions in the expansion problem, and M is the mass of the atom. In analyzing the expansion of a magnetized plasma we can replace v_0 in this formula by the initial Alfvén velocity $v_A = H_0/2(\pi N_0 M)^{1/2}$, where H_0 is the initial magnetic field strength and N_0 is the initial plasma density.

To examine the gain properties of an expanding plasma one must solve the population relaxation and temperature equations together with the gas-dynamic equations. A problem of this type under conditions of inertial expansion was first solved in [122], but the population kinetics was not analyzed there.

The relaxation equations for the electronic level populations in the inertial stage of a uniform expansion scheme, including the change in the density, take the form [44] (cf. Eq. (3.2))

$$\frac{dN_n}{dt}=\Gamma_n-N_n\frac{\mu}{t}, \tag{11.1}$$

where μ is the dimensionality of the expansion, i.e., $\mu = 0, 1, 2$, and 3 for a motionless plasma and for plane, cylindrical, and spherical expansion geometries, respectively. The heat balance equations are also somewhat more complicated than Eqs. (3.8) and (3.9):

$$\frac{3}{2}N_e\frac{dT_e}{dt}=-N_eT_e\frac{\mu}{t}+\frac{3}{2}T_e\sum_{n=1}^{n_1}\Gamma_n+Q_{\text{inel}}-Q_{\Delta T}; \tag{11.2}$$

$$\frac{3}{2}\frac{dT}{dt}=T\frac{\mu}{t}+\frac{Q_{\Delta T}}{N(t)}. \tag{11.3}$$

Here

$$N=N_0\,(t_0/t)^{\mu} \tag{11.4}$$

is the total concentration of heavy particles and $t_0 = x_0/v$, where x_0 is the initial linear dimension.

The quantity Q_{inel} [see Eq. (3.10)] can be evaluated with the formula [122] $Q_{\text{inel}} = E^*\beta_{Ne}$, where E^* is an effective estimate of the energy released upon recombination of a single electron: for $E > E^*$, collisional recombination is predominant, while for $E < E^*$, the energy of a recombining electron leaves the volume as radiation.

The explicit appearance of t on the right-hand sides of the relaxation equations limits the domain of applicability of the constant-sink approximation. The condition for its applicability,

$$N_e/N_n\gg 1+\mu\tau_{N_e}/t, \tag{11.5}$$

is more stringent than the condition $N_u \ll N_e$ for slow (compared to the expansion time t) recombination. When it is not obeyed, a substantial portion of the excited atoms leave the volume under consideration, having not changed their state during recombinational relaxation [123]. This effect is similar to a freezing of the degree of ionization of the plasma [122] such that the free electrons cannot recombine during rapid expansion.

The relaxation of a xenon plasma over the electronic levels as it expanded in a nozzle was studied in [124]. Equation (11.1) was solved for the 13 lowest excited states taking reabsorption into account. An analysis of the populations showed that in an optically dense plasma the minimum values of $\delta_{ab} \equiv (N_a/g_a)/(N_b/g_b)$ for the Xe(7p) → Xe(5d) and Xe(4*f*) → Xe(5d) transitions lie at the center of the flow channel and are roughly the same in both cases at $(\delta_{ab})_{\min} \simeq 1/3$. For a transparent plasma, they take their minima at the outlet of the nozzle. Numerical calculations [125] on a nitrogen plasma heated by an arc discharge and then expanded through a nozzle showed that recombination leads to a significant overpopulation of the upper levels. The possibility was noted of obtaining a population inversion in the N atom at a wavelength of $\lambda = 1745$ Å.

An inverted population of electronic states in an expanding plasma has been observed in a number of experiments with mixtures in nozzles. An inversion in levels of atomic hydrogen was observed in [98] in an expanding argon–hydrogen plasma jet. Measurements [126] on plasma mixtures of helium and argon with methane as they recombined upon expansion through a nozzle after being dissociated into atoms and ionized in an arc indicate an inversion in the $6S_{1/2}-5P_{3/2}$ transition of the carbon atom (λ = 2478.6 Å). The possibility of creating an inverted population in plasma jets was studied experimentally in [127, 128]. An inversion in levels of helium and hydrogen was observed in an accelerated plasma jet in [128].

(b) Compressed Plasma Lasers. Rapid compression of a plasma leads to strong deviation in the electronic level populations of ions from their thermodynamic equilibrium values. Energy is efficiently delivered to a plasma in this fashion in various devices using the pinch effect. A linear (Z) pinch is produced in a cylindrical vessel, most often filled with a light gas, by means of a longitudinal breakdown accompanied by a high current in the plasma (up to about 10^6 A) and an azimuthal magnetic field produced by this current which compresses the plasma column toward the axis of the vessel. If a straight plasma column is compressed by an external longitudinal field, the result is a θ-pinch in which a dense, high-temperature plasma is produced with record high free-electron densities ($N_e > 10^{17}$ cm^{-3}). Population inversions in a number of ion levels have been observed in spectroscopic studies of pinch compressed plasmas. Both in Z- and θ- pinches lasing has been obtained in argon and xenon plasmas [129-134]. In a linear pinch [132] lasing occurred after a delay relative to the beginning of the current pulse during the stage of increasing plasma compression and ceased after the pinch column began to break up. The peak power was about 1 kW. Lasing was observed on Ar II (the most intense transition was $4s^2P_{1/2} \rightarrow 4p^2P_{3/2}$, λ = 4765 Å, with an active region of diameter $d_{las} \approx 0.4$-0.5 cm and a plasma column diameter $d_{col} \approx 0.3$ cm) as well as on Ar III (the most intense line was $4p^3P_2 \rightarrow 4s^3S_1^0$, λ = 3511 Å, with an active region of diameter $d_{las} = 1.2$ cm and a column diameter $d_{col} = 1.0$ cm) transitions. The diameter of the chamber was D = 2.0 cm and the compression lifetime was $\tau_{com} \approx 3.0 \cdot 10^{-7}$ sec.

The kinetics of the processes in a linear pinch argon ion laser were examined in [135]. With the substitution

$$x_{nl} = N_{nl}(Nf_{nl})^{-1}, \qquad f_{nl} = g_{nl}\exp(-E_{nl}/T_e), \tag{11.6}$$

where N is the density of heavy particles, the equations for the populations in a compressible plasma reduce to the form of Eq. (3.2) which corresponds to the population kinetics during free relaxation. With the formulas of [57] [see Eq. (4.11)] analytic expressions were obtained for $N_{nl}(t)$. It was shown that the population inversion is created by an ionization flux over the levels of Ar II and is due to rapid growth of the electron temperature as the plasma is compressed. The radiative mechanism controls the emptying of the lower working level. Thus, as in ordinary ionization lasers using Ar II, reabsorption limits the volume and density of the active medium.

Therefore, even though pinches make it possible to rapidly deliver energy at high densities to a plasma, an efficient laser has not yet been built using them. It is appropriate in this regard to analyze the possibility of producing an intensely recombining plasma with a pinch. We note that in principle recombination can occur in the plasma compression phase as well as in the expansion phase. In a multiply charged ion plasma bremsstrahlung energy loss by the free electrons sharply inhibits any rise in their temperature, while their density increases during compression. Then we might expect recombination of a hard-to-ionize component in a rapidly pinch compressed plasma mixture. In such a situation recombination filling of the upper working levels and collisional depopulation of the lower working levels would be realized [135].

(c) Plasma Jets. MHD Lasers. Along with expanding and compressed plasmas there is also interest in uniform plasma jets, which act as a time scan of a pulsed population inversion, for cw lasers. To produce an inversion the plasma jet must be subjected to some additional interaction.

The effect on an electrical pulse applied perpendicular to the plasma stream was discussed in [136]. Since the duration of the inversion may be 10^{-7}-10^{-6} sec, the pumping rates must be supersonic and transitions between comparatively high atomic levels may be used.

Another possibility for a laser is to pass a plasma jet through a magnetic field using an apparatus such as an MHD converter (magnetohydrodynamic converter of thermal energy into elecrical) [137]. The physical kinetics in this case is complicated; both an ionization and a recombination regime are possible. The ionization regime will occur because as a thermally equilibrium plasma flows in a magnetic field an electric field is induced in the plasma and the electron component may be selectively heated [138]. A recombination regime may develop when nonequilibrium plasma jets are used. Without dwelling on details, we note only the possibility in principle of obtaining both electrical and coherent optical energy from MDH installations.

CHAPTER V

PLASMA CHEMICAL LASERS

12. Chemical Depopulation of Lower Atomic Levels

When the electrons in a dense plasma are substantially supercooled, it is difficult to find processes which could compete with the recombination flux in filling the upper discrete levels. Thus, as opposed to chemical lasers, in which the chemical energy of the atoms, ions, and radicals must excite the vibrational levels of the molecules, the main purpose of chemical reactions in the active medium of a plasma chemical laser is to empty the lower working level. It seems that in this way it might be possible to achieve a population inversion even in transitions to the ground state of atoms. Two directions can be distinguished here. One of these, plasma lasers using dissociating (disintegrating) molecules, is attracting ever more attention at this time. Such molecules can be formed in a molecular gas under the influence of an electrical discharge, an electron beam, or a laser beam. The lower term of these molecules disintegrates and an inversion is formed automatically. The other direction has to do with the removal of atoms in certain states due to a chemical reaction. We shall first deal briefly with the second approach.

The fundamental difficulty in analyzing the population kinetics of the electronic levels during a reaction that takes place in a given mixture is the lack of research on the basic plasma chemical processes (i.e., the absence of data on the rates of ionic reactions) as well as on electronically excited atoms and molecules.† Thus, at this stage only crude estimates will be meaningful, beginning with the simplest models and verifying the conclusions experimentally. The populations of hydrogen atoms in a supercooled chemically active plasma have been estimated in [54]. The effect of losses of recombined H(n) atoms from the ground (n = 1) and excited states due to a chemical reaction was taken into account in the same way as reactions (8.2): the quantity $\alpha(1 + \alpha n)$ was added to the diagonal element of the relaxation matrix, where n is the principal quantum number and α is a constant which determines the loss rate of un-

† Reactions in an atomic–molecular plasma can take place at rates comparable with the electron relaxation times. Here nonadiabatic transitions, which have large cross sections [139-140], play an important role.

excited atoms, H(1), due to the chemical reaction (α is an estimate of the dependence of this rate on the excitation).

These estimates showed that in the range of plasma parameters of interest for laser applications, the population kinetics for $\alpha \sim 1$ are practically independent of the value of α. For definiteness the calculations were done in detail for $\alpha = 0.2$. The graphs of Fig. 9 show that for $\alpha \sim 10^8$-10^9 sec^{-1} an inversion occurs on the first two transitions of the Balmer series H(3) → H(2) and H(4) → H(2). The possibility of amplification of the Lyman line radiation is of interest. Since in an analysis of such a problem the principal role in the population of the ground level is played by spontaneous radiative transitions, it was assumed in the time-independent problem that

$$N_1 = \frac{1}{\alpha} \sum_{n=2}^{n_1} A_{1n} N_n. \tag{12.1}$$

For $N_e = 10^{16}$ cm^{-3} and $T_e = 0.4$ eV an inversion is realized on the H(2) → H(1) transition for $\alpha > 3 \cdot 10^9$ sec^{-1}; $\alpha = 5 \cdot 10^9$ sec^{-1} corresponds to a gain coefficient of $\varkappa_{21} = 5 \cdot 10^{-2}$ sec^{-1}.

Chemical reactions which might ensure rapid removal of hydrogen from the ground state include the following:

$$\begin{gathered} H + F + HF \to MF + HF, \qquad H + F + F_2 \to HF + F_2, \\ H + Cl + HCl \to HCl + HCl, \qquad H + Cl + Cl_2 \to HCl + Cl_2. \end{gathered} \tag{12.2}$$

Their rate constants are about 10^{-29} cm^6/sec [141]. Thus, a sufficient value of α for a Lyman

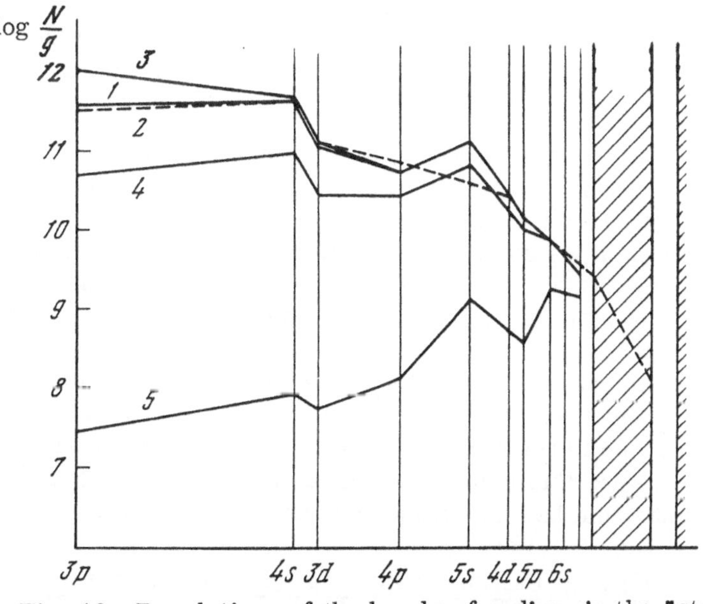

Fig. 12. Populations of the levels of sodium in the "stationary-sink" regime for $T_e = 0.1$ eV and $N_e = 10^{14}$ cm^{-3}. 1) Calculating the collisional transition rates according to the equations of [41]; 2) using the formulas of [15]; 3) including the reabsorption of radiation [41]; 4, 5) including the effect of a chemical reaction $\alpha = 10^8, 10^9$ sec^{-1}.

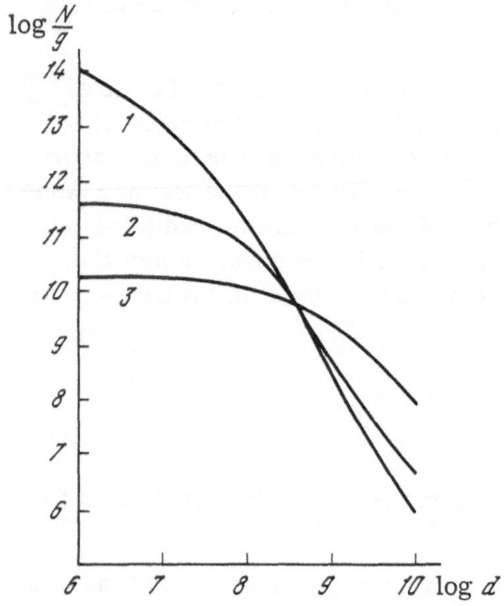

Fig. 13. Dependence of the level populations of Na on the rates of the chemical reaction for $T_e = 0.1$ eV and $N_e = 10^{14}$ cm^{-3}. Curve 1 corresponds to the 3s level; curve 2, to 3p; curve 3, to 5p.

transition laser is reached at concentrations $N_{Cl} \sim 3 \cdot 10^{19}$ cm^{-3} and $N_{Cl_2} \sim 3 \cdot 10^{20}$ cm^{-3} in a mixture of chloride and hydrogen chloride (or fluorine and hydrogen fluoride).

Losses of recombined sodium atoms due to a chemical reaction were modeled in an analogous fashion in [49]. Here reactions of the form

$$M + Na^* + Cl \rightarrow (NaCl)^* + M \qquad (12.3)$$

were considered. Their rate constants and those of reactions (12.2) are obviously close. The results of some calculations to estimate the values of α at which an inversion develops in the Na atom relative to the ground state are shown in Figs. 12 and 13. Here an inversion is achieved at smaller values of α than for hydrogen. An estimate of the unsaturated gain coefficient for $N_e = 10^{14}$ cm^{-3} and $T_e = 0.1$ eV yields $\varkappa_{3p,2s} = 1.5 \cdot 10^{-2}$ cm^{-1}.

We now briefly explain the significance of these estimates. We shall consider a gaseous mixture of Cl_2 and NaCl. We subject it to thorough ionization. Due to thermal dissociation of the unstable Cl_2 molecule and the molecular NaCl ions (see Section 10) [$Cl_2 \rightarrow Cl + Cl$, $(NaCl)^+ \rightarrow Na^+ + Cl$] and to rapid cooling of the electrons in collisions with heavy particles, three-body recombination, $Na^+ + e + e \rightarrow Na^* + e$, takes place, followed by relaxation of Na* to the ground state at the same time as this atom is lost due to reaction (12.3). Relaxation of Na* from the upper excited levels in collisions with free electrons takes place over very small characteristic times close to the elastic collision times for free electrons. It is clear that the chemical reaction has practically no effect on the populations of such levels. By choosing a concentration of Cl such that the loss due to reaction (12.3) would be faster than transitions into the Na(1) state from the excited levels, we obtain an inversion on transitions into this state. An advantage of such an arrangement would be high efficiencies and high output powers.

13. Plasma Lasers Using Disintegrating (Dissociating) Molecules. A Scheme for Filling an Excited State. Conditions for Amplification

In an analysis of the prospects for lasing on molecular electronic transitions in a recombining plasma, "disintegrating" molecules attract greatest attention, that is, molecules with a dissociating (or disintegrating) electronic ground state. The idea of lasing on photodissociating transitions to the ground state of such molecules was considered many years ago by Houtermans and was published by him in a note [142]. As a medium with an "automatic" inversion he suggested the use of a gaseous discharge plasma of inert and alkaline earth elements, molecules of which are stable only in excited electronic states.

Lasing with disintegrating molecules is of interest at this time for the following reasons:

1. In many dissociating molecules the working transition lies in the vacuum ultraviolet.
2. The closeness of the upper working level to the ionization energy for such molecules leads us to expect high powers and efficiencies.
3. The unusually wide (for an active medium) homogeneous line corresponding to the spontaneous transition leads us to anticipate both obtaining quasimonochromatic frequency-tunable coherent radiation and building a laser with extremely short pulses.

At the same time, there are considerable difficulties in obtaining lasing because of the large linewidth of photodissociating transitions ($\Delta\omega \sim 10^{15}$ sec^{-1}). The large linewidth leads to a correspondingly small cross section for the photodissociation transition. Thus, despite the comparative simplicity of ensuring a population inversion of "disintegrating" molecules, the gain coefficient required for lasing is reached here only at very high populations of the upper working level. To obtain such populations the gas must have a high density and a low temperature (since the depth of the terms is small even for the excited electronic states of these molecules), which suggests a preference for the plasma principle.

(a) Relaxation Channels. The recombination scheme for a laser using dissociating molecules proposed in [143, 144] consists of the following. A dense atomic gas is ionized by a pulse from a transverse field or by a source of penetrating particles and forms a supercooled plasma. In a dense plasma with supercooled electrons two recombination channels are important, both of which populate the first excited bound electronic state $X_2(2)$ of the molecule (Fig. 14).

The first recombination channel looks this way. The recombination flux of electrons over the excited states of the atoms X^* (see Section 1) ensures that electron collisions are mainly of the second type. Relaxation over the excited atomic levels X(m) continues right to the first excited state X(2). If the energy of this state is sufficiently high that deexcitation by

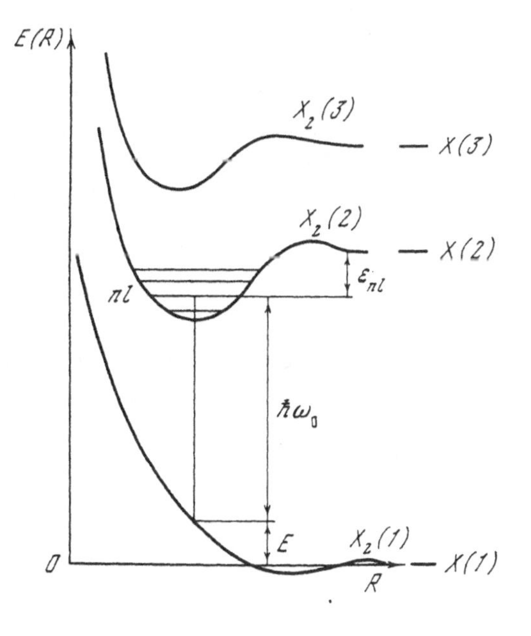

Fig. 14. Term diagram of an X atom and a dissociating X_2 molecule.

electrons in the reaction

$$X(2)+e\rightarrow X(1)+e$$

is unlikely and the volume of dense plasma is large enough that radiation from the $X(2)\rightarrow X(1)$ transition is trapped, then recombination from the excited states of the atom "turn" toward formation of an excited molecule $X_2(2)$:

$$X(2)+X(1)+X(1)\rightarrow X_2(2)+X(1). \tag{13.1}$$

The other channel is relaxation over the excited electronic states of the molecule $X_2(m)$. Such relaxation begins with the molecular ion X_2^+ being produced by conversion of the atomic ion:

$$X^{+}+X(1)+X(1)\rightarrow X_2^{+}+X(1). \tag{13.2}$$

The atomic and molecular relaxation channels may mix, for example, due to the reactions

$$X_2^{+}+e\rightleftarrows X(m)+X(1),$$
$$X_2(m)+e\rightleftarrows X(m)+X(1)+e;$$

however, this does not change the essence of the matter. The molecular relaxation channel in an intensely recombining dense plasma need not lead to appreciable energy losses from the source or to increased heating of the gas.

(b) Ideal Case. Let the lowest excited electronic state $X_2(2)$ decay solely due to dissociative transitions into the ground (dissociating) state,

$$X_2(2nl)\rightarrow X_2(1,\ E)+\hbar\omega_0=X(1)+X(1)+E+\hbar\omega. \tag{13.3}$$

Here E is the energy of the atoms when the molecule disintegrates, n and l are the vibrational and rotational quantum numbers of the bound excited electronic state, and ω_0 is the frequency of the emitted light. If in this ideal scheme we neglect photoabsorption as well, then the gain coefficient may be evaluated using the formula

$$\varkappa=\frac{\lambda_0}{4}\frac{AN_2^{X_2}}{\Delta\omega}=\frac{\lambda_0^2}{4}\frac{\nu_i}{\Delta\omega}N_1^X, \tag{13.4}$$

where $\lambda_0=c/2\pi\omega_0$ is the wavelength, $\Delta\omega$ is the width of the line, $N_2^{X_2}$ is the concentration of $X_2(2)$ molecules, N_1^X is the concentration of gas atoms in the ground state, and ν_i is the rate of inelastic collisions with fast particles (for example, beam electrons), also including inelastic collisions of secondary electrons. Assuming, for example,† that $\lambda_0 = 1500$ Å and $\Delta\omega = 1$ eV, we find that to attain a gain coefficient of $\varkappa > 10^{-2}$ cm^{-1} it is necessary to achieve sufficiently high concentrations,

$$N_2^{X_2}>3\cdot 10^{23}/A,\ \text{sec}^{-1}. \tag{13.5}$$

This condition is equivalent to requiring an ionization source of sufficient power. For an electron beam

$$N_1^X(\text{cm}^{-3})\,j\,(\text{A/cm}^2)>5\cdot 10^{20}/(\sigma_p/\pi a_0^2) \tag{13.6}$$

† The parameters correspond roughly to the dimer Xe_2.

must be satisfied. Here σ_H is the cross section for interaction of beam electrons with gas atoms (in Bohr units, $a_0 = 0.5 \cdot 10^{-8}$ cm^{-1}). Condition (13.6) is not strict for the beams presently available; however, it is true only if $\nu_H N_1^X = AN_2^{X_2}$, that is, when the ideal scheme is applicable where each act of ionization and excitation of an X atom is accompanied by formation of a molecule and then by its radiative decay. The latter event is not always true. If the principal relaxation channels actually "lead" to an X_2 state, then it may not only decay radiatively.

Evidently, the most dangerous "parasitic" reaction at this time is assumed to be a variant of the Penning effect, ionization during collisions of excited particles,

$$X_2(2) + X_2(2) \to X_2^+ + X(1) + X(1) + e. \tag{13.7}$$

Introducing the rate of this reaction, $q_P = \langle \sigma_P v \rangle$, we may write, for the population of the upper working level $X_2(2)$ [145]

$$N_2^{X_2} = \frac{A}{q_P}\left(\sqrt{1 + N_1^X \nu \frac{q_P}{A^2}} - 1\right). \tag{13.8}$$

From this it follows that for

$$N_1^X \nu_q \ll A^2$$

reaction (13.7) is unimportant and the concentration $N_2^{X_2}$ increases in proportion to νN_1^X. With further increases in the intensity of ionization $N_2^{X_2}$ increases to values on the order of some critical value $N_{cr} = A/q_P$, after which the growth is inhibited.

For

$$N_1^X \nu q \gg A^2$$

we have

$$N_2^{X_2} = \sqrt{\frac{\nu}{q_P}} N_1.$$

If we take $\nu_H = 10^3$ sec^{-1} and assume the molecular parameters to be close to those of a xenon molecule [146] ($A \sim 10^7$ sec^{-1} and $q_P \sim 10^{-10}$ cm$^3\cdot$sec^{-1}), then the gain coefficient increases linearly with the gas pressure up to values of about 100 atm and may then reach about 0.1 cm^{-1}. At the same time, it is evidently impossible to achieve lasing on a working transition $X_2(2) \to X_2(1)$ for a molecule with small values of the Einstein coefficient $A < 10^4$ sec^{-1}.

(c) Conditions for Amplification. The light amplified by dissociative transitions is mainly absorbed due to inverse (photoassociative) transitions and to photoionization of the upper working state $X_2(2)$ of the molecule. The problem of amplifying light on dissociative transitions taking photoassociation into account was discussed in [142] where the expression

$$\varkappa = \sigma_{\text{diss}}^{\text{ph}}(nl, E)\left[N_{2nl}^{X_2} - (2l+1)\left(\frac{2\pi\hbar^2}{\mu T}\right)^{3/2} \exp\left(-\frac{E_{nl} - \hbar\omega_0}{T}\right)(N_1^X)^2\right] \tag{13.9}$$

was found for the gain coefficient. Here E_{nl} and $N_{2nl}^{X_2}$ are the energy and population of the $N_2(2nl)$ state, $\sigma_{\text{diss}}^{\text{ph}}(nl, E)$ is the photodissociation cross section [reaction (13.3)], E is the dissociation energy, T is the gas temperature, and μ is the reduced mass of the molecule. In this expression the first term in the square brakets takes photodissociative processes into account;

the second, photoassociative processes. Absorption of light due to photoionization may be included by putting the factor

$$(1-\sigma^{ph}/\sigma^{ph}_{diss}),$$

where σ^{ph}_{ion} is the photoionization cross section, ahead of the first term in the square brackets of Eq. (13.9). Usually† $\sigma^{ph}_{ion} \ll \sigma^{ph}_{diss}$, so photoionization may be neglected. We now briefly consider the role of photoassociation. The maximum possible population of the $X_2(nl)$ state is determined by the Boltzmann formula to be

$$N^{X_2}_{nl}=(2l+1)\left(\frac{2\pi\hbar}{\mu T}\right)^2 \exp\left(\frac{\varepsilon_{nl}}{T}\right)N^X_2 N^X_1, \quad (13.10)$$

where N_2^X is the population of the X(2) state, $\varepsilon_{nl} = E_2^X - E^X_{2nl}$ is the energy depth of the term, and E_2^X is the excitation energy of the X(2) state (see Fig. 14).

Substituting Eq. (13.10) into Eq. (13.9), we obtain

$$\varkappa=\left(N^X_2 - N^X_1 \exp -\frac{\Delta E}{T}\right)\left(\frac{2\pi\hbar^2}{\mu T}\right)N^X_1 \sum_{nl} \sigma^{ph}_{diss}(nl, E)(2l+1),$$

where $\Delta E = \varepsilon_{nl} + E = E_2^X - \hbar\omega_0$. Thus, $\varkappa > 0$ for

$$T<\frac{\Delta E}{\ln(N^X_1/N^X_2)}. \quad (13.11)$$

Since the population N_1^X of the ground state of the atom is usually several orders of magnitude greater than the population N_2^X of the excited state, it is clear that we must have $T \ll \Delta E$. As a rule $\Delta E \sim 1$ eV, so to achieve amplification it is necessary that the gas be fairly cold ($T \sim 0.1$ eV).

In connection with condition (13.11) we note that to achieve a population inversion it is not sufficient that the expansion time be small compared to the lifetime of the excited state, as is often assumed.

14. Problems in Making Plasma Lasers Using Dissociating Molecules

First we shall briefly enumerate the other methods of achieving lasing on dissociating molecules discussed in the literature. In [147] optical pumping of the $X_2(2)$ state through a photoassociation reaction was proposed. This idea is difficult to realize, first of all, because of the small probability of such a process (for details, see [145]). More realistic is another method: optical pumping of the excited X(2) state of an atom so that later, during the collisional association reaction (13.1), an $X_2(2)$ molecule will be formed. However, this approach is impractical from an energetic standpoint. It would be rather more useful in studies of vibrational–rotational relaxation of molecules [148].

A number of authors, proceeding from the principle of the gas laser, have suggested using a plasma with superheated electrons formed in the rising front of a heating field [149] or in an ordinary gaseous discharge [150, 151] (for example, in an arc [150]). However, despite the optimistic estimates, made according to an idealized scheme in which each act of excita-

† This is because the wave function of the ejected electron is "smeared" over a larger energy interval than the wave function of the nuclei.

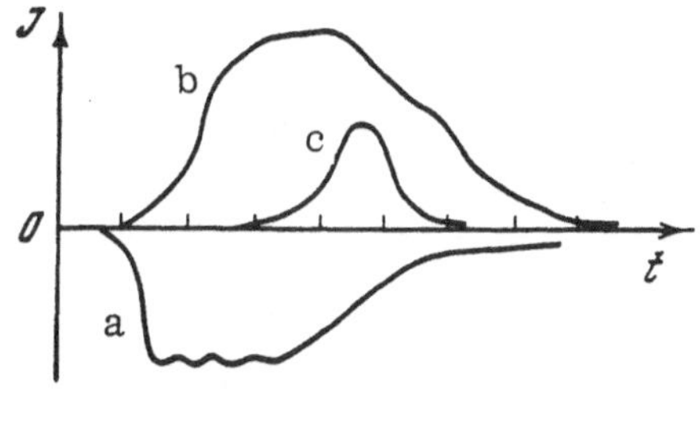

Fig. 15. Dynamics of the development [153] of the pumping (a), spontaneous emission (b) and laser output (c) pulses. Time scale: 20 nsec/div.

tion of an atom to the X(2) state is accompanied by emission of a "photodissociation" quantum, numerous attempts to build a laser in the ionization regime have failed (see, for example, [151]). These failures were predetermined by the fact that such an idealized scheme is less realistic in the ionization regime. The ionizing electron flux does not "get stuck" in the X(2) and $X_2(2)$ states as it moves "upward" through the excited levels (for details, see [145]). There are other reasons as well. For example, in an arc discharge the gas may be overheated and condition (13.11) will not hold.

During the past year a number of reports have at last appeared [146, 152-54] on the construction of relativistic electron beam plasma lasers using dissociating molecules. Thus, in [146, 153] lasing was observed on a molecular transition (λ = 1730 Å) in dense (about 30 atm) xenon ionized by an electron beam with an electron energy U of about 1.5 MeV, a current density j of about 200 A/cm^2, and a pulse duration of $\tau \sim$ 50 nsec. An examination of the data of [153] shows that the medium begins to amplify weakly shortly before the beam interaction ceases. Maximum gain comes in the afterglow and appears later than the peak in the spontaneous emission. It is clear that lasing takes place in the recombination regime (Fig. 15). The deterioration in the conditions for amplification on disintegrating molecules while the beam is acting cannot be explained by excitation of the lower working level (as takes place among atoms; cf. Section 4, paragraph a). It remains to suppose that the weak amplification while the beam was acting in the experiments of Wayne and Gerardo is due to the high temperature of the electrons in the plasma that has been created. Turning off the beam leads to rapid cooling of the electrons (over times τ_{Te} ; see Section 10, paragraph a) and intensified recombination and gain. Comparatively recently [152, 155] the possibility of lasing in the recombination regime using the same transition of xenon was examined theoretically and then realized experimentally. This group of researchers has also been able to obtain lasing in high-pressure krypton and xenon–argon mixtures.

Using the results of Section 13 we shall examine some schemes for plasma lasers using dissociating molecules. We shall evaluate the conditions for lasing on the transition (13.3) in the afterglow of a short ionization pulse starting with the simplest concepts of the decay behavior of a dense supercooled inert gas plasma. Atomic ions X^+ disappear through three-body recombination, $X^+ + e + e \rightarrow X^* + e$, and conversion, $X^+ + X + X \rightarrow X_2^+ + X$, while molecular ions disappear through dissociative recombination. The latter process takes place much faster than conversion. The loss rates of the X^+ ions and electrons are the same and equal the rates of filling the lowest excited state $X_2(2)$ of the molecule. The decay of the level $X_2(2)$, which is fairly close to the continuum, takes place due to a spontaneous transition ($2 \rightarrow 1$) and collisional ionization by the Penning effect,

$$X_2(2) + X_2(2) \rightarrow \begin{cases} X_2^+ + X(1) + X(1) + e, \\ X^+ + X(1) + X(1) + X(1) + e. \end{cases} \tag{14.1}$$

Thus, the kinetics of the population $N_2(t)$ of the $X_2(2)$ level is described by the equations

$$dN_e/dt = -\beta N_e^3 - kN^2 N_e + qN_2^2, \quad dN_2/dt = \beta N_e + kN^2 N_e - 2qN_2^2 - AN_2, \tag{14.2}$$

which involve the coefficients of three-body recombination β, conversion k, and Penning ionization q. These equations are not complete because β and k depend on the time through the temperature of the free electrons. At large gas densities N the entire recombination flux of electrons flows through the $X_2(2)$ level. The relaxation of its population, near the threshold for self-excitation at high free electron densities $N_e(t)$, is the "narrowest" part of this flux. Thus, $N_e(t)$ follows $N_2(t)$ in a quasistationary manner; that is, we can set $dN_e(t)/dt = 0$ in Eq. (14.2) and solve the differential equation

$$\frac{dN_2}{dt}(t) = -qN_2(t)^2 - AN_2(t) \tag{14.3}$$

together with the algebraic equation

$$\beta N_e(t)^3 + kN^2N_e(t) = qN_2(t)^2. \tag{14.4}$$

The solution of Eq. (14.3), $N_2(t) = [(N_{20}^{-1} + q/A)e^{At} - q/A]^{-1}$, falls monotonically from its initial value $N_2(0) = N_{20}$. We now estimate N_{20} assuming that all the energy in the ionizing pulse is absorbed by the electrons, which rapidly exchange it among themselves and, much more slowly, with the heavy particles. Assuming that the energy (ε) distribution of the free electrons falls linearly with increasing ε to zero at $\varepsilon = \varepsilon_{ex}$ (ε_{ex} is the energy of the lowest excited atomic level), we write

$$N_e(0)(J + \varepsilon_{ex}/3) + N_{20}\varepsilon_{ex} = w,$$

where J is the ionization energy of the atom and w is the energy density stored in the plasma upon ionization. Neglecting three-body recombination in Eq. (15.4) when t = 0 we find

$$N_{20} = \frac{kN^2}{2q}(\sqrt{1+\mu} - 1), \qquad \mu \equiv \frac{4q}{kN^2}\frac{w}{I}.$$

When the pumping level is low ($\mu \ll 1$),

$$N_{20} = \frac{w}{J}\left(1 - \frac{\mu}{4}\right).$$

Then if the energy W stored by the entire medium upon ionization is specified, the cavity gain is practically independent of the length l of the active medium even for t = 0; that is,

$$\exp[\varkappa(0)\,l] - R = \exp\left(\frac{A\lambda}{4\Delta\omega}\frac{W}{SI}\right) - R.$$

Here $\varkappa(0)$ is the unsaturated linear gain coefficient at t = 0; R is the radiant energy loss coefficient, S is the cross-sectional area of the active medium, and $\Delta\omega$ and λ are the bandwidth and wavelength of the laser radiation. For small losses ($R \lessapprox 0.1$) the condition for lasing takes the form

$$\frac{W}{S} < U_T, \quad U_T = 4R\frac{\Delta\omega}{A}\frac{J}{\lambda^2}.$$

The threshold population $N_2^T = 2R(\Delta\omega/A)(1/\lambda^2 l)$ corresponds to a characteristic decay time $\tau_I = (A + N_2^T q)^{-1}$. If the relative amount δ by which the self-excitation threshold is exceeded is small, then for the duration θ of the lasing process we find $\theta = \delta\tau_I$.

We now write down the conditions for pulsed lasing in a dense inert gas:

1. The ratio of the aborbed energy to the cross-sectional area must exceed a threshold value, i.e., $W/S \geq U_T(1 + \delta)$. Taking $\delta = 1$ for R = 0.1, we have $W/S \geq 0.3(\Delta\omega/A) \times (I/\lambda^2)$. For lasing on Xe_2, for example, we obtain $W/S \geq 10^{-3}$ J/cm^2.

2. The density w of absorbed energy must not overheat the gas. It is shown in [143, 144] that the condition for an inversion on the transition (2.1) fails for a gas temperature $T_L \sim 0.1$ eV. Thus, we must have $W/Sl < w_L$, where the limiting energy density is given by $w_L = c_v N/2$ (atm) T_L, c_v is the heat capacity of the gas "under normal conditions," and N (atm) is its density in units of $3 \cdot 10^{19}$ cm^{-3}. Taking Xe_2 as an example we find w_L (J/cm^2) = 2N (atm).

3. To avoid substantial losses it is appropriate to make the duration of the ionization pulse less than τ_L. For observation of lasing it is sufficient that the risetime of the ionizing pulse not exceed τ_I. As an example, for lasing on Xe_2 $\tau_L \sim 10^{-8}$ sec.

Chapter VI

SHORT-WAVELENGTH PLASMA LASERS

15. Vacuum Ultraviolet Lasers

The possibility of efficient lasing in the vacuum ultraviolet (VUV) and x-ray range is presently a cause of heightened interest in plasma lasers. A whole series of articles (see, for example [156-170]) is devoted to the general problems of building sources of coherent radiation in the short-wavelength range (VUV, x-ray, and gamma radiation). We shall therefore limit ourselves to questions having direct bearing on plasma laser research which have not been reflected completely or accurately enough in these articles.

A plasma medium for short-wavelength lasers has some characteristic features which distinguish it from the preceding considerations. This plasma will contain multiply charged ions and have a high density and electron temperature. The nature of the relaxation processes is also different from the usual (see Chapter II). The role of relaxation processes increases, while that of collisional processed decreases (for hydrogenlike atoms by z^4 and z^{-4} times, respectively) and autoionization also becomes important. Below we shall consider Be and He ions as examples of active media. They are clearly not optimal but nevertheless serve as illustrations of the prospects for VUV plasma lasers.

After plasma lasers on a number of alkaline earth ionic transitions [100-102] were tested, the analysis of the conditions for making an active medium of a recombining plasma made up of doubly ionized Be and alkaline earth metal ions became of immediate interest [91]. The wavelength λ of one of the intended laser transitions Be II(3S) → Be II(2P), equal to 1776 Å, is at the long wavelength boundary of the VUV region. (Here the light can be transmitted through a lithium fluoride window and it is still possible to count on comparatively good mirrors.) The lower working level Be II (2P) is close to the ground state and is well depopulated by collisions with the cold plasma electrons. Calculations showed that an inversion on the 3s → 2p transition occurs only for sufficiently large electron densities ($N_e \gtrsim 10^{15}$ cm^{-3}), when collisional depopulation is no longer less effective than radiative repopulation of the Be II (2p) level. For parameters attainable at present in gaseous discharge devices without extreme difficulty (for example, in the afterglow of a high-voltage pulse in an inert gas) the gain coefficients are sufficient for lasing. Thus, for $N_e = 10^{15}$ cm^{-3} $T_e = 0.2$ eV, and $N_{++}/N_e = 10^{-2}$ ($\beta N_e^2 = 10^8$ sec^{-1}), $\varkappa = 0.03$ cm^{-1}.

The recombination of a helium plasma with a substantial density of He^{++} was computed numerically in [89]. The first transition of the Balmer series of the hydrogenlike ion, He II (3)→ He II (2), was chosen as the working level. The lower working level is emptied as neutral helium atoms are ionized (Penning effect):

$$\mathrm{He\,II}\,(n) + \mathrm{He\,I}\,(1) \rightarrow 2\,\mathrm{He\,II}\,(1) + e. \quad (15.1)$$

According to [89] for $T \sim 0.5$ eV and $n = 2$ we have $q_2 = \langle\sigma_2 v\rangle \sim 10^{-9}$ cm^3/sec for the rate constant of this reaction. From this we find for a helium atom ground state density of $N_{HeI} \sim 10^{19}$ cm^{-3} that the rate of depopulation of the lower working level by reaction (15.1) exceeds the rate of radiative decay of this level of the He II ion ($A_{21} = 8 \cdot 10^9$ sec^{-1}). At the same time, calculations [46] made neglecting reaction (15.1) and reabsorption of resonance radiation from He II showed that radiative deexcitation in a transparent plasma ensures a population inversion on the $3 \rightarrow 2$ transition over a wide range of variation of the parameters of the recombining plasma. This means that for sufficiently high neutral helium densities $N_{HeI} > 10^{19}$ cm^{-3} an inversion over the $3 \rightarrow 2$ transition should be conserved even for complete reabsorption of the resonance radiation. This is indeed confirmed by a detailed calculation. The system of relaxation equations including reaction (15.1) was solved on a computer for various parameters: $N_e = 10^{15}$–10^{17} cm^{-3}, $N_{HeI} = 3 \cdot 10^{18}$–$10^{20}$ cm^{-3}, $T_e = T = 0.2$–1 eV, $N_{HeIII}/N_{HeII} = 0.01$. Figure 16 shows the time variation of the populations of the levels of He II with principal quantum numbers $n = 2$ and 3 for $N_e = 10^{16}$ cm^{-3}, $T_e = 0.3$ eV, and various values of N_{HeI}. For free decay of the plasma an inverted population is realized for $N_{HeI} = 3 \cdot 10^{18}$ cm^{-3}. Figure 17 shows the time dependence of the gain coefficient for this transition ($\lambda = 1641$ Å). It is clear that for den-

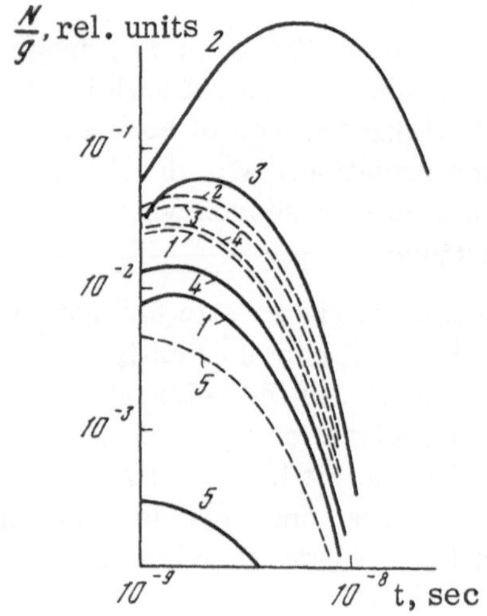

Fig. 16. Populations of the n = 2 and 3 levels as a function of time for various helium densities: 1) neglecting absorption; 2) $N_{He} = 0$; 3) $N_{He} = 3 \cdot 10^{18}$ cm^{-3}; 4) $N_{He} = 10^{19}$ cm^{-3}; 5) $N_{He} = 10^{20}$ cm^{-3}. Smooth curve, n = 2; dashed curve, n = 3.

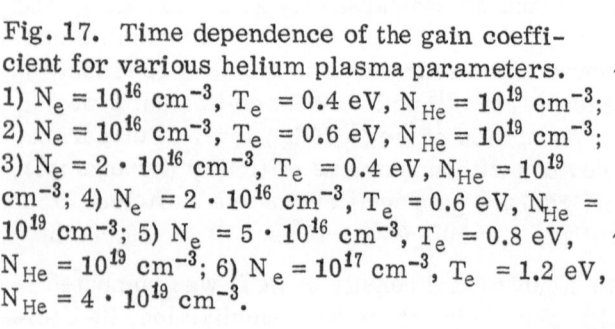

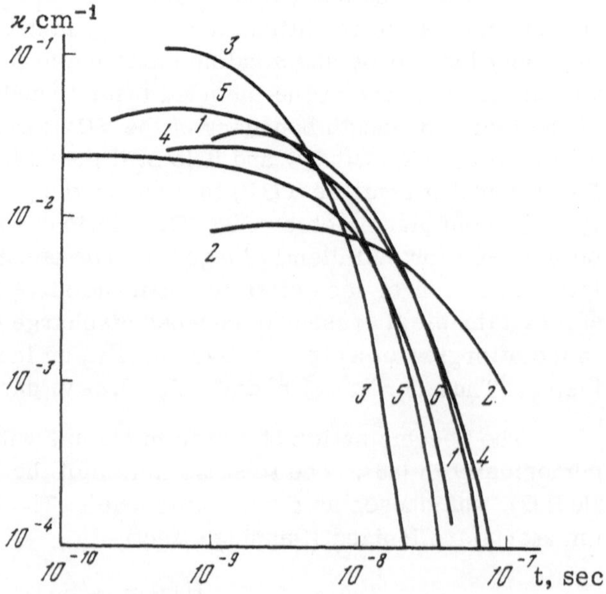

Fig. 17. Time dependence of the gain coefficient for various helium plasma parameters. 1) $N_e = 10^{16}$ cm^{-3}, $T_e = 0.4$ eV, $N_{He} = 10^{19}$ cm^{-3}; 2) $N_e = 10^{16}$ cm^{-3}, $T_e = 0.6$ eV, $N_{He} = 10^{19}$ cm^{-3}; 3) $N_e = 2 \cdot 10^{16}$ cm^{-3}, $T_e = 0.4$ eV, $N_{He} = 10^{19}$ cm^{-3}; 4) $N_e = 2 \cdot 10^{16}$ cm^{-3}, $T_e = 0.6$ eV, $N_{He} = 10^{19}$ cm^{-3}; 5) $N_e = 5 \cdot 10^{16}$ cm^{-3}, $T_e = 0.8$ eV, $N_{He} = 10^{19}$ cm^{-3}; 6) $N_e = 10^{17}$ cm^{-3}, $T_e = 1.2$ eV, $N_{He} = 4 \cdot 10^{19}$ cm^{-3}.

sities $N_e \sim 10^{16}$-10^{17} cm^{-3} and $N_{\text{He I}} \sim 10^{19}$ to $4 \cdot 10^{19}$ cm^{-3}, and temperatures $T_e = T \sim 0.4$-1.2 eV, the value of $\varkappa_{32}$ remains at a reasonable level ($\varkappa > 10^{-2}$ cm^{-1}) for a laser over a time interval $\tau \sim 10^{-8}$ sec. Here the cooling time of the electrons after a short ionization pulse is much less than the lifetime of the inversion. We note that electron heating during recombination of the He III is insignificant so that during this time interval only a small fraction of the electrons recombine.

16. The X-Ray Laser Problem

In discussions of ways of making x-ray lasers it is most often proposed that a dense plasma of multiply charged ions be used as the active medium. An important part of such articles is the analysis of specific difficulties in this wavelength range, of which there are three:

1. The ratio of the probabilities of induced and spontaneous transitions is very small here. In fact, $B_{ab}/A_{ab} \sim \omega_{ab}^{-3}\Delta\omega_{ab}^{-1}$, while the linewidth, $\Delta\omega_{ab}$, is usually proportional to the frequency ω_{ab} of the transition.
2. Here there exist no good mirrors with correspondingly high Q cavities. If use of reflection from crystals at the Bragg angle [171-174] for resonators is problematical in the hard x-ray range, it is not at all promising in the range $\lambda \sim 10$-100 Å.
3. The increased probability of spontaneous (radiative and radiationless) transitions with higher ionization states of the ions leads to a sharp speedup in the relaxation.

Besides these objective difficulties, the lack of research on the relaxation processes involved is certainly a hindrance in analyzing the conditions for x-ray amplification. Even in a plasma of relatively simple composition it is necessary to take into account the existence of ions of different charges, multielectron excitation, and spontaneous radiationless autoionization (Auger effect). In a high-density plasma a certain role may be played, not only by pairwise quasimolecular processes (for example, charge exchange and the Penning effect), but also by various collective particle interactions. Meanwhile, in a number of theretical papers there are separate considerations of the conditions for amplification by such a plasma of x-radiation: Methods are proposed for delivering a high density of energy to the medium, prescriptions are given for choosing the chemical composition of the active medium, etc. At this stage it is perhaps most relevant to formulate several questions, the answers to which would aid in choosing reasonable directions for experimental research.

(a) The Power Source for an X-Ray Laser. The energy density in the active medium required for efficient amplification of x-rays greatly exceeds the usual level achieved in gaseous discharges. Even in the ideal case of selective filling of the upper working level X(b), the minimum energy $E_{\min}$ per excited z-fold ion of the element X is very large, with

$$E_{\min} = E_{b1} + \sum_{\zeta=0}^{z-1} I(\zeta),$$

where E_{b1} is the excitation energy of the X_z ion and $I(\zeta)$ is the ionization energy of a ζ-fold ion. The smallness of the ratio of the radiative probabilities B_{ab}/A_{ab} and the absence of mirrors require the production of large populations N_b in the upper level with a sufficient length of active medium. The energy expenditures must be still higher than this because of the impossibility of purely selective pumping of the X_z(b) level and because of the large contribution of nonresonance channels (in particular, photoionization of excited X_z ions and ions X_ζ with lower charge). The requirement of high energy density with more rapid relaxation leads in general to the need for high power sources [162, 165]. In the steady-state regime an increase in the frequency by an order of magnitude will require an increase in the power by four orders of magnitude.

In the problem of designing an x-ray laser one should not aim for general schemes. If, for example, we take into account the possibility of using metastable states the above estimates of the power are greatly excessive. As an illustration of "slow" delivery of energy, we point out the possibility of storing energy in the metastable state of the He_2 molecule with subsequent pulsed transfer of the excitation to the resonance level and effective amplification of radiation ($\lambda \sim 600$ Å).

(b) The Principle of Creating an Inverted Population. Composition of the Medium. In pulsed ionization lasers the inversion is produced as the external ionizing field rises. Progress toward shorter wavelengths is here associated with a need to reduce the duration of the field pulse risetime. In a note [175] it has been shown that even with idealization of the properties of a dense gas ionized by an extremely fast rising ideal pulse, lasing cannot be achieved at wavelengths shorter than a critical value $\lambda_{min} \sim 400$ Å. Including the real gas properties will evidently increase λ_{min} substantially.

Thus, it is appropriate to discuss the possibility of x-ray amplification in a recombining plasma. Here it is possible to rely on the comparatively long afterglow stage of the pulse or on quasistationary nonequilibrium conditions as external ionizing particles act on the plasma. The problem of electron cooling becomes one of the most important questions. In fact the heat capacity of a multiply ionized plasma is determined by the free electrons and no longer by the heavy particles (as in the weakly ionized medium of visible or UV plasma lasers). Under these conditions it is necessary to evaluate the role of electron bremsstrahlung and to compare the rate of cooling of electrons due to adiabatic expansion of the plasma [87] with the loss of fast electrons to the "wall" [86].

The choice of the chemical composition of the active medium must be coordinated with the result of such estimates and with calculations of recombination for filling the upper and emptying the lower working level. Work published until recently has dealt with charge exchange for filling the upper working level [157], charge exchange for emptying the lower level by removal of an inner electron [159], a recombination–relaxation scheme for hydrogenlike ions [160], and then an attempt [162] to combine the approaches of [159, 160] by including charge exchange during recombination of hydrogenlike ions. In [163] a mechanism for creating an x-ray pumping inversion due to the difference in the probabilities of removing inner and outer electrons from ions was discussed. The estimates given in these papers cannot yet serve as a basis for a recommendation on the choice of chemical composition, the charge of the amplifying ions, or the density of the active medium of an x-ray laser. Taking into account the required high energy density, a maximum value must be obtained for the ratio of the energy of an output photon to the energy $E_b(z)$ used to produce a single laser ion X_z. Hence, the suggestion of [156] that multiply charged ions from isoelectronic sequences (starting from transitions on which lasing has been obtained with gaseous ion lasers in the visible and UV) be used for an x-ray laser is unjustified. This is because it is unreasonable to attempt to set up a laser at wavelengths of hundreds of angstroms on 20- to 30-fold ions.

(c) Conditions for Energy Input. Sources which would permit delivery of high density energy to a plasma over short times at present primarily include pulsed lasers and high-current electron accelerators. The characteristics of a laser plasma, produced by focusing a pulsed laser on a solid target, have been studied in many articles (see, for example, [176-179]. Light pulses at powers of about 10^{11} W and duration $\leqslant 10^{-10}$ sec have been obtained from neodymium lasers. Upon focusing to a spot of diameter $\sim 10^{-2}$ cm (an incident light intensity of about 10^{15} W/cm^2) a field of about 10^6-10^7 V/cm is produced in the vaporized multiply ionized medium. Commensurate plasma characteristics are obtained with a powerful electron beam with electron energies of several megaelectron volts and currents of hundreds of kiloamperes [180-181]. A laser plasma has been considered as an active medium for a laser in [158, 162, 165, 182, 183], and the plasma produced by an electron beam, in [184]. The kinetics of the interaction of light and fast electrons with solids is still little studied. It is possible that collective effects will have an important role here [185].

In order to obtain an inverted population in X_z ions without metastable states, the characteristic cooling time of the plasma must be much less than the relaxation time of the density of X_{z+1} ions. In order to rely on cooling by adiabatic expansion of the plasma into a vacuum or by loss of fast electrons to the walls, it is necessary to achieve a minimal transverse dimension of the active region. Maximum transverse focusing of the input energy also aids efficient escape from the medium of resonance radiation from the X ion (so a resonance state of this ion can be used as a lower working level), reduces competing emission of transverse modes, and reduces the source power requirements. The presence of a high-density gas in front of the target makes focusing difficult. The use of charge exchange of ions in the laser plasma with the surrounding neutral ballast gas (instead of three-body recombination $X_{z+1} + e + e \rightarrow X_z^* + e$) is made difficult by other factors as well. At temperatures still low enough for efficient three-body recombination of X_{z+1} ions the plasma electrons will ionize the ballast gas, making it unfit for charge exchange.

There are now a number of prescriptions for building an x-ray plasma laser. Along with experiments, an important part of the solution of this problem will be a detailed study of the kinetics of a plasma with multiply charged ions.

LITERATURE CITED

1. P. J. Berger and D. C. Smith, Appl. Phys. Lett., 21:167 (1972).
2. L. I. Gudzenko and L. A. Shelepin, Zh. Éksp. Teor. Fiz., 45:1445 (1963).
3. Gas Lasers [in Russian], Mir, Moscow (1968).
4. D. Bates, ed., Atomic and Molecular Processes, Academic Press (1962).
5. Ya. B. Zel'dovich and Yu. P. Raizer, The Physics of Shock Waves and High-Temperature Hydrodynamical Phenomena [in Russian], Nauka, Moscow (1966).
6. B. M. Smirnov, Atomic Collisions and Elementary Processes in Plasmas [in Russian], Atomizdat, Moscow (1968).
7. L. M. Biberman, V. S. Vorob'ev, and I. T. Yakubov, Usp. Fiz. Nauk., 107:353 (1972).
8. D. R. Bates, A. E. Kingston, and R. W. P. McWhirter, Proc. Roy. Soc., A267:297 (1963).
9. D. R. Bates and A. E. Kingston, Planet. Space Sci., 11:1 (1963).
10. R. W. P. McWhirter and A. J. Hearn, Proc. Phys. Soc., 82:641 (1963).
11. S. T. Belyaev and G. N. Budker, in: Plasma Physics and the Problem of Controlled Thermonuclear Reactions, Izd. AN SSSR (1958), Vol. 3.
12. A. V. Gurevich, Geomagn. Aeron., 4:10 (1964).
13. L. P. Pitaevskii, Zh. Éksp. Teor. Fiz., 42:1326 (1962).
14. H. R. Griem, Plasma Spectroscopy [Russian translation], Atomizdat, Moscow (1969).
15. I. I. Sobel'man, Introduction to the Theory of Atomic Spectra [in Russian], Fizmatgiz, Moscow (1963).
16. B. N. Glennon and W. L. Wiese, Atomic Transition Probabilities, N.B.S. (U.S.A.) Monograph No. 50 (1962).
17. W. L. Wiese, M. W. Smith, and B. M. Glennon, Atomic Transition Probabilities, N.B.S. (U.S.A.) Report NSRDS-4 (1966), (VII, 1968).
18. D. R. Bates and A. Damgaard, Ph. P. Trans. Roy. Soc., 242:101 (1949).
19. A. Burgess and M. J. Seaton, Rev. Mod. Phys., 30:992 (1958).
20. H. Bethe and E. Salpeter, Quantum Mechanics of One- and Two-Electron Atoms, Springer-Verlag (1957).
21. L. M. Biberman, Dokl. Akad. Nauk SSSR, 59:659 (1948).
22. T. Holstein, Phys. Rev., 72:1212 (1947).

23. V. Ya. Beldre, in: Electron–atom Collisions [in Russian], Zinatne, Riga (1965), p. 3.
24. M. R. U. Rudge, Rev. Mod. Phys., 40:564 (1968).
25. B. L. Moiseiwitsch and S. I. Smith, Rev. Mod. Phys., 40:238 (1968).
26. M. J. Seaton, Atomic Physics, New York (1969), p. 295.
27. A. Burgess, D. L. Hammer, and I. A. Jully, Philos. Trans. Roy. Soc., A226:225 (1970).
28. G. J. Schulz, Atomic Physics, New York (1969), p. 321.
29. A. Burgess and J. C. Percival, in: Advances in Atomic and Molecular Physics (1968), Vol. 4.
30. L. Vriens, in: Case Studies in Atomic Collision Physics, Vol. 1, North-Holland (1969).
31. L. J. Kieffer, and G. H. Dunn, Rev. Mod. Phys., 38:1 (1968).
32. L. J. Kieffer, JILA Int. Center Report No. 6 (1969), p. 7.
33. I. L. Beigman and L. A. Vainshtein, Izv. Akad. Nauk SSSR, Ser. Fiz., 27:1018 (1963).
34. J. E. Elwort, Z. Naturforsch., 7a:432 (1952).
35. L. M. Biberman, Yu. P. Toropkin, and K. S. Ulyanov, Zh. Tekh. Fiz., 32:827 (1962).
36. R. Post, High-Temperature Plasmas and Controlled Thermonuclear Reactions [Russian translation], IL (1961).
37. R. Post, Int. Summer Course in Plasma Phys., N18 (1960).
38. H. V. Drawin, Collision and Transport Cross Sections, EUR-CEA-FC-385 (1966).
39. G. D. Alkhazov, Zh. Tekh. Fiz., 40:97 (1970).
40. D. H. Samson and L. B. Golden, Astron. J., 161:321 (1970).
41. L. A. Vainshtein, I. L. Sobel'man, and E. A. Yukov, Cross Sections for Excitation of Atoms and Ions by Electrons [in Russian], Nauka, Moscow (1973).
42. E. M. Cherkasov, Candidate's Dissertation, Physics Institute, Academy of Sciences of the USSR (1969).
43. L. P. Presnyakov, A. D. Ulantsev, and L. A. Shelepin, Opt. Spektrosk., 24:677 (1968).
44. L. I. Gudzenko, S. S. Filippov, and S. I. Yakovlenko, Preprint FIAN No. 156 (1968); Zh. Prikl. Spektrosk., 13:357 (1970).
45. E. Hinnov and J. G. Hirschberg, Phys. Rev., 125:795 (1962).
46. Yu. K. Zemtsov, Candidate's Dissertation, Institute of Nuclear Physics, Moscow State University (1971).
47. B. F. Gordiets, L. I. Gudzenko, and L. A. Shelepin, Zh. Éksp. Teor. Fiz., 55:942 (1968).
48. A. V. Potapov and L. E. Tsvetkova, Teplofiz. Vys. Temp., 9:182 (1971).
49. L. I. Gudzenko, V. V. Evstigneev, S. S. Filippov, and S. I. Yakovlenko, Institute of Applied Mechanics Preprint No. 36 (1973).
50. V. A. Abramov, Teplofiz, Vys. Temp. 3:23 (1965).
51. D. W. Norcross and P. M. Stone, J. Quant. Spectrosk. Radiat. Transfer, 8:655 (1968).
52. B. F. Gordiets, I. A. Dymova, and L. A. Shelepin, Zh. Prikl. Spektrosk., 15:205 (1971).
53. B. F. Gordiets, L. I. Gudzenko, and L. A. Shelepin, Zh. Prikl. Mekh. Tekh. Fiz., No. 5, p. 116 (1968); No. 8, p. 971 (1968).
54. L. I. Gudzenko, V. V. Evstigneev, Yu. I. Syts'ko, S. S. Filippov, and S. I. Yakovlenko, Institute of Applied Mechanics Preprint No. 63 (1971).
55. I. S. Berezin and N. P. Zhidkov, Computational Techniques [in Russian], Nauka, Moscow (1966), Vol. 2.
56. V. S. Vorob'ev, Zh. Éksp. Teor. Fiz., 51:327 (1966).
57. S. A. Reshetnyak and L. A. Shelepin, Zh. Prikl. Mekh. Tekh. Fiz., No. 4, p. 18 (1972).
58. B. F. Gordiets, A. I. Osipov, E. V. Stupochenko, and L. A. Shelepin, Usp. Fiz. Nauk, 108:665 (1972).
59. C. K. N. Patel, in: Lasers, London (1966), Vol. 2.
60. G. G. Petrash, Usp. Fiz. Nauk, 105:645 (1971).
61. A. V. Eletskii and B. M. Smirnov, Gas Lasers [in Russian], Atomizdat, Moscow (1971).
62. V. F. Kitaeva, A. I. Odintsov, and N. N. Sobolev, Usp. Fiz. Nauk, 99:361 (1969).
63. V. I. Donin, Zh. Éksp. Teor. Fiz., 62:1648 (1972).

64. I. Corog, RCA Rev., 32:88 (1971).
65. Laser Focus, 6:16 (1970).
66. W. W. Simmons and R. S. Witte, IEEE J. Quant. Electron., 6:648 (1970).
67. Laser Focus, 6:5 (1970).
68. P. L. Rubin, N. N. Sobolev, and V. N. Faizulaev, Proceedings of the Conference on Electronic Techniques. Gas Lasers [in Russian], Vol. 2 (18), Institute of Electronics, Moscow (1970), p. 19.
69. E. J. Gordon, E. F. Labuda, and W. B. Bridges, Appl. Phys. Lett., 4:178 (1964).
70. W. R. Bridges, R. N. Chester, et al., Proc. IEEE, 59:724 (1971).
71. W. B. Bridges and G. N. Mercer, IEEE J. Quant. Electron., QE-5:476 (1969).
72. W. B. Bridges and R. N. Chester, Appl. Opt., 4:573 (1965).
73. J. R. Fendley, IEEE J. Quant. Electron., QE-4:627 (1968).
74. W. Wolinsky, Electronika, 13:355 (1972).
75. M. Gallardo, M. Garavaglia, A. A. Tagliaferri, and E. G. Luesma, IEEE J. Quant. Electron., QE-6:745 (1970).
76. D. T. Hodes, Appl. Phys. Lett., 17:11 (1970).
77. J. Opt. Soc. Am., 60:845 (1970).
78. T. G. Giallorensi and S. A. Ahmed, IEEE J. Quant. Electron., QE-7:11 (1971).
79. M. B. Klein and W. T. Silfast, Appl. Phys. Lett., 18:482 (1971).
80. W. K. Schuebel, IEEE J. Quant. Electron., QE-7:39 (1971).
81. W. T. Walter, IEEE J. Quant. Electron., QE-4:355 (1968).
82. W. T. Walter and J. S. Deech, Appl. Phys. Lett., 11:97 (1967).
83. A. A. Isaev, M. A. Kazaryan, and G. G. Petrash, Pis'ma Zh. Éksp. Teor. Fiz., 16:40 (1972).
84. L. I. Gudzenko and L. A. Shelepin, Dokl. Akad. Nauk SSSR, 160:1296 (1965).
85. L. I. Gudzenko, A. T. Matachun, and L. A. Shelepin, Zh. Tekh. Fiz., 37:833 (1967).
86. B. F. Gordiets, L. I. Gudzenko, and L. A. Shelepin, Zh. Tekh. Fiz., 36:1622 (1966).
87. L. I. Gudzenko, S. S. Filippov, and L. A. Shelepin, Zh. Éksp. Teor. Fiz., 51:1115 (1966).
88. L. I. Gudzenko and S. I. Yakovlenko, Zh. Éksp. Teor. Fiz., 59:1863 (1970).
89. L. I. Gudzenko, Yu. K. Zemtsov, and S. I. Yakovlenko, in: Proc. of the Third All-Union Conf. on Low-Temperature Plasma Physics [in Russian], Izd. MGU (1971), p. 256.
90. L. I. Gudzenko and S. I. Yakovlenko, Kratk. Soobshch Fiz., No. 7, p. 3 (1970).
91. L. I. Gudzenko, V. V. Evstigneev, and S. I. Yakovlenko, Kratk. Soobshch. Fiz., No. 9, p. 23 (1973); Preprint FIAN No. 186 (1973).
92. L. I. Gudzenko, Yu. K. Zemtsov, and S. I. Yakovlenko, Pis'ma Zh. Éksp. Teor. Fiz., 12:244 (1970).
93a. S. I. Yakovkenko, Institute of Atomic Energy Preprint No. 2174 (1972).
93b. L. I. Gudzenko and S. I. Yakovlenko, Dokl. Akad. Nauk SSSR, 207:1085 (1972).
94. L. I. Gudzenko, M. V. Nezlin, and S. I. Yakovlenko, Zh. Tekh. Fiz., 43:1931 (1973).
95. L. I. Gudzenko, Yu. I. Syts'ko, and S. I. Yakovlenko, FIAN Preprint No. 70 (1973).
96. L. I. Gudzenko and S. I. Yakovlenko, Kratk. Soobshch Fiz., No. 2, p. 3 (1974).
97. S. I. Yakovlenko, Candidate's Dissertation, Moscow State University (1973).
98. V. M. Gol'farb, E. V. Il'ina, I. G. Kostygova, G. A. Luk'yanov, and V. A. Silant'ev, Opt. Spektrosk., 20:1085 (1966).
99. K. Bockasten, T. Lundholm, and O. d'Andrade, J. Opt. Soc. Am., 56:1260 (1966).
100. E. L. Latush, V. S. Mikhalevskii, and M. F. Sem, Opt. Spektrosk. 34:214 (1973).
101. E. L. Latush and M. F. Sem, Kvant. Élektron. No. 3, p. 66 (1973).
102. E. L. Latush and M. F. Sem, Zh. Éksp. Teor. Fiz., 64:2017 (1973).
103. B. M. Smirnov and O. V. Firsov, Pis'ma Zh. Éksp. Teor. Fiz., 2:478 (1965).
104. M. Biondi, Phys. Rev. 93:1136 (1954).
105. R. M. Pirton and G. R. Fowles, Phys. Lett., 29A:654 (1969).

106. L. I. Gudzenko, Yu. K. Zemtsov, and S. I. Yakovlenko, Kratk. Soobshch. Fiz., No. 12, p. 3 (1971).
107. S. Kobayashi et al., IEEE J. Quant. Electron. QE-9:699 (1966).
108. R. K. Leonov, E. D. Protsenko, and Yu. M. Satunov, Opt. Spektrosc., 21:243 (1966).
109. D. Ya. Dudko, Yu. P. Korchevoi, and V. I. Lukashenkov, Opt. Spektrosk., 34:33 (1973).
110. E. E. Antonov, Yu. I. Korchevoi, and V. I. Lukashenko, Proc. 11th Int. Conf. on Phenomena in Ionized Gases, Prague (1973), Contributed Papers, p. 33.
111. V. L. Granovskii, Electrical Currents in Gases [in Russian], Nauka, Moscow (1971).
112. B. F. Zharov, V. K. Malinovskii, Yu. S. Neganov, and G. M. Chumak, Pis'ma Zh. Éksp. Teor. Fiz., 16:219 (1972).
113. R. T. Hodgson and R. W. Dreyfus, Phys. Lett., A38:213 (1972).
114. R. W. Dreyfus and R. T. Hodgson, Appl. Phys. Lett., 20:195 (1972).
115. N. G. Basov, É. M. Belenov, V. A. Danilychev, O. M. Kerimov, A. S. Podsosonnyi, and A. F. Suchkov, Kratk. Soobshch. Fiz., No. 5, p. 44 (1972).
116. N. G. Basov, E. M. Belenov, V. A. Danilychev, O. M. Kerimov, I. B. Kovsh, and A. F. Suchkov, Pis'ma Zh. Éksp Teor. Fiz., 14:421 (1971).
117. D. Yu. Zaroslov, E. K. Karlova, N. V. Karlov, G. P. Kuz'man, and A. M. Prokhorov, Pis'ma Zh. Éksp. Teor. Fiz., 15:665 (1972).
118. Yu. V. Tkach, Ya. B. Fainberg, L. I. Bolotin, Ya. Ya. Bessarab, N. P. Gadetskii, I. I. Magda, and A. V. Sidel'nikova, Zh. Éksp. Teor. Fiz., 62:1702 (1972).
119. V. G. Averin, A. I. Karchevskii, and G. V. Yurina, Zh. Éksp. Teor. Fiz., 63:85 (1972).
120. C. B. Collins, A. J. Cunningham, and B. W. Johnson, Proc. 11th Int. Conf. on Phenomena in Ionized Gases, Prague (1973), Contributed Papers, p. 31.
121. K. Thom and R. Schneider, AIAA J., 10:400 (1972).
122. N. M. Kuznetsov and Yu. P. Raizer, Zh. Prikl. Mekh. Tekh. Fiz., No. 4, p. 10 (1965).
123. L. N. Gudzenko, Yu. I. Syts'ko, S. S. Filippov, and S. I. Yakovlenko, Institute of Applied Mechanics Preprint No. 37 (1973).
124. E. L. Stupitskii and G. I. Kozlov, Institute of Applied Mechanics Preprint No. 13 (1972).
125. S. W. Bowen and C. Park, AIAA J., 9:493 (1971).
126. S. W. Bowen and C. Park, AIAA J., 10:522 (1972).
127. V. M. Gol'farb, I. V. Il'ina, I. E. Kostygova, and G. A. Luk'yanov, Opt. Spektrosk., 27:204 (1969).
128. V. M. Goldfarb, E. V. Ilyina, G. L. Lukyanov, and V. V. Sachin, Proc. 11th Int. Conf. on Phenomena in Ionized Gases, Prague (1973), Contributed Papers, p. 16.
129. V. M. Likhachev, M. S. Rabinovich, and V. M. Sutovskii, Pis'ma Zh. Éksp. Teor. Fiz., 5:55 (1967).
130. A. N. Vasil'eva, V. M. Likhachev, and V. M. Sutovskii, Zh. Tekh. Fiz., 39:341 (1969).
131. J. S. Hitt and W. T. Haswell, IEEE J. Quant. Electron. QE-2:60 (1966).
132. V. M. Sutovskii, Trudy FIAN, 56:66 (1971).
133. R. Illingworth, J. Phys. D (Appl. Phys.), 3:924 (1970).
134. R. Illingworth, J. Phys. D (Appl. Phys.), 5:686 (1972).
134a. A. Papayoanou and J. Gumeiner, J. Appl. Phys., 42: 1914 (1971).
135. S. A. Reshetnyak and L. A. Shelepin, Preprint FIAN No. 108 (1973).
136. A. S. Biryukov and L. A. Shelepin, Zh. Tekh. Fiz., 43:355 (1973).
137. L. I. Gudzenko, V. N. Kolesnikov, N. N. Sobolev, and L. A. Shelepin, Magnitn. Gidrodinam., 1:54 (1965).
138. MHD Energy Conversion; A Handbook [in Russian], Fizmatgiz, Moscow (1963).
139. E. E. Nitikin and S. Ya. Umanskii, Dokl. Akad. Nauk SSSR, 196:145 (1971).
140. F. McTaggart, Plasma Chemistry in Electrical Discharges, American Elsevier (1967).
141. V. N. Kondrat'ev, Rate Constants of Gaseous Phase Reactions [in Russian], Nauka, Moscow (1971).
142. F. G. Houtermans, Helv. Phys. Acta, 33:933 (1960).

143. S. I. Yakovlenko, Institute of Atomic Energy Preprint No. 2174 (1972).
144. L. I. Gudzenko and S. I. Yakovlenko, Dokl. Akad. Nauk SSSR, 207:1085 (1972).
145. L. I. Gudzenko, I. S. Lakoba, and S. I. Yakovlenko, Preprint FIAN No. 1 (1974).
146. J. B. Gerardo and A. W. Jonson, Conf. on Laser Eng. and Appl., Washington (1973), Digest, p. 29.
147. V. L. Borovach and V. S. Zuev, Zh. Éksp. Teor. Fiz., 58:1794 (1970).
148. A. A. Belyaeva, R. B. Dushin, E. V. Nikiforov, Yu. B. Predchetenskii, and L. D. Shcherba, Dokl. Akad, Nauk SSSR, 198:1117 (1971).
149. M. M. Mkrtchan and V. T. Platonenko, Pis'ma Zh. Éksp. Teor. Fiz., 17:28 (1973).
150. C. V. Heer, Phys. Lett., A31:160 (1970).
151. R. J. Carbone and M. M. Litvak, Solid State Res. Lincoln Lab., M.I.T., No. 2, p. 21 (1964); No. 4, p. 21 (1965).
152. P. W. Hoff, J. C. Swingli, and C. K. Phodes, Appl. Phys. Lett., 23:245 (1973).
153. A. Wayne and J. J. B. Gerardo, Proc. 11th Int. Conf. on Phenomena in Ionized Gases, Prague (1973), Contributed Papers, p. 164.
154. O. J. Bragly, M. D. R. Hutchinson, and H. Koster, Opt. Commun., 7:187 (1973).
155. E. V. George and C. K. Phodes, Appl. Phys. Lett., 23:139 (1973).
156. A. G. Molchanov, Usp. Fiz. Nauk, 106:165 (1972).
157. B. M. Smirnov, Pis'ma Zh. Éksp. Teor. Fiz., 6:565 (1967).
158. L. I. Gudzenko and L. A. Shelepin, in: Problems in the Physics of Low-Temperature Plasmas [in Russian], Nauka i Tekhnika, Minsk (1970), p. 478.
159. L. I. Gudzenko and L. A. Shelepin, Pis'ma Zh. Éksp. Teor. Fiz., 7:246 (1968).
160. B. F. Gordiets, L. I. Gudzenko, and L. A. Shelepin, Zh. Prikl. Mekh. Tekh. Fiz., No. 5, p. 115 (1966).
161. L. P. Presnyakov and V. P. Shevel'ko, Pis'ma Zh. Éksp. Teor. Fiz., 13:286 (1971).
162. A. V. Vinogradov and I. I. Sobel'man, Zh. Éksp. Teor. Fiz., 63:2113 (1972).
163. M. A. Dugney and P. M. Rentzepis, Appl. Phys. Lett., 10:350 (1967).
164. M. H. Reilly, R. C. Elton, R. W. Wayrand, and P. A. Andrews, Gamma Ray and Vacuum UV Lasers, NRL Report No. 7412 (1972).
165. B. Lax and A. H. Guenther, Appl. Phys. Lett., 21:361 (1972).
166. M. J. Bernstein and G. G. Comisar, J. Appl. Phys., 41:729 (1970).
167. P. Jagob and A. Carillon, Phys. Lett., A36:167 (1971).
168. R. V. Khokhlov, Pis'ma Zh. Éksp. Teor. Fiz., 15:580 (1972).
169. D. Marcus, Proc. IEEE, 51:849 (1963).
170. V. I. God'danskii and Yu. N. Kagan, Zh. Éksp. Teor. Fiz., 64:90 (1973).
171. W. L. Bond, M. A. Dugnay, and P. M. Rentzepis, Appl. Phys. Lett., 10:216 (1967).
172. B. Okkers, Phillips Res. Rep. 18:413 (1963).
173. R. D. Deslatter, Appl. Phys. Lett., 12:133 (1967).
174. R. M. J. Cotterill, Appl. Phys. Lett., 12:403 (1968).
175. Yu. V. Afanas'ev, É. M. Belenov, and I. N. Knyazev, Kratk. Soobshch. Fiz., No. 1, p. 59 (1970).
176. Yu. V. Afanas'ev, É. M. Belenov, O. N. Krokhin, and I. A. Poluéktov, Zh. Éksp. Teor. Fiz., 63:121 (1972).
177. T. T. Suboi and K. Terao, Jpn. J. Appl. Phys., 9:401 (1971).
178. S. Witkowski, Naturwissenschaften, 57:211 (1970).
179. C. Jamansky, Phys. Lett., 38A:495 (1972).
180. G. A. Mesyats, Yu. I. Bychkov, and V. V. Kremnov, Usp. Fiz. Nauk, 107:201 (1972).
181. M. S. Rabinovich and I. A. Kossyi, Usp. Fiz. Nauk, 101:83 (1970).
182. J. Peyrand and N. Peyrand, J. Appl. Phys., 43:2993 (1972).
183. L. I. Gudzenko, S. D. Kaitmazov, A. A. Medvedev, and E. I. Shklovskii, Pis'ma Zh. Éksp. Teor. Fiz., 9:561 (1969).
184. R. B. Baksht, Yu. I. Bychkov, and G. A. Mesyats, Kvant. Élektron., No. 3, p. 89 (1972).
185. S. D. Fanchenko, At. Énerg., 32:444 (1972).

THE KINETICS OF PLASMA AND ROTATIONAL TRANSITION LASERS†

S. A. Reshetnyak

Quasistationary population distributions are derived and used to examine rotational kinetics and the kinetics of recombination and ionization in plasmas. The possibility of increasing the efficiency of lasers in the far infrared and submillimeter ranges is studied. Lasers using low-temperature plasmas of both simple and complex (mixture of elements) chemical composition are examined. The basic relationships for moving plasma lasers (plasma-dynamic and pinch discharge lasers) are calculated. The kinetics of multiply charged ion plasmas is considered. The theory of high-current discharges is used to evaluate the feasibility of using them as powerful light sources.

INTRODUCTION

The modern theory of gas and plasma lasers is based on molecular and atomic kinetics. The process of creating a population inversion is itself the result of rapid relaxation among atoms and molecules. If the analysis of this process immediately after the first lasers were made was limited to two or three levels, then by the early 1960's it was already obvious that many levels would have to be included [1, 2]. The multilevel approach is being developed for the analysis of atomic kinetics. Bates, Kingston, and McWhirter [3] analyzed the population balance equations for the levels of the hydrogen atom and proposed the constant-sink approximation, according to which all time derivatives of the level populations (except of the first level) on the left-hand side of the balance equations are assumed to equal zero. For a long time this approximation has been the basis of the kinetic analysis of recombining plasma lasers.

The diffusion approach was developed in [4-6]. According to it the motion of an electron over the levels during recombination and ionization is regarded as diffusion in energy space. A review of the state of atomic kinetics is given in [6, 7]. In the last decade there has been a rapid development in the molecular kinetics of vibrational processes (see [8]). Population distributions over the vibrational levels of molecules have been found under the most varied of nonequilibrium conditions for diatomic and multiatomic molecules, for single-component gases and gaseous mixtures, and for molecular dissociation and recombination with thermal, electrical, and laser pumping of energy into the vibrational degrees of freedom.

The applicability of quasiequilibrium distributions in vibrational kinetics depends on differences among the characteristic relaxation times of a given physical system. A distribution is established over a given set of levels (degree of freedom) and will depend on certain param-

† This article includes material from a dissertation for the degree of Candidate of Physical and Mathematical Sciences. The research adviser was L. A. Shelepin.

eters. The introduction of partial vibrational temperatures for various vibrational modes of multiatomic molecules [9] and the use of the Treanor distribution [10] and distributions depending on the concentrations of the gases [11] have been highly effective. The operating mechanisms of most lasers using vibrational–rotational transitions have been analyzed on the basis of these models. Quasiequilibrium distributions appear to be important not only for vibrational levels. Thus, in the theory of nuclear magnetic resonance it has been useful to introduce the spin temperature and to examine quasiequilibrium states of spin–spin subsystems [12].

The purpose of the present article is twofold. The first is to extend the method of quasiequilibrium distributions and to examine those stages of the kinetics which can be described analytically. Here we obtain analytic distribution functions for the populations of the rotational states of molecules and the electronic states of atoms, and for the charge states of ions. Second, on the basis of these distribution functions some specific questions are raised about the kinetics of plasma lasers and of lasers using rotational transitions.

A characteristic feature of quasistationary distributions in plasmas is the fact that the level populations are described by two (or several) parameters. This makes it possible to solve complicated problems of physical kinetics in plasma lasers, in particular, to analyze moving plasma lasers. Here we analyze relaxation processes during compression and expansion of a plasma. The kinetics of lasers using a pinch discharge [13] and of plasma-dynamic lasers [14] will be studied.

Here we shall not review the literature as that has already been done [7, 15] and shall only present original results. The basic contents of the present article are found in [16-23]. The article is made up of four chapters. In the first chapter distribution functions for the rotational levels of molecules are obtained and far infrared lasers are analyzed. The second chapter deals with the quasistationary population distribution functions of the electronic levels in a low-temperature plasma. The use of these distribution functions in plasma lasers employing mixtures is analyzed. The third chapter is devoted to a study of the physical kinetics of a moving plasma laser (plasma-dynamic and pinch discharge lasers). The fourth chapter discusses some problems in the kinetics of a multiply charged ion plasma and in the use of a high-current self-pinching discharge plasma as a high-power light source. As opposed to the case of low-temperature plasmas, however, our results are by no means a complete picture and are rather like fragments of a future theory.

CHAPTER I

ROTATIONAL KINETICS AND FAR-INFRARED (FIR) MOLECULAR LASERS

This chapter deals with a number of questions about quasistationary rotational level population distributions. These distributions make it possible to study the kinetics of the formation of an inverted population and to analyze lasers in the far infrared region of the spectrum.

Analytic distribution functions for the populations of the rotational levels of diatomic molecules in the presence of a source of rotationally excited molecules are found on the basis of the stochastic Fokker–Planck equation. Lasing mechanisms in multiatomic molecules are studied. Quasistationary rotational energy distribution functions are obtained in the adiabatic region of the spectrum in the single-quantum approximation (discrete model).

1. Rotational Kinetics of Diatomic Molecules

At the present time lasing in the far infrared and submillimeter spectral regions has been obtained on a large number of transitions of a whole series of molecules (H_2O, D_2O, HCN,

BrCN, ICN, H_2CO, H_2S, SO_2, COS, CH_3F, CH_3OH; see, for example, [24-26]). However, as opposed to the near infrared, the development of lasers in this spectral region has come much slowly, despite their obviously important applications (biological effects, molecular structure, radiospectroscopy, atmospheric studies). The powers obtained so far in the continuous regime are still negligible, the mechanism for producing the population inversion is often unknown, and in many cases the transition lines are not even known.

The inadequate efficiency of present submillimeter lasers is due to three basic difficulties which arose during their development. The first is related to the high rates of energy pumping required to produce a disequilibrium state. The corresponding laser transitions are either purely rotational or vibrational–rotational in multiatomic molecules. Then, in order to have appreciable radiative probabilities, sufficient for the required gain coefficients, dipole molecules are used. The characteristic times for both rotational and vibrational relaxation of such molecules are very small, so high pumping rates are needed. The second difficulty is due to the absence of information on the excitation rates of the individual rotational levels by various mechanisms (chemical, electrical, etc.) and to our completely inadequate knowledge of the radiative and collisional transition probabilities. This greatly inhibits choice of specific materials for lasers. The third difficulty in the way of building efficient submillimeter lasers may be regarded as the undeveloped state of the physical kinetics of the processes associated with the creation of a population inversion (especially rotational relaxation). Knowledge of the kinetics provides a fundamental basis for seeking new materials and transitions for increasing the efficiency of existing lasers.

In diatomic molecules lasing in the far infrared takes place on rotational transitions. We shall examine the basic equations describing purely rotational relaxation with a positive source of rotationally excited molecules approximated by a rigid rotator model.

We shall assume that a system of such rotators is located in a temperature bath with translational temperature T. Then the behavior of the relaxing system may be described by the balance equations for the populations N_j of the rotational levels:

$$\frac{dN_j}{dt}=\sum_i [P(i, j) N_i - P(j, i) N_j] + Q_j, \tag{1.1}$$

where P(i, j) is the probability per unit time of a transition from the i-th to the j-th rotational level due to collisions of rigid rotators with gas particles and Q_j is the power of the source that populates level j.

However, because of a lack of reliable published data on P(i, j) and the need in many cases to take multiquantum transitions into account it seems more convenient to analyze rotational relaxation on the basis of a stochastic equation and to assume that the energy spectrum of the rotational states is continuous. Then Eq. (1.1) takes the form

$$\frac{\partial n}{\partial t}=\int_0^\infty d\varepsilon' [P(\varepsilon', \varepsilon) n(\varepsilon', t) - P(\varepsilon, \varepsilon') n(\varepsilon, t)] + q(\varepsilon, t). \tag{1.2}$$

Here $n(\varepsilon, t)d\varepsilon$ is the number of molecules with rotational energy in the interval $(\varepsilon, \varepsilon + d\varepsilon)$, and $P(\varepsilon', \varepsilon)d\varepsilon' d\varepsilon$ is the probability per unit time of a transition in the continuous spectrum from the interval $(\varepsilon', \varepsilon' + d\varepsilon')$ to the interval $(\varepsilon, \varepsilon + d\varepsilon)$. The source density distribution $q(\varepsilon, t)$ is defined in an analogous manner.

By integrating Eq. (1.2) over ε from zero to infinity and using the normalization $\int_0^\infty n(\varepsilon, t)\, d\varepsilon = N$ we obtain

$$\int_0^\infty q(\varepsilon, t)\, d\varepsilon = \frac{dN}{dt}, \tag{1.3}$$

where N is the total number of rigid rotators per unit volume.

The transformation from the populations of discrete levels to a density distribution also presupposes the inverse procedure:

$$N_j = \int_{\varepsilon'_j}^{\varepsilon''_j} n(\varepsilon, t)\, d\varepsilon, \qquad Q_j = \int_{\varepsilon'_j}^{\varepsilon''_j} q(\varepsilon, t)\, d\varepsilon, \tag{1.4}$$

where

$$\varepsilon'_j = \frac{1}{2}(\varepsilon_j + \varepsilon_{j-1}), \qquad \varepsilon''_j = \frac{1}{2}(\varepsilon_j + \varepsilon_{j+1}), \qquad \varepsilon_j = B_e j(j+1),$$

and B_e is the rotational constant of the molecule. Comparing Eq. (1.4) with the equilibrium rotational level distribution function,

$$N_{je} = N_0 g_j \exp\left(-\frac{\varepsilon_j}{kT}\right), \qquad g_j = 2j + 1, \qquad N_0 = \frac{B_e}{kT} N, \tag{1.5}$$

it is easy to see that the equilibrium density distribution has the form

$$n_e(\varepsilon) = N_0 G(\varepsilon) \exp\left(-\frac{\varepsilon}{kT}\right), \qquad G(\varepsilon) = \frac{g_j}{d\varepsilon/dj} = B_e^{-1}. \tag{1.6}$$

The method of deriving the stochastic Fokker–Planck equation for rotational relaxation has been discussed in [27]. When there is a positive source of rotationally excited molecules this equation takes the form

$$G(\varepsilon) \exp\left(-\frac{\varepsilon}{kT}\right) \frac{\partial \Phi}{\partial t} = \frac{\partial}{\partial \varepsilon}\left[G(\varepsilon) \exp\left(-\frac{\varepsilon}{kT}\right) D \frac{\partial \Phi}{\partial \varepsilon}\right] + q(\varepsilon, t), \tag{1.7}$$

where Φ is related to the density distribution n in the following way:

$$n(\varepsilon, t) = G(\varepsilon) \exp(-\varepsilon/kT)\, \Phi(\varepsilon, t). \tag{1.8}$$

The coefficient D(ε) in Eq. (1.7) is defined as the product of the frequency of collisions between rigid rotators and gas particles Z and the average of the square of the change in energy of the rotator during collisions, $\langle \Delta\varepsilon^2 \rangle$:

$$D(\varepsilon) = \frac{Z}{2} \langle \Delta\varepsilon^2 \rangle, \qquad \langle \Delta\varepsilon^2 \rangle = Z^{-1} \int_0^\infty P(\varepsilon, \varepsilon')(\varepsilon' - \varepsilon)^2\, d\varepsilon. \tag{1.9}$$

Equation (1.7) must be supplemented by some specific boundary conditions. Taking the diffusion flux to be zero at the point ε = 0, we have

$$D \frac{\partial \Phi}{\partial \varepsilon}\bigg|_{\varepsilon=0} = 0. \tag{1.10}$$

To this boundary condition we must add the requirement that the density distribution be bounded at infinity.

Depending on the relationship between the collision time between the rotator and the gas particles and the rotational period of the rigid rotator, the rotational spectrum is divided into nonadiabatic and adiabatic regions. In the nonadiabatic region the time for collisions between the rotator and gas particles $\tau_{coll} \sim (2\pi/a)(\mu/2kT)^{1/2}$ is less than the rotational period of the former, $\tau_{rot} \sim (1/2\varepsilon)^{1/2}$; that is,

$$\left(\frac{kT}{\varepsilon}\right)^{1/2} \gg \frac{2\pi}{a}\left(\frac{\mu}{I}\right)^{1/2} = \xi. \tag{1.11}$$

Here a^{-1} is the characteristic size of the interaction region, μ is the reduced mass of the colliding particles, I is the moment of inertic of the rotator, ε is the rotational energy of the rotator, and T is the translational temperature of the gas. When $(\varepsilon/kT)^{1/2} \gg \xi^{-1}$ the range of values of ε is adiabatic.

There is a significant advantage to using Eq. (1.7), rather than Eq. (1.1), as the basis for analyzing relaxation since the need to know the individual rotational transition probabilities no longer exists. The role of these probabilities in Eq. (1.7) is played by the the coefficient D which effectively takes multiquantum transitions into account as well. Solution of Eq. (1.1) including such transitions is exceedingly difficult.

The coefficient D was calculated in [27] for both the nonadiabatic and the adiabatic regions of the rotational spectrum and has a very complicated dependence on ε. In the calculation it was assumed that the rigid rotator interacts with a structureless particle through a modified Morse potential with a potential well of depth V_0. On the basis of the results of [27] in the limiting case $V_0 \to 0$ we can obtain a simple analytic expression for D over a wide range of energies:

$$D(\varepsilon) = 16Z\left(\frac{3}{a}\right)^2 \frac{\mu}{\xi^{-2} I} \varepsilon^2 \int_0^\infty \frac{\exp(-x)\,dx}{\mathrm{sh}^2\left[\xi\sqrt{\frac{\varepsilon}{kTx}}\right]} \approx 16Z\left(\frac{3}{a}\right)^2 \frac{\mu}{I} kT\varepsilon \exp\left\{-\frac{2}{3}\xi\sqrt{\frac{\varepsilon}{kT}}\right\}, \tag{1.12}$$

where β is a parameter that indicates the deviation of the potential from spherical.

This form of D makes it possible to substantially simplify the analysis of rotational relaxation. In the following we analyze two cases of interest for laser applications: relaxation of an initial perturbation in the rotational distribution function (for the nonadiabatic region of the spectrum) and a quasistationary nonequilibrium distribution with a positive source of rotationally excited molecules (over the entire spectrum). An analysis of the relaxation of an initial nonequilibrium rotational level distribution of the molecules is of interest in studies of the inverted population in cases where the pumping pulse duration is less than the characteristic rotational relaxation time. Such an analysis is easily done for the nonadiabatic region of rotational energies $(\varepsilon/kT)^{1/2} \ll \xi^{-1}$, where the expression for D is especially simple,

$$D = \alpha\varepsilon, \quad \alpha = 16Z\left(\frac{\beta}{a}\right)^2 \frac{\mu}{I} kT. \tag{1.13}$$

The stochastic equation (1.7) reduces to the usual diffusion equation with a diffusion coefficient D which depends linearly on the coordinate ε.

We shall consider relaxation of a nonequilibrium initial distribution in the form of a Boltzmann distribution with an additional number N^* of particles at level j_0 with energy ε_0:

$$\Phi(\varepsilon, 0) = N_0 + G^{-1} N^* \exp\left(\frac{\varepsilon_0}{kT}\right) \delta(\varepsilon - \varepsilon_0). \tag{1.14}$$

A solution of Eq. (1.7) including Eqs. (1.13) and (1.14) when $\varepsilon \ll kT$ (which is almost always satisfied in the nonadiabatic portion of the spectrum) can be found using the Laplace–Carson transform. A second-order ordinary differential equation results on transforming Eq. (1.7) from its original form. By an appropriate change of variables this equation may be reduced to a Bessel equation whose linearly independent solutions will be modified Bessel functions of zeroth order I_0 and the MacDonald function K_0. Including the boundary condition for Eq. (1.7) and the asymptotic behavior of these functions at infinity, we find that the solution of the problem in the region $0 < \varepsilon < \varepsilon_0$ will be $C_1 I_0$ and $C_2 K_0$ in the region $\varepsilon > \varepsilon_0$. To find the constants C_1 and C_2 we first use the continuity conditions on the solution at the point ε_0 and, second, integrate the original differential equation for the transform in the neighborhood of the point ε_0 from $\varepsilon_0 - \varepsilon_1$ to $\varepsilon_0 + \varepsilon_1$ with later transition to the limit $\varepsilon_1 \to 0$. As a result of changing from the transform to the original equation, which was done with the aid of tables [28], the following expression was found for n:

$$n(\varepsilon, t) = \exp\left(-\frac{\varepsilon}{kT}\right)\left[\frac{N_0}{B_e} + \frac{N^*}{at}\exp\left(-\frac{\varepsilon_0 + \varepsilon}{at}\right) I_0\left(\frac{2\sqrt{\varepsilon_0 \varepsilon}}{at}\right)\right]. \qquad (1.15)$$

Relaxation of a nonequilibrium initial distribution has also been discussed in [29]. There, with a linear dependence of the diffusion coefficient on the energy, a general solution of Eq. (1.7) was found for a deviation of the distribution function from equilibrium (for an arbitrary value of the ratio ε/kT) given by

$$\Delta n = n - n_e = \exp\left(-\frac{\varepsilon}{kT}\right)\sum_n C_n L_n\left(\frac{\varepsilon}{kT}\right)\exp\left(-\frac{nt}{\tau_R}\right),$$
$$C_n = \int_0^\infty \Delta n(\varepsilon, 0) L_n\left(\frac{\varepsilon}{kT}\right)\frac{d\varepsilon}{kT}, \quad \tau_R = \frac{kT}{a}, \qquad (1.16)$$

where L_n is a Laguerre polynomial and $\Delta n(\varepsilon, 0)$ is the (arbitrary) initial deviation of the distribution function from equilibrium.

We shall now show that Eq. (1.16) transforms to Eq. (1.15) in the case of a δ-shaped initial distribution. To do this we substitute $\Delta n(\varepsilon, 0) = N^*\delta(\varepsilon - \varepsilon_0)$ in Eq. (1.16) and use the properties of Laguerre polynomials.† We therefore have

$$\Delta n = \frac{N^*}{kT(1-\Theta)}\exp\left(-\frac{\varepsilon}{kT} - \frac{\Theta}{1-\Theta}\cdot\frac{\varepsilon + \varepsilon_0}{kT}\right) J_0\left(\frac{2\sqrt{\varepsilon\varepsilon_0}}{kT}\cdot\frac{\Theta}{1-\Theta}\right),$$
$$\Theta = \exp\left(-\frac{t}{\tau_R}\right). \qquad (1.17)$$

Expanding the exponent in the equation for Θ in a series and limiting oursleves to the first two terms of this expansion, we obtain Eq. (1.15).

It follows from Eq. (1.15) that the characteristic relaxation time of the initial nonequilibrium distribution is τ_R. Over the very short time τ_R, a δ-shaped initial distribution transforms to a Boltzmann distribution, so the inverted population on the j_0 and $j_0 - 1$ levels disappears.

This is an indication of the difficulty of using the nonadiabatic portion of the spectrum for lasing. A second difficulty in using this regime is the necessity of creating substantial nonequilibrium populations over a short time, less than τ_R. Thus, it is of more practical in-

† $\sum_{n=0}^{\infty}\frac{n!}{\Gamma(n+\alpha+1)} L_n^{(\alpha)}(x) L_n^{(\alpha)}(y) z^n = \frac{(xyz)^{-\alpha/2}}{1-z}\cdot\exp\left(-z\frac{x+y}{1-z}\right) J_\alpha\left(\frac{2\sqrt{xyz}}{1-z}\right), \quad |z| < 1.$

terest to analyze rotational relaxation and the conditions for existence of an inversion with a slowly varying source of excited molecules acting over a time interval greater than τ_R. Let a source of power $q(\varepsilon)$ which is constant or slowly varying over a time τ_R act at some point ε_0 in the rotational spectrum. For concreteness we shall further assume that this source produces rotationally excited molecules as a result of a chemical reaction.† After a time of order τ_R following activation of the source a quasistationary distribution is established over the rotational levels. This distribution is characterized by a Boltzmann distribution with a time-varying total number of particles and a distortion determined by the power of the source.

The physical considerations on which the existence of such a distribution is based are similar to those for the case of sources which distort the distributions over the vibrational or translational degrees of freedom of molecules [8, 30, 31]. Thus, for a slowly varying source after some time of order τ_R the solution to Eq. (1.7) may be found in the form

$$\Phi(\varepsilon, t) = N_0(t) + F(\varepsilon). \qquad (1.18)$$

Substituting Eq. (1.18) into Eq. (1.7) we obtain an equation for the function F which describes the distortion in the Boltzmann distribution n_e:

$$G \exp\left(-\frac{\varepsilon}{kT}\right)\dot{N}_0 = \frac{\partial}{\partial \varepsilon}\left[G \exp\left(-\frac{\varepsilon}{kT}\right) D \frac{\partial F}{\partial \varepsilon}\right] + q(\varepsilon). \qquad (1.19)$$

Integrating Eq. (1.19), we have

$$F(\varepsilon) = \int_0^{\varepsilon} D^{-1}(\varepsilon')\left[kT\dot{N}_0\left(\exp\left(\frac{\varepsilon'}{kT}\right) - 1\right) - B_e \exp\left(\frac{\varepsilon'}{kT}\right)\int_0^{\varepsilon'} q(\varepsilon'')\, d\varepsilon''\right] d\varepsilon'. \qquad (1.20)$$

For simplicity, in the following we shall analyze the case of a δ-shaped pumping distribution $q(\varepsilon) = Q\,\delta(\varepsilon - \varepsilon_0)$. Then in the region $0 < \varepsilon < \varepsilon_0$, F takes the form

$$F(\varepsilon) = \frac{B_e Q}{\alpha} f\left(\frac{\varepsilon}{kT}\right), \quad f\left(\frac{\varepsilon}{kT}\right) = \alpha \int_0^{\varepsilon} \frac{\exp\left(\frac{\varepsilon'}{kT}\right) - 1}{D(\varepsilon')}\, d\varepsilon' \quad (Q = N). \qquad (1.21)$$

Equations (1.20) and (1.21) are quite general. In the nonadiabatic and adiabatic spectral regions they can be written in a simple form. In the case of Eq. (1.21) we have

$$f(x) = x, \quad x = \varepsilon/kT$$

for the nonadiabatic region, and

$$f(x) = x^{-1} \exp(x + 2/3\xi x^{1/2})$$

for the adiabatic region.

We note that this analysis demonstrates the great similarity between rotational and vibrational relaxation. As rotational analogies of vibrational–translational and vibrational–vibrational exchange (VT and VV processes) one may identify RT processes (exchange of rotational and translational energy) and RR processes (rotational exchange). The probabilities

† The results obtained are easily generalized to the case of excitation with a constant total number of molecules (for example, with optical or electrical pumping).

of RT and VT processes are similar and are given by the integral in Eq. (1.12). Also similar are the distribution functions over rotational and vibrational levels in the presence of sources of excited molecules. Thus, an entire set of results on vibrational relaxation [8] may be transferred to rotational relaxation in the adiabatic part of the spectrum, where according to Eq. (1.12) single-quantum transitions predominate and their probabilities are determined from Eqs. (1.9) and (1.12).

We now turn to an analysis of a population inversion ΔN over rotational transitions of a diatomic molecule. Going to the discrete description in Eqs. (1.18) and (1.21) and assuming that the source acts on level j_0, we have

$$\Delta N=\frac{N_{j_0}}{2j_0+1}-\frac{N_{j_0-1}}{2j_0-1}=\frac{B_e}{kT}\exp\left(-\frac{\varepsilon_{j_0}}{kT}\right)\left[Q\tau_R\varphi_{j_0}-N\left(\exp\left(\frac{\Delta\varepsilon}{kT}\right)-1\right)\right],$$
$$\varphi_{j_0}=f(\varepsilon_{j_0})-f(\varepsilon_{j_0-1})\exp\left(\frac{\Delta\varepsilon}{kT}\right),\quad \Delta\varepsilon=\varepsilon_{j_0}-\varepsilon_{j_0-1}. \tag{1.22}$$

Further analysis of the population inversion is easy once the time variation of the chemical pumping Q is given. For nonbranching chain reactions it is often possible to assume that $Q=At^m$ and $N=\int_0^t Qdt=\frac{A}{m+1}t^{m+1}$. Using these expressions we find from Eq. (1.22) that the lifetime of the inversion t_0 and the time to reach its maximum value t_{max} are given by

$$t_0=\frac{m+1}{m}t_{max},\quad t_{max}=\tau_R\frac{m\varphi_{j_0}}{\exp(\Delta\varepsilon/kT)-1}. \tag{1.23}$$

In the nonadiabatic and adiabatic portions of the spectrum these expressions simplify to

$$t_{max}=m\tau_R,\quad (\varepsilon_{j_0}/kT)^{1/2}\ll\xi^{-1},$$
$$t_{max}=\frac{2}{3}\xi\tau_R\frac{m}{j_0+1}\left(\frac{kT}{\varepsilon_{j_0}}\right)^{1/2}\left[\exp\left(\frac{\Delta\varepsilon}{kT}\right)-1\right]^{-1}\exp\left[\frac{\varepsilon_{j_0}}{kT}+\frac{2}{3}\xi\left(\frac{\varepsilon_{j_0}}{kT}\right)^{1/2}\right],\quad \left(\frac{\varepsilon_{j_0}}{kT}\right)^{1/2}\gg\xi^{-1}. \tag{1.24}$$

From Eqs. (1.23) and (1.24) it is clear that in a quasistationary pumping regime the lifetime of an inversion in the nonadiabatic region is on the order of the rotational relaxation time τ_R, while in the adiabatic region, it is much greater than τ_R. As an example we can examine the possibility of obtaining a population inversion on transitions of the HCl molecule as rotationally excited HCl molecules are formed in the reaction $H_2 + Cl_2 \rightarrow HCl^*$ when initiated by light. When the mixture is at a pressure of about 1 torr, a temperature of about 300°K, and the chlorine concentration $[Cl] \simeq 5\cdot 10^{20}t$ [32], estimates of the maximum inversion on the $j_0 = 15 \rightarrow j_0 - 1 = 14$ transition yield $\Delta N \lessdot 10^{11}$ cm^{-3}. This corresponds to a pump power of $Q \sim 3\times 10^{20}$ mole/sec$\cdot$cm^3.

Using $A_{j_0,j_0-1} \simeq 2\Delta\varepsilon^3\tilde{\mu}^2/3\hbar^4c^3$ (c is the speed of light) for the probability A_{j_0,j_0-1} of the radiative transition $j_0 \rightarrow j_0 - 1$, the gain coefficient $\varkappa_{j_0,j_0-1}$ for the case of purely rotational transitions may be written in the forms

$$\varkappa_{j_0,j_0-1}\ (\text{cm}^{-1})\simeq 4.4\cdot 10^{-15}\sqrt{\frac{M}{T}}\tilde{\mu}^2\Delta N \tag{1.25a}$$

for Doppler broadening and

$$\varkappa_{j_0,j_0-1}\ (\text{cm}^{-1})\simeq 1.3\cdot 10^{-10}\frac{\tilde{\mu}^2B_ej_0}{\Delta\nu_{coll}}\Delta N \tag{1.25b}$$

for collisional broadening. Here M is the mass of the molecule (atomic units); $\tilde{\mu}$ is the dipole moment of the molecule (Debyes), B is the rotational constant (cm^{-1}), and $\Delta\nu_{coll}$ is the collisional linewidth (sec^{-1}). Since in the quasistationary regime $\Delta N \propto B_e$, it follows from Eq. (1.25) that molecules with large dipole moments $\tilde{\mu}$ and large rotational constants B_e will be most favorable for lasing.

For the above examples with $\Delta N \leqslant 10^{11}$ cm^{-3} we have $\varkappa_{15,14} \leqslant 3 \cdot 10^{-4}$ cm^{-1}. The laser power estimated from the flux of molecules passing through level j_0 must be of order < 0.3 W/cm^{-3} with a pulse length of about 10^{-3} sec. Since the rate of formation of molecules in the chemical reaction with a particular rotational state is much less than the total rate of formation of HCl used in these estimates, the actual laser parameters (gain coefficient and power) would be much less. Nevertheless, these estimates show the possibility in principle of obtaining lasing on rotational transitions in nonbranching chain reactions.

In the case of branching chain reaction the time dependence of the pumping rate in the initial period of the reactions can often be written in the form Q = A exp (st). Substituting this in Eq. (1.22), it is easy to determine that a population inversion among the levels exists and increases in time if

$$s > \frac{\exp(\Delta\varepsilon/kT) - 1}{\tau_R \varphi_{j_0}}. \tag{1.26}$$

When the inequality is opposite to this there is no inversion in the quasistationary regime. It follows from Eq. (1.26) that the conditions for lasing become much less stringent for pumping in the adiabatic region of the spectrum.

Calculations in [33] show that rotationally excited HF molecules are produced over the j_0 = 6-12 levels in the reaction $F + H_2 \rightarrow HF^* + H$ at T = 300°K, that is, in the region of the spectrum where $(\varepsilon_{j_0}/kT)^{1/2} \gtrsim \xi^{-1}$. This is an indication that lasing is also really possible in this spectral region ($\lambda \sim$ 20-40 μ) using the branching chain reaction of $H_2 + F_2$. Estimates show that lasing on HF is also possible by directly mixing H_2 with F atoms. In fact, considering the ($v = 1$, $j_0 = 9$) → ($v = 1$, $j_0 - 1 = 8$) transition (wavelength $\lambda \sim 28$ μ) and using the rate constants computed in [33] for the individual vibrational and rotational states of the reaction $F + H_2 \rightarrow HF^* + H$, with Eqs. (1.22) and (1.25) it is possible to estimate the resulting population inversion ΔN and gain coefficient $\varkappa_{9,8}$. For T = 500°K and taking the initial concentrations to be $[H_2]_0 \sim 10^{15}$ cm^{-3} and $[F]_0 \sim 3.8 \cdot 10^{14}$ cm^{-3} we find that after a time $\tau_x \sim K[H_2]_0 \sim 10^{-5}$ sec (where $K = 3.9 \cdot 10^{-11}$ cm^3 sec is the overall rate constant for the reaction) there is still an inversion over this transition of about $1 \cdot 10^{11}$ cm^{-3}. Then $\varkappa_{9,8} \sim 6 \cdot 10^{-4}$ cm^{-1} and the possible output power is $W \sim 4 \cdot 10^{-2}$ W/cm^3. Lasing in this manner could evidently be achieved in the cw regime as with the usual continuous gas-feed chemical laser using vibrational transitions of HF. The flows of F and H_2 must be mixed over a time less than τ_x and the pumping rate for a working zone of length l must be $\sim l/\tau_x$.

Pulsed lasing on rotational transitions of HCl and HF has been realized in [34, 35]. Optical pumping of HF was used in [34], and chemical pumping as a result of the electrochemical action of a current pulse in the mixtures $CF_4 + H_2$ and $CH_3Cl + Cl_2$ was used in [35]. The experimental results of [35] are in qualitative agreement with the above-described mechanism for lasing in the adiabatic region.

2. Lasing on Transitions of Multiatomic Molecules (Far Infrared and Submillimeter Ranges)

Lasing on transitions of multiatomic molecules in the far infrared and submillimeter wavelength ranges may be achieved both on purely rotational and on vibrational–rotational transitions. We shall first examine purely rotational relaxation. For symmetric and slightly symmetric tops the structure of the rotational levels is specified by the projection k along with the quantum number j:

$$\varepsilon_{jk} = B_e j(j+1) + \Delta B_e k^2, \tag{1.27}$$

where B_e and ΔB_e depend on the moments of inertia of the molecule. For flattened tops ΔB is negative and for elongated tops it is positive. The relative smallness of the probabilities of radiative transitions between levels with different k should be noted (for a symmetric top, dipole transitions with $\Delta k \neq 0$ are forbidden). This indicates that there are apparently two relaxation times for symmetric and slightly asymmetric tops: one, more rapid, within a vibrational mode having a given k and the other, slower, between various rotational modes with different values of k (here there is a definite analogy with the relaxation of vibrational energy between different modes of multiatomic molecules). Thus, in principle, it is possible to create a population inversion among levels with different k. Then, for a slightly asymmetric top the radiative transition probability may be sufficiently large for the inversion to lead to lasing (the wavelengths of the transitions lie in the millimeter range).

We shall briefly discuss rotational relaxation for this type of system with a positive source of rotationally excited molecules and various rates of formation of these molecules in modes with different k.

Regarding the rotational spectrum as continuous, the relaxation of the molecules may be treated as diffusion of the particles over j levels and over k levels (with different diffusion coefficients). Limiting ourselves to the quasistationary regime and assuming for simplicity that the source acts only on the level with $j = j_0$ and $k = k_0$ it is possible, on the basis of the results of the previous section, to find the populations of the levels and determine the population inversion in the transition $(j_0, k_0) \rightarrow (j_0, k_0 - 1)$ of an elongated top as a function of the source power Q:

$$N_{j_0, k_0} - N_{j_0, k_0-1} = \frac{Q}{a_k}\left[f(\varepsilon_{0, k_0}) \exp\left(\frac{-j_{0}, k_0}{kT}\right) - f(\varepsilon_{0, k_0-1}) \exp\left(-\frac{\varepsilon_{j_0, k_0-1}}{kT}\right) + \right.$$

$$\left. + \frac{a_k}{a_j} f(\varepsilon_{j_0, k_0} - \varepsilon_{0, k_0}) \exp\left(-\frac{\varepsilon_{j_0, k_0} - \varepsilon_{0, k_0}}{kT}\right)\right] - N_{00}\left[\exp\left(-\frac{\varepsilon_{j_0, k_0-1}}{kT}\right) - \exp\left(-\frac{\varepsilon_{j_0, k_0}}{kT}\right)\right]. \tag{1.28}$$

Here $f(\varepsilon_{jk})$ are functions describing the distortion of the Boltzmann distribution over the levels j and k and are found from Eq. (1.21); α_j and α_k are parameters analogous to Eq. (1.13) and characterize the times of relaxation within and between modes, respectively. We note that since $\alpha_k < \alpha_j$ and practically always

$$\exp\left(-\frac{\varepsilon_{j_0, k_0-1}}{kT}\right) - \exp\left(-\frac{\varepsilon_{j_0, k_0}}{kT}\right) \simeq \exp\left(-\frac{\varepsilon_{j_0, k_0}}{kT}\right)\frac{\Delta B_e k_0^2}{kT} \ll 1$$

[which substantially reduces the term with N_{00} in Eq. (1.28)] it is easier to obtain an inversion between levels with different k. Practical construction of a laser requires nonuniform pumping in various rotational modes with specified k. Since k is related to the shape of the molecule, this kind of pumping may be done by sorting the molecules in a quadrupole electric field in addition to the standard excitation methods. With the aid of such sorting, beam lasers have been built using formaldehyde, H_2CO (λ = 4.12 mm), and hydrogen sulfide, H_2S (λ = 1.78 mm). To increase the power of such lasers it is possible to use electric field sorting together with optical, chemical, or electrical pumping for producing an additional disequilibrium in k.

In many multiatomic molecules a large number of transitions between levels corresponding to various vibrational modes are in the far infrared and submillimeter wavelength range.

Because of the need to have sufficiently high radiative transition probabilities for lasing in this range, it is appropriate to use dipole molecules. In such molecules the relaxation rates are high; in a number of cases the rate of vibrational relaxation is close to that of rotational relaxation. However, even here in the first approximation the vibrational kinetics may often be considered separately from the rotational kinetics and may be analyzed on the basis of the usual mechanism of partial vibrational temperatures [36].

A population inversion develops in such systems (as in CO_2 lasers [36]) due to a divergence in the vibrational temperatures of different modes. We shall briefly consider the demands on these temperatures. Modeling the vibrational modes of the molecules with harmonic oscillators having quanta of energy E_i, the conditions for an inversion over the transition $mE_1 \to nE_2$ may be written in the form

$$\exp\left(-\frac{mE_1}{kT_1}-\frac{E_{rot}(1)}{kT_{rot}}\right) > \exp\left(-\frac{nE_2}{kT_2}-\frac{E_{rot}(2)}{kT_{rot}}\right),$$
$$\frac{mE_1}{kT_1}-\frac{nE_2}{kT_2} < \frac{E_{rot}(2)-E_{rot}(1)}{kT_{rot}} = \frac{mE_1-nE_2-h\nu}{kT_{rot}}. \tag{1.29}$$

Here T_1 and T_2 are the vibrational temperatures of the modes, $E_{rot}(1)$ and $E_{rot}(2)$ are the energies of the rotational levels, and T_{rot} is the rotational temperature. The energy of the working transition, $h\nu$, is usually close to $mE_1 - nE_2$. If $T_2 \sim T_{rot} \sim T$, the condition (1.29) is satisfied for the following divergence of T_1 from T:

$$\frac{T_1}{T} > \frac{mE_1}{mE_1 - h\nu}. \tag{1.30}$$

Since usually $E_1 \gtrsim 1000\ cm^{-1}$ and in this frequency range $h\nu \sim 100\ cm^{-1} \ll E_1$, the required divergence between T_1 and T may be small. This is an advantage held by lasers in the submillimeter range of the spectrum over, for example, the well-known CO_2 laser, where it is necessary that $T_a/T \sim T_a/T_D > 1.4$ (T_a and T_D are the temperatures of the asymmetric and deformed modes) to produce a population inversion. One shortcoming, however, is that because of the small energy defect, $\Delta E = mE_1 - nE_2 \sim h\nu$, the rate of relaxation of the upper laser level mE_1 due to vibrational exchange between modes in the $mE_1 \to nE_2$ channel may be large. In the limiting case when this rate exceeds the rates of the other processes the vibrational temperatures T_1 and T_2 will be related by

$$\frac{mE_1}{T_1}-\frac{nE_2}{T_2}=\frac{mE_1-nE_2}{T}$$

and no inversion occurs on the $mE_1 \to nE_2$ transition. In this case, however, an inversion is possible between other levels wE_2 and vE_1. Then it is easy to show that the difference between T_1 and T must satisfy

$$\frac{T_1}{T} > E_1 \frac{\frac{m}{n}w - v}{\left(\frac{m}{n}w - v\right)E_1 + h\nu}. \tag{1.31}$$

To analyze the prospects for lasing between these levels it is also important to know the radiative probabilities. At present it is difficult to calculate them. In certain cases it is possible to isolate individual pairs of levels that are convenient for lasing, where due to a number of random factors (for example, mixing of the rotational states of the vibrational levels of different modes) the radiative probabilities increase anomalously (see, for example, [24]).

This mechanism for lasing on the vibrational–rotational transitions of multiatomic molecules may lead to high powers. In this regard optimization of the water vapor laser seems promising. At this time the $\nu_1 \rightarrow 2\nu_2$ and $\nu_3 \rightarrow 2\nu_2$ transitions of H_2O (wavelengths $\lambda \simeq 30$-$120\ \mu$) are among the most powerful laser transitions in a pulsed regime [24, 37] (according to [37] the pulse power is about 15 kW). At the same time the cw power obtained is small. To increase it substantially a transverse discharge with a high feed rate (pumpthrough) of H_2O vapor might be used. In a water vapor discharge there is partial dissociation of H_2O into OH and O and excitation of the ν_1 and ν_3 vibrations of H_2O and OH by the electrons. Since the frequency $\nu_{OH} = 3558\ cm^{-1}$ is close to the ν_1 ($3652\ cm^{-1}$) and ν_3 ($3756\ cm^{-1}$) frequencies of H_2O, excited OH may serve as an additional reservoir of vibrational energy for the ν_1 and ν_3 modes of H_2O, thereby improving the conditions for formation of an inverted population. Additional pumping may also be provided by recombination,

$$OH + H + M \longrightarrow H_2O^* + M.$$

It is most convenient to use an $H_2O + H_2$ mixture with an excess of H_2 rather than pure H_2O vapors in the pumpthrough regime. Because of the extremely efficient excitation of vibrations in H_2 by electrons (at an electron energy of about 2 eV the cross section for the process $H_2\ (v = 0) + e \rightarrow H_2\ (v = 1) + e$ is $\sigma \sim 3 \cdot 10^{-17}\ cm^2$) and the energy exchange reaction $H_2\ (v = 1) + H_2O \rightarrow H_2\ (v = 0) + H_2O\ (v_3 = 1)$, it is possible to realize substantial energy pumping and differences between the vibrational and gas temperatures in H_2O. Since $\nu_{H_2} = 4170\ cm^{-1} > \nu_3\ (H_2O)$, rapid nonresonance vibrational exchange between H_2 and the asymmetric $\nu_3\ (H_2O)$ mode leads to an additional growth in T_3 [37] and, therefore, to an increase in the gain coefficient and output power. An increase in the pulsed output power from H_2O has been observed experimentally when H_2 was added to the discharge [38, 39].

We now make some quantitative estimates of some system parameters as well as of the power and efficiency of a cw laser using a mixture of $H_2O + H_2$ operating on the $\nu_3 \rightarrow 2\nu_2$ transition in a design with a transverse discharge and rapid pumpthrough. This apparatus may be similar to that used in [40] for lasing on CO_2 transitions. Let an $H_2O : H_2$ mixture with partial pressures in the ratio 1 : 3 and a total pressure of about 1 torr flow through a discharge zone of length $l \approx 3$ cm at a speed V. The case where the molecules spend a time equal to the relaxation time $\tau_{00\nu_3}$ of the upper laser level (that is, the $(00\nu_3)$ H_2O vibrations) in the discharge is optimum for the lasing regime (cf. [41]). Taking $\tau_{00\nu_3} \sim 10^{-5}$ sec [39], we find that the speed V of the gas flow must be $V \sim l/\tau_{00\nu_3} \sim 3 \cdot 10^5$ cm/sec, i.e., supersonic.

For a cross section for electron impact excitation of H_2 vibrations of $\sigma \sim 3 \cdot 10^{-17}\ cm^2$ [42], an electron density $N_e \sim 3 \cdot 10^{13}\ cm^{-3}$ will be required to obtain a vibrational temperature $T_{H_2} \sim 6000°K$ (in H_2O T_3 would then be of the same order of magnitude). Such values of N_e correspond to current densities of about $140\ A/cm^2$, which are quite attainable. We note that if the ν_1 and ν_3 vibrations of H_2O are excited more effectively by the electrons than H_2 then the requirements on the current densities are reduced. At these densities we obtain a photon flux $Q = \langle\sigma v\rangle_{H_2} N_e N_{H_2}$ (v is the electron velocity) of about 10^{21} quanta/cm^3 · sec. The maximum possible power W at a frequency $\nu = 118\ cm^{-1}$ will be $W \simeq h\nu Q \approx 2\ W/cm^3$ and the maximum efficiency of the system will equal $\nu/\nu_{H_2} \sim 1/40$. Of course, the actual efficiency will be much less due to possible excitation of the deformed mode of H_2O and to the various channels for dissipation of the energy in the ν_2, ν_3, and ν_{H_2} modes. Nevertheless, these estimates demonstrate the effectiveness of this pumping arrangement. We now mention, without evaluating them, three more concepts for producing a population inversion in H_2O and H_2.

(a) Combustion $H_2 + O_2 \rightarrow H_2O$ with subsequent supersonic escape through a nozzle and use of the energy stored in H_2 (surplus H_2). This arrangement may be combined with a transverse discharge.

(b) A cw chemical laser using a H_2O + HF mixture similar to lasers using CO_2 + HF or CO_2 + DF mixtures.

(c) Mixing ozone, O_3, with a flow of H_2O ($H_2O + H_2$) through a discharge zone. In the resulting mixture of gases, H_2O, H_2, O, O_2, O_3, OH, and H, the reaction

$$O_3 + H \rightarrow OH^* + O_2$$

takes place. A substantial portion of the energy of this exothermic reaction goes into exciting vibrations of OH (up to the ninth level). Because of nonresonance vibrational exchange, energy from OH* will be transferred to the ν_1 and ν_3 vibrations of H_2O molecules.

We note that when ozone is mixed with an H_2 flow passing through a discharge zone we might expect lasing on OH as well. Here it is desirable to obtain as great a dissociation of H_2 as possible since the $OH^* + H_2 \rightarrow OH + H_2^*$ channel is detrimental here.

On the whole, a study of the basic mechanisms for creating a population inversion in submillimeter and FIR molecular lasers has disclosed a number of unused possibilities.

Despite the high relaxation rates and the rigid requirements on the pumping rate the efficiency of lasers using the schemes proposed here may be increased considerably, both for rotational and for vibrational–rotational transitions.

3. Rotational Energy Distribution Functions and Rotational Transition Lasers (Discrete Model)

It has already been noted previously that the comparatively low efficiency of existing lasers using purely rotational transitions is due to the high rate of rotational relaxation, which then places considerable demands on the pump. In the adiabatic region of the rotational spectrum the relaxation rate is much less than in the nonadiabatic region. The kinetics is in many respects similar to the kinetics of vibrational relaxation. Single-quantum transitions play the basic role. The most important processes are rotational–translational and rotational–rotational exchange and exchange of rotations between different rotators (RT, RR, and RR' processes, respectively, which have direct analogies in the forms of VT, VV, and VV' vibrational processes [31]). This analogy can be shown to extend even to the probabilities of the corresponding transitions. In the following we examine two limiting cases in which relaxation is determined by rotational–translational processes and by rotational–rotational exchange, respectively.

We now examine the first case, rotational relaxation of diatomic molecules that are a small impurity in an inert gas. We shall seek the rotational level population distribution function in the adiabatic region of the rotational spectrum, where the rotational period of the diatomic molecules (modeled as rigid rotators) is much less than the collision time with inert gas particles. The rotational quantum number which divides the spectrum of the rotator into adiabatic and nonadiabatic regions is defined by the equality

$$j^* = \frac{\pi\alpha\hbar}{B_e}\left(\frac{kT}{\mu}\right)^{1/2}, \qquad (1.32)$$

where B_e is the rotational constant of the diatomic molecule, μ is the reduced mass of the colliding particles, T is the translational temperature of the gaseous mixture, α^{-1} is the characteristic dimension of the potential well for the interaction of colliding particles, and k is Boltzmann's constant.

We shall assume that the distance between neighboring rotational levels is on the order of or greater than the gas temperature T and, therefore, the most probable processes are inelastic collisions of a rotator with inert gas particles, leading to single-photon transitions in the region of the rotational spectrum with $j > j^*$. Then the population balance equations for the rotational levels have the form

$$dN_j/dt = P_{j+1,\,j}N_{j+1} - P_{j,\,j+1}N_j + P_{j-1,\,j}N_{j-1} - P_{j,\,j-1}N_j + Q_j, \tag{1.33}$$

where N_j is the population of the j-th rotational level, $P_{j+1,j}$ is the frequency of transitions from the level j + 1 to level j due to collisions of a rigid rotator with inert gas particles, and Q_j is the power of the source for pumping rotationally excited molecules to the level j.

Summing the equations over j subject to the normalization condition $\sum N_j = N$ yields

$$\sum Q_j = \frac{dN}{dt}.$$

Let a source of rotationally excited molecules of constant power Q = dN/dt act on the level $j_0 > j^*$. By a time on the order of the rotational relaxation time, after the source is turned on a quasistationary level population distribution similar to Eq. (1.18) will have been established. We now give an expression for the quasistationary distribution function without justifying it, referring the reader to Chapter II where the solution of a system of relaxation equations similar to Eq. (1.33) will be considered in detail:

$$N_j = g_j \exp\left(-\frac{E_j}{kT}\right)\left(N_0 + \beta_j \frac{dN_0}{dt} - \mu_j Q\right) \quad (j > j^*); \tag{1.34}$$

$$\beta_j = \sum_{m=j^*+1}^{j} \frac{\exp\left(\frac{E_m}{kT}\right)}{g_m P_{m,\,m-1}} \cdot \sum_{n=j^*}^{m-1} g_n \exp\left(-\frac{E_n}{kT}\right); \tag{1.35}$$

$$\mu_j = \sum_{m=j_0+1}^{j} \frac{\exp\left(\frac{E_m}{kT}\right)}{g_m P_{m,\,m-1}}, \qquad \mu_j = 0 \ (j \leqslant j_0), \tag{1.36}$$

where $g_j = 2j + 1$ and $E_j = B_e j(j + 1)$ are the statistical weight and energy of the j-th rotational level. In deriving Eq. (1.34) it was assumed that the group of states with $j \le j^*$ is in equilibrium with the translational degrees of freedom. From Eqs. (1.34)-(1.36) it is clear that the deviation of the distribution function from equilibrium increases with j all the way up to the j_0-th rotational level.

Differentiating the normalization condition $\sum_j N_j = N$ with respect to time and assuming the distortion in the Boltzmann function to be stationary, we obtain

$$\frac{dN_0}{dt} = \frac{B_e}{kT} Q. \tag{1.37}$$

The parameter which controls the kinetics of the relaxation of a quasistationary distribution to a Boltzmann distribution is N_0. It may be related† to the temperature Θ_{j*} between the levels $j^* + 1$ and j^*. Setting $\Theta_{j*} = k(T + \Delta\Theta)$, with $\Delta\Theta \ll T$ and $\Delta\Theta/T \ll kT/E_{j^*+1,j^*}$, we find

$$N_0 = \frac{QB_e}{E_{j^*+1,\,j^*}} \cdot \frac{T}{\Delta\Theta} \cdot \frac{2j^*+1}{2j^*+3} \cdot \frac{\exp\left(E_{j^*+1,\,j^*}/kT\right)}{P_{j^*+1,\,j^*}}. \tag{1.38}$$

Since the temperatures between neighboring rotational levels is determined by the ratio f_{j+1}/f_j, they and, therefore, the character of the quasistationary distribution are independent

† By the temperature between two neighboring levels j + 1 and j we shall mean the quantity

$$\Theta_j = \frac{kT}{1 - \frac{kT}{E_{j+1,\,j}} \ln \frac{f_{j+1}}{f_j}}, \qquad E_{j+1,\,j} = E_{j+1} - E_j.$$

of the power of the pumping source Q. In addition, if the frequencies of the rotational transitions are represented in the form $P_{j+1,j} = P_0 f(j) \sim N_A f(j)$, then the ratio f_{j+1}/f_j is also independent of the density of the inert solute N_A and is of the same order of magnitude as the rotational transition probabilities.

For a specific calculation of the temperatures Θ_j the probabilities of an inelastic collision between a rigid rotator and a structureless particle (atom) in the adiabatic region of the rotational spectrum were computed in a way similar to that used for a harmonic oscillator, for example, in [43, 44]. In doing this it was assumed that a collision of a rigid rotator with a structureless particle takes place in a plane and with an interaction potential $V(r, \theta) \sim \exp(-\alpha r)(1 + \beta \cos \theta)$, where r is the distance between the center of mass of the rigid rotator and the atom, θ is the angle between the axis of the rotator and a line joining the center of mass of the rotator and the atom, and β is a parameter indicating the deviation of the potential from spherical.

Carrying out some calculations analogous to those in [43, 44], we obtain

$$P_{j,\,j-1} \sim \omega_{j,\,j-1}^2 \int_0^{\infty} \exp\left(-\frac{2\pi\omega_{j,\,j-1}}{\alpha v} - \frac{\mu v^2}{2kT}\right) v^3 dv \sim \omega_{j,\,j-1}^3 \exp\left(-3\gamma_{j,\,j-1}^{2/3}\right),$$

$$\gamma_{j,\,j-1} = \frac{\pi\omega_{j,\,j-1}}{\alpha}\left(\frac{\mu}{2kT}\right)^{1/2},$$

where v is the relative velocity of the colliding particles and $\hbar\omega_{j,j-1} = E_{j,j-1}$.

We shall examine the rotational relaxation of an HI molecule in an inert gas (He) with a translational temperature, T = 300°K, $j^* = 15$, and $\alpha^{-1} = 2$ Å, if the source of rotationally excited HI molecules acts on the $j_0 = 20$ level. The results of some calculations of the temperatures between neighboring rotational levels lying above j_0 for several values of the parameter $\Delta\Theta$ are shown in Fig. 1 in the form of graphs of the temperatures Θ_j as functions of the rotational quantum number j. It is clear that when a continuously acting source is turned on, strong deviations in the distribution function from its equilibrium form will occur for relatively large values of $\Delta\Theta$. Calculations show that in the quasistationary stage a state of absolute inversion of the populations of the j_0 and $j_0 - 1$ levels is possible if $\Delta\Theta > 0.5$°K.

The general character of the curves shown in Fig. 1 will be conserved for other diatomic molecules. In particular, an inverted population can also be produced in the HCl molecule.

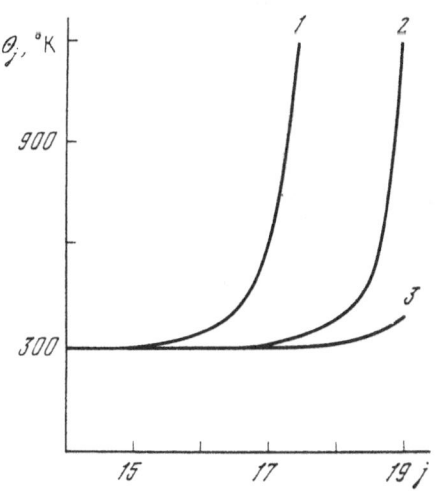

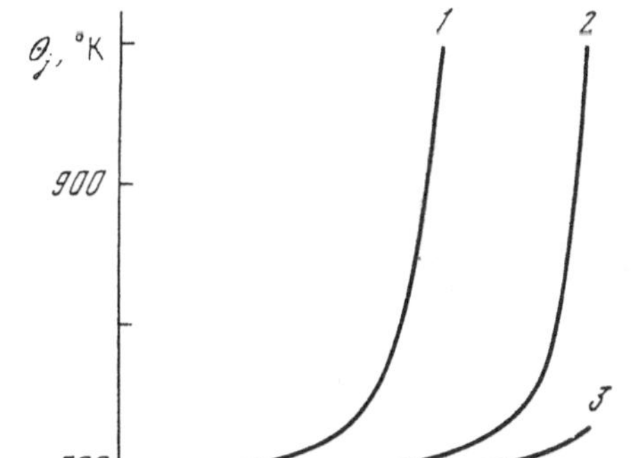

Fig. 1. The temperatures between neighboring rotational levels as a function of j when a source of rotationally excited molecules acts on the j_0 = 20 level. Curves 1, 2, and 3 correspond to $\Delta\Theta$ = 4.9, 0.24, and $4.6 \cdot 10^{-2}$°K.

Neglecting the populations of rotational levels with $j > j_0$ in the normalization condition, we find that an inversion in the populations of the j_0 and $j_0 - 1$ levels is possible if

$$Q > \frac{N}{R + \frac{B_e}{kT} S}, \tag{1.39}$$

where

$$R = \frac{\beta_{j_0} \exp(-E_{j_0}/kT) - \beta_{j_0-1} \exp(-E_{j_0-1}/kT)}{\exp(-E_{j_0-1}/kT) - \exp(-E_{j_0}/kT)},$$

and

$$S = \sum_{j=j^*+1}^{j_0} g_j \exp\left(-\frac{E_j}{kT}\right) \beta_j.$$

Substituting the initial density of rigid rotators instead of N in Eq. (1.39) it is possible to compute the threshold pumping source power for obtaining an absolute inversion in the populations of the j_0 and $j_0 - 1$ levels. With Eq. (1.39) we can also calculate the time beginning with which the inversion disappears. To do this we must set $N = N(0) + Qt$ in Eq. (1.39).

We now turn to the case of rotational–rotational exchange. Let a gas consisting of rigid rotators be in equilibrium at a translational temperature T_0. Then, over times shorter than the rotational relaxation time the translational temperature changes to some value T_1. We assume that the exchange of rotational quanta, beginning with some quantum number j_1, is more probable than rotational–translational relaxation. Then a Treanor distribution [10] is established fairly rapidly over the upper levels:

$$N_j = N_1 (2j+1) \exp\left(-\frac{E_j}{kT_1} + \gamma j\right), \quad j > j_1. \tag{1.40}$$

It the final temperature $T_1 > T_0$, then the stored rotational energy $E_{\text{rot}} = \sum_{j>j_1} E_j N_j$ corresponding to the Treanor distribution (1.40) is less than the equilibrium accumulation corresponding to a temperature T_1, and the parameter γ is therefore negative. For $T_1 < T_0$, on the other hand, γ is positive. It follows from Eq. (1.40) that an absolute inversion in the populations of the $j + 1$ and j levels is achieved for $\gamma > (2Be_e/kT_1)(j + 1)$ and is therefore possible only upon cooling a gas of rigid rotators.

As opposed to exchange of quanta between anharmonic oscillators in vibrational kinetics [45], the temperatures between neighboring rotational levels decrease with increasing quantum numbers j when the gas is cooled. In this respect, a rigid rotator is analogous to an "inverted" anharmonic oscillator.

We shall assume that until and after the translational temperature of the gas is changed, the stored rotational energy and the average number of rotational quanta $V_{\text{rot}} = \sum_{j>j_1} jN_j$ in the region $j > j_1$ do not change. Then going from a sum to an integral in the expressions for E_{rot} and V_{rot} we obtain an equation for finding the parameter γ:

$$\sqrt{\frac{T_0}{T_1}} f\left(j_1 \sqrt{\frac{B_e}{kT_0}}\right) = \varphi\left(\frac{\gamma}{2} \sqrt{\frac{kT_1}{B_e}}\right), \tag{1.41}$$

$$f(x) = \frac{1 + x^2}{x + \frac{\sqrt{\pi}}{2} [1 - \Phi(x)] \exp(x^2)},$$

$$\varphi(x) = x + \frac{1 + x^2 + \sqrt{\pi}x\,[1 + \Phi(x-a)]\exp(x-a)^2}{a + x + \frac{\sqrt{\pi}}{2}(1 + 2x^2)\,[1 + \Phi(x-a)]\exp(x-a)^2},$$

where $\Phi(x)$ is the error integral and $a = j_1(Be/kT_1)^{1/2}$.

An analysis of Eq. (1.41) shows that the root of the equation lies in the region $\gamma > 0$ for $T_1 < T_0$ and in the region $\gamma < 0$ for $T_1 > T_0$. Thus, upon specifying the values $j_1 = 10$, $T_1 = 300°K$, and $B_e = 15.2°K$ (HCl molecule), it is possible to calculate the initial temperatures up to which a gas of rigid rotators must be heated and then rapidly cooled to room temperature in order to obtain an absolute population inversion in the rotational levels 10-15. It appears that the gas must first be heated to a temperature $T_0 \gtrsim 2300°K$ to obtain this inversion. Gas-dynamic methods may be used to cool the gas rapidly. The basic difficulty with this scheme is the short cooling times which are determined by the rate of rotational relaxation.

We note that these quasistationary rotational level population distribution functions may be used, not only in the design of molecular lasers employing rotational transitions, but also in a number of problems in molecular kinetics, in particular, the theory of chemical reactions.

CHAPTER II

QUASISTATIONARY POPULATION DISTRIBUTION FUNCTIONS IN LOW-TEMPERATURE PLASMA

An analytic expression is obtained for the population distribution function in an atomic plasma in the form of a series of successive time derivatives of the population of the ground state. The first approximation involves using the well-known "stationary sink" method. The quasistationary distribution obtained for hydrogen is in good agreement with numerical computations of recombination and ionization, and those obtained for lithium, helium, and argon are in qualitative agreement with the numerical computations. Quasistationary distribution functions in a plasma made up of a mixture of elements are discussed taking charge exchange into account.

1. The Level Population Distribution Function for Hydrogen

Determining the population distribution function for the discrete levels in an atomic plasma is important for problems in ionization, recombination, plasma radiation, and the design and optimization of lasers using atomic transitions. Its role is especially emphasized by the results of molecular kinetics, where quasistationary distributions over the vibrational levels of molecules have recently been obtained. These are the Treanor distribution [10] and its various generalizations. The simple analytic form of these distributions has made it possible to ascertain a number of new phenomena in vibrational kinetics, and they have found wide application in the theory of lasers [32, 45, 46].

In atomic kinetics the situation is more complicated. Here there are two approaches. One of these is based on a numerical solution of the population balance equations, where simplified equations have often been solved numerically in the "stationary sink" approximation (see, for example, [3, 47-49]).

The other approach [6, 50-54] is based on a consideration of the motion of an electron over the levels as a probabilistic process similar to Brownian motion in energy space and described by the Fokker–Planck equation. Vorobev et al. [52-54] have succeeded in preserving the actual discrete structure of the energy spectrum using a modified, finite difference, diffusion approximation for the Fokker–Planck equation. If the first approach is inconvenient because it is cumbersone, the second will be completely adequate for the upper levels, but inaccurate for the lower levels.

We shall examine the principles for solving the equations of atomic kinetics using the example of a relaxing hydrogen plasma. We introduce some simplifying assumptions. For the free electrons we use a Maxwellian distribution function with a constant electron temperature T_e. Of the elementary collision processes we include only electron–atom collisions of the first and second kinds and neglect the unlikely interatomic and atom–ion collision processes, as well as ionization, and radiative and three-body recombination. In addition, the plasma will be taken to be spatially uniform.

These assumptions make it possible to write the system of balance equations for the discrete levels of the hydrogen atom in the following way:

$$\frac{dN_n}{dt} = N_e \sum_{n'} [V(n', n) N_{n'} - V(n, n') N_n], \qquad n = 1, 2, \ldots, \tag{2.1}$$

where N_n is the population of the n-th level of the hydrogen atom, $V(n, n') = \langle v\sigma_{n,n'}\rangle$ is the averaged (over the electron distribution) product of the electron velocity and the cross section for inelastic atom–electron collisions in which the atom shifts from state n to state n', and N_e is the electron density.

In a plasma with a Maxwellian electron velocity distribution the following relation exists between the excitation and deexcitation rates:

$$V(n, n') g_n \exp(-E_n/kT_e) = V(n', n) g_{n'} \exp(-E_{n'}/kT_e), \tag{2.2}$$

where g_n and E_n are the statistical weight and energy of the n-th level.

Equation (2.2) means that the direct and reverse processes take place in the same way and that, with the passage of time and a dominant role for inelastic collisions of atoms with electrons in the plasma, a Boltzmann level population distribution evolves with an electron temperature of T_e. In the following the relation (2.2) between the rates of the direct and reverse processes will be called the "principle of detailed balance."

We assume that the most probable inelastic atom–electron collisions are those in which the principal quantum number n changes by unity (single-quantum approximation). Then the system of population balance equations takes the form

$$dN_n/dt = V(n+1, n) N_{n+1} N_e - V(n, n+1) N_n N_e + V(n-1, n) N_{n-1} N_e - V(n, n-1) N_n N_e. \tag{2.3}$$

Let the degree of ionization of the plasma at the initial time not correspond to the temperature of the free electrons. We shall examine the relaxation of such a nonequilibrium plasma for a constant electron temperature T_e on the basis of the population balance equations (2.3). To do this we introduce the new variable

$$\tau = \int_0^t N_e(t')\, dt'$$

and sum Eq. (2.3) form 1 to m. As a result, using the principle of detailed balance, we have

$$N_{m+1}=\frac{g_{m+1}}{g_m}\exp\left(-\frac{E_{m+1,\,m}}{kT_e}\right)N_m+\frac{1}{V\,(m+1,\ m)}\sum_{n=1}^{m}\frac{dN_n}{d\tau}, \tag{2.4}$$

where $E_{m+1,m} = E_{m+1} - E_m$.

From Eq. (2.4) it is easy to see that the population of some level m may be written in the form of a series of derivatives of the population of the first level (the ground state of the hydrogen atom),

$$N_m=\sum_{i=0}^{m-1}\alpha_m^i\frac{d^{(i)}N_1}{d\tau^i}. \tag{2.5}$$

To find the α_m^i we substitute Eq. (2.5) in (2.4) and equate the coefficients of time derivatives of the same order on both sides of Eq. (2.4). Changing the order of summation in the second term according to the rule $\sum_{n=1}^{m}\sum_{i=1}^{n}\to\sum_{i=1}^{m}\sum_{n=i}^{m}$, we find the following recurrence relations for the unknown coefficients:

$$\alpha_{m+1}^i=\frac{g_{m+1}}{g_m}\exp\left(-\frac{E_{m+1,\,m}}{kT_e}\right)\alpha_m^i+\frac{1}{V\,(m+1,\ m)}\sum_{n=i}^{m}\alpha_n^{i-1}. \tag{2.6}$$

Expanding these relations, we obtain

$$\alpha_m^0=g_m\exp\left(-\frac{E_{m,\,1}}{kT_e}\right),\qquad g_m=m^2; \tag{2.7}$$

$$\alpha_n^i=\sum_{m=i+1}^{n}\frac{g_n\exp\left(-E_{n,m}/kT_e\right)}{g_mV\,(m,\ m-1)}\sum_{k=i}^{m-1}\alpha_k^{i-1}; \tag{2.8}$$

$$\alpha_{m+1}^m=\prod_{n=1}^{m}V^{-1}(n+1,\ n). \tag{2.9}$$

Equations (2.7)-(2.9) make it possible to compute any coefficient α_m^i in the expansion (2.5).

This distribution function (2.5) describes the process of relaxation to a Boltzmann distribution with an electron temperature T_e. In fact, with the passage of time $N_m \to N_1 g_m \times \exp(-E_{m1}/kT_e)$. The deviation from a Boltzmann distribution is given by the terms with first, second, and so on, derivations of the population of the ground state. The parameter which controls the recombination or ionization kinetics will be the population of the ground state.

Over a time on the order of the collision time between free electrons the populations of the upper discrete levels with $n \geq n_0$ come into equilibrium with the continuous spectrum and obey the Saha equation

$$N_n=K\,(T_e)\,N_e^2g_n\exp\left(-E_{n,1}/kT_e\right), \tag{2.10}$$

where $K(T_e) = (2\pi\hbar^2/mkT_e)^{3/2}\exp(R/kT_e)$ and R is the ionization potential of hydrogen. Joining the solution of Eq. (2.5) to the Saha distribution for the upper levels and using the normalization condition

$$\sum_{n=1}^{n_0}N_n=N-N_e \tag{2.11}$$

(N is the density of heavy particles in the plasma) we obtain a differential equation of order n_0 for the population of the ground state N_1. To solve it, concrete initial conditions must be specified (in this case, the initial distribution function).

In a majority of cases of practical interest, however, it is not necessary to solve such an equation. The fact is that recombination (and ionization) takes place in a way such that the atoms begin to "forget" their initial conditions with the passage of time. In other words, terms with high derivatives may be important only in the initial stage of the process and insignificant in the final state. Thus, in the quasistationary relaxation state of the solution of (2.5) we can limit ourselves to the first derivative term and call the distribution quasistationary.

We now consider the relation of this quasistationary distribution to the often used "stationary sink" approximation. To do this we now turn to the problem of finding the distribution function by iteration as the most effective method and show that use of this method leads to the exact solution of Eq. (2.5).

We seek a population distribution function in the form

$$N_n = g_n \exp(-E_n/kT_e) f_n \qquad (E_1 = 0). \tag{2.12}$$

We shall call the quantity f_n the reduced population of the n-th level. Substituting Eq. (2.12) in Eq. (2.3) and using the principle of detailed balance, we have

$$df_n/d\tau = V(n,\ n+1)(f_{n+1} - f_n) + V(n,\ n-1)(f_{n-1} - f_n). \tag{2.13}$$

We introduce the notation

$$f_n = f_1 + \sum_{m=2}^{n} a_m. \tag{2.14}$$

We now substitute Eq. (2.14) on the right-hand side of Eq. (2.13) and expand the resulting recurrence relation for a_n. Thus,

$$\begin{gathered} a_m = \frac{\exp(E_m/kT_e)}{g_m V(m,\ m-1)} \sum_{k=1}^{m-1} g_k \exp\left(-\frac{E_k}{kT_e}\right) \frac{df_k}{d\tau}, \\ f_n = f_1 + \sum_{m=2}^{n} \frac{\exp(E_m/kT_e)}{g_m V(m,\ m-1)} \sum_{k=1}^{m-1} g_k \exp\left(-\frac{E_k}{kT_e}\right) \frac{df_k}{d\tau}. \end{gathered} \tag{2.15}$$

This equation is nothing but another way of writing the system of equations (2.3).

In order to find the distribution function we set $f_k = f_1$ on the right-hand side of Eq. (2.15) as the first step in the iteration (this corresponds to a Boltzmann test distribution function). Then

$$f_n = f_1 + \beta_n^1 \frac{df_1}{d\tau}, \qquad \beta_n^1 = \alpha_n^1/\alpha_n^0. \tag{2.16}$$

In the second step we again substitute the resulting distribution, Eq. (2.16), on the right-hand side of Eq. (2.15), and so on. At the (m − 1)-st step

$$f_m = \sum_{i=0}^{m-1} \beta_m^i \frac{d^{(i)} f_1}{d\tau^i}, \qquad \beta_m^i = \alpha_m^i/\alpha_m^0. \tag{2.17}$$

From this it is clear that Eq. (2.17) coincides exactly with Eq. (2.5) since for $E_1 = 0$, $f_1 = N_1$ and $N_m = \alpha_m^0 f_m$.

From an analysis of Eq. (2.15) one may conclude that the quasistationary distribution (2.16) is valid only when the time derivatives of the reduced populations equal the time derivatives of the ground state; that is

$$df_k/dt = df_1/dt. \tag{2.18}$$

For small recombination temperatures $T_e \sim 0.05$-0.5 eV, the time derivatives of the populations of the excited levels $dN_n/dt = g_n \exp(-E_n/kT_e)(df_1/dt)$ are exceedingly small compared to dN_1/dt; that is, in solving the system (2.3) we may assume $dN_n/dt = 0$ $(n \neq 1)$. Thus, the quasistationary distribution function for low electron temperatures is equivalent to the stationary-sink approximation.

We now consider recombination and ionization of a hydrogen plasma at relatively high free electron densities, when collisional transitions predominate. The level population distribution function is conveniently characterized by the temperatures Θ_n between neighboring levels, which are defined by

$$\frac{N_{n+1}}{N_n} = \frac{g_{n+1}}{g_n} \exp\left(-\frac{E_{n+1,\,n}}{\Theta_n}\right).$$

In the case of recombination of a plasma whose degree of ionization exceeds the equilibrium level corresponding to the free electron temperature $T_e = 0.05$-0.5 eV, we find, on using the expression for the populations [Eq. (2.12)] and Eq. (2.16)], that

$$\Theta_n = \frac{T_e}{1 + \frac{T_e}{E_{n+1,\,n}} \ln \frac{1 + \varkappa\beta_n^1}{1 + \varkappa\beta_{n+1}^1}}, \tag{2.19}$$

where

$$\beta_n^1 = \sum_{m=2}^{n} \frac{\exp(E_m/kT_e)}{g_m V(m,\, m-1)}; \qquad \varkappa = \frac{1}{N_e} \frac{d}{dt} \ln N_1.$$

The parameter $\varkappa$ can be related to the temperature between the first and second levels as follows:

$$\varkappa = 4V(2,\, 1)[\exp(-E_2/k\Theta_1) - \exp(-E_2/kT_e)]. \tag{2.20}$$

The temperatures were calculated for fixed $T_e = 0.1$ eV and various values of the parameter Θ_1 using Eq. (2.19). The probabilities of electron–atom collisions of the first and second kinds were taken from expressions based on Bethe's formula (see, for example, [47]):

$$V(m+1,\, m) = 1.73 \cdot 10^{-7} \sqrt{\frac{R}{kT_e}} \left[\frac{m(m+1)}{2m+1}\right]^3 \frac{m^2}{2m+1} U\left(\frac{E_{m+1,\,m}}{kT_e}\right),$$

$$U(x) = 1 + x \exp(x) \operatorname{Ei}(-x),$$

where $\operatorname{Ei}(-x)$ is the exponential integral.

The results of the calculation are shown in Fig. 2 as plots of Θ_n versus n for various Θ_1. It is interesting to note that when Θ_1 is reduced to the value determined by the condition $\exp(-E_2/\Theta_1) \gg \exp(-E_2/T_e)$, $\Theta_1 < 0.102$ and the values of the other temperatures remain practically constant and are given by the equation

$$\Theta_n = \frac{T_e}{1 + \frac{T_e}{E_{n+1,\,n}} \ln \frac{\beta_n^1}{\beta_{n+1}^1}} \qquad (n \neq 1).$$

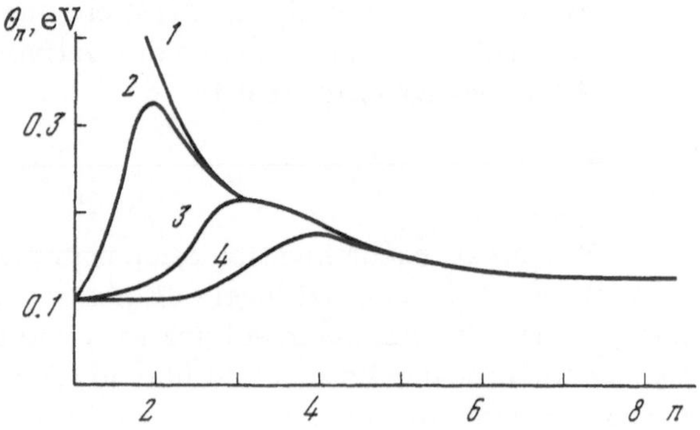

Fig. 2. Quasistationary distributions of the temperatures over the levels of a recombining hydrogen plasma. Curve 1 corresponds to values of the parameter $\theta_1 > 0.102$ eV, and curves 2, 3, and 4 to values of the parameter $\varkappa = 1.12 \cdot 10^{-52}$, $1.12 \cdot 10^{-57}$, and $1.12 \cdot 10^{-59}$.

In the stationary-sink approximation [47] the distribution obtained by numerical computation is also independent of the temperature between the first and second levels and coincides with the analytic solution.

During recombination an equilibrium is established fairly rapidly between the electrons in the upper levels and the free electrons. Then atoms accumulate in the first level while the distribution remains unchanged at the upper levels (curve 1, Fig. 2). Later, as soon as the temperature Θ_1 comes close to equilibrium, there is a drop in the temperature between the second and third levels and atoms accumulate in the second level while the distribution remains unchanged at the higher levels (curves 2 and 3), and so on.

Thus, the recombination process may be regarded as a sequential filling of levels, beginning with the first.

We shall now consider the problem of ionization in a hydrogen plasma whose degree of ionization at the initial time is less than the equilibrium value corresponding to a free electron temperature of $T_e = 5$–10 eV. Unlike recombination, here the parameter $\varkappa$ is negative since the population of the ground state decrease to some equilibrium value. The temperature between the first and second levels rises to the electron temperature.

The results of a calculation of the temperatures for $T_e = 5$ eV and various values of the parameter Θ_1 are shown in Fig. 3. According to the resulting distributions, at first, when the electron temperature differs greatly from the temperature between the first and second levels, ionization takes place rather strongly. At this state of ionization the contribution of multi-quantum transitions at the upper levels is important. Thus, curve 1 of Fig. 3 (where only the single-quantum transitions are taken into account) does not tend sufficiently rapidly toward the electron temperature. Beginning at the time when the difference between the electron temperature and the temperature between the first and second levels becomes small, the populations of

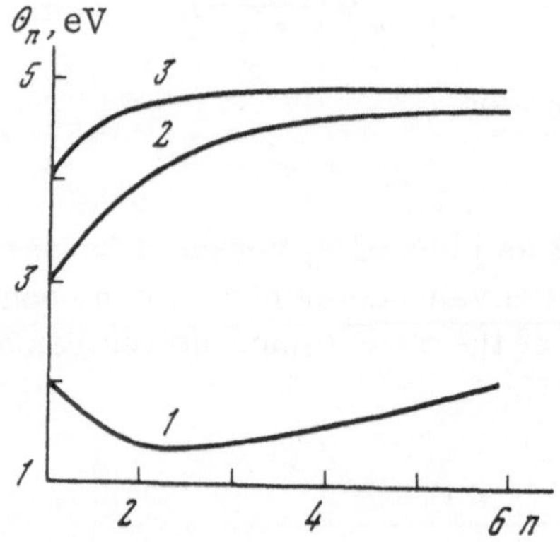

Fig. 3. The distribution of the temperatures over the levels in the hydrogen plasma ionization problem.

the excited levels are mainly determined by the population of the next lower level (stepwise ionization). Then the temperatures between neighboring upper levels successively relax to the electron temperature. The character of this relaxation of the distribution to equilibrium in the stepwise ionization stage is illustrated in curves 2 and 3 of Fig. 3.

The quasistationary distribution function of Eq. (2.16) was obtained in the single-quantum approximation (only transitions between neighboring levels were taken into account). This approximation is not satisifed for the upper levels, which obey the Saha distribution of Eq. (2.10). To obtain a unique distribution we join Eqs. (2.10) and (2.16) and use the normalization condition (2.11) to obtain

$$N_1 + \frac{\beta^1_{n_0}}{N_e}\frac{dN_1}{dt} = K(T_e)\,N_e^2; \tag{2.21}$$

$$Z\frac{dN_1}{dt} = -\frac{dN_e}{dt}, \qquad Z = \sum_{n=1}^{n_0} g_n \exp\left(-\frac{E_n}{kT_e}\right) \tag{2.22}$$

[Eq. (2.22) is obtained by differentiating Eq. (2.11) with N = const].

Differentiating Eq. (2.21) with respect to τ and using Eq. (2.22), we find the conditions under which we may neglect the term with i = 2 in the expansion (2.5):

$$\left|\frac{d^{(2)}N_1}{d\tau^2}\Big/\frac{dN_1}{d\tau}\right| = \frac{2K(T_e)\,ZN_e + 1}{\beta^1_{n_0}} \ll \frac{\alpha^1_{n_0}}{\alpha^2_{n_0}}\,. \tag{2.23}$$

In the case of recombination at small electron temperatures ($T_e \sim 0.05$-0.5 eV) this inequality may be simplified since

$$\alpha^1_{n_0}/\alpha^2_{n_0} \simeq V(2,\,1), \qquad Z = 1, \text{ and } 2K(T_e)\,N_e \gg 1.$$

Thus,

$$N_e \ll V(2,\,1)\,\frac{\beta^1_{n_0}}{2K(T_e)}\,. \tag{2.24}$$

Therefore, the quasistationary distribution function, as does the stationary-sink approximation, correctly reflects the recombination process for small T_e, but not for very high electron densities.

In the case of ionization the condition for neglecting the second derivatives transforms to

$$\frac{1}{\beta^1_{n_0}} \ll \frac{\alpha^1_{n_0}}{\alpha^2_{n_0}}\,. \tag{2.25}$$

This inequality limits the electron temperature from below. For T_e = 5 eV and n_0 = 9 it is satisfied with a reserve of one or two orders of magnitude.

Matching the quasistionary distribution function to the Saha distribution for the upper levels is analogous to specifying the relaxation flux from the continuous spectrum to the discrete levels. For recombination almost always $K(T_e)N_e^2 \ll N_1$, and from Eqs. (2.21) and (2.22) we find the recombination coefficient for a hydrogen plasma,

$$\gamma = -\frac{1}{N_e^2}\frac{dN_e}{dt} = \frac{K(T_e)\,N_e}{\beta^1_{n_0}}\,. \tag{2.26}$$

From this equation it is clear that the characteristic recombination time is $\tau_R \sim \beta^1_{n_0}/K(T_e)N_e^2$.

We now turn to the ionization problem in the case $K(T_e)N_e^2 \ll N_1$. This means that the populations of the upper levels, beginning with $n = n_0$, are small compared with the lower level populations and that the ionization flux is unidirectional and controlled by the flux from the $(n_0 - 1)$-st level to the n_0-th level. In the following we shall refer to this as the "absolutely absorbing wall" approximation.

Thus, from Eq. (2.21) we have

$$N_1 = N_1(0)\exp(-\tau/\beta^1_{n_0}),$$

where $N_1(0)$ is the initial value of the population of the ground state. Keeping Eq. (2.11) in mind and including the fact that $d\tau/dt = N_e$, we find that

$$N_0 = \varepsilon N_1, \qquad \varepsilon = Z - \frac{\sum\limits_{n=1}^{n_0} g_n \exp(-E_n/kT_e)\,\beta^1_n}{\beta^1_{n_0}} \simeq 1; \tag{2.27}$$

$$N_1 = \frac{N}{\varepsilon + \frac{N_e(0)}{N_1(0)}\exp\left(Nt/\beta^1_{n_0}\right)}; \tag{2.28}$$

$$N_e = \frac{N}{1 + \varepsilon\frac{N_1(0)}{N_e(0)}\exp\left(-Nt/\beta^1_{n_0}\right)}, \tag{2.29}$$

where $N_e(0)$ is the initial value of the electron density. The characteristic ionization time is $\tau \sim \beta^1_{n_0}/N$.

These calculations show that the coefficient $\beta^1_{n_0}$ has an important effect on the recombination and ionization. Thus, we shall dwell in more detail on its behavior as a function of the boundary level n_0. Let us now put the probabilities based on the Bethe formula in the expression for $\beta^1_{n_0}$ and restrict ourselves to examining small recombination temperatures such that

$$\sum_{k=1}^{n_0} g_k \exp(-E_k/kT_e) = 1:$$

$$\beta^1_{n_0} = \frac{10^7}{1.73}\left(\frac{kT_e}{R}\right)^{1/2}\exp\left(\frac{R}{kT_e}\right)\sum_{m=2}^{n_0}\frac{(2m-1)^4}{[m(m-1)]^5}\,\frac{\exp(-R/kT_e m^2)}{U(E_{m,\,m-1}/kT_e)}. \tag{2.30}$$

Analyzing the expression under the summation sign in Eq. (2.30), it is clear that it has a sharp maximum in m since for small m the first terms in the sum are small due to the smallness of the exponential factor (because of the large distance between neighboring energy levels), and for large m they are small because of the increase in the transition probabilities $V(m, m-1)$. The term with $m = m^*$, which is the main contributor to the sum in Eq. (2.30), corresponds to the so-called bottleneck in the energy level spectrum of the hydrogen atom. The bottleneck has a strong effect on the entire recombination process.

By studying the expression under the summation at the extremum in m it is possible to find the location of the bottleneck in the energy level spectrum of hydrogen in two limiting cases of the function $U(x)$. For $x \ll 1$, $U(x) = 1$, and $x = E_{m,\,m-1}/kT$, m^* is a root of the equation

$$\frac{R}{kT_e} = \frac{m^2(12m^2 - 12m + 5)}{2(2m-1)(m-1)}, \qquad m^* \simeq \sqrt{\frac{R}{3kT_e}}.$$

For $x \gg 1$, $U(x) = x^{-1}$, m^* is a root of the equation

$$\frac{R}{kT_e} = \frac{m^2(18m^2 - 18m + 7)}{2(2m-1)(m-1)}, \qquad m^* \simeq \frac{1}{3}\sqrt{\frac{2R}{kT_e}}.$$

A computer calculation of the sum in Eq. (2.30) showed that over the temperature interval 0.1 eV ≤ kT_e ≤ 1.2 eV the factor $\beta^1_{n_0}$ is well approximated by an analytic expression of the form

$$\beta^1_{n_0} = (1.4 \pm 0.2) \cdot 10^6 T_e^4 (\text{eV}) \exp\left(\frac{R}{kT_e}\right). \tag{2.31}$$

The lower the electron temperature, the higher the bottleneck m = m* lies, which divides the energy level spectrum into two parts. The population distribution function for the energy levels lying below the bottleneck deviates strongly from equilibrium and

$$\beta^1_m \simeq \frac{\exp(E_m/kT_e)}{g_m V(m, m-1)}.$$

For the energy levels lying above the bottleneck ($\beta^1_n \simeq \beta^1_{n_0}$) the distribution is almost a Boltzmann distribution and is in equilibrium with the continuous spectrum. Then $\beta^1_{n_0}$ is independent of the choice of n_0. This implies that the boundary level n_0 must be chosen so that it lies above the botteneck. Otherwise the results of the calculation would depend on the choice of n_0. In the ionization problem, at high electron temperatures the sum (2.30) is determined (as to order of magnitude) by the first term. In this regard we may say that the bottleneck lies between the ground and first excited states.

Thus, these distributions describe a rather wide range of phenomena. This is also true of the distribution which takes only the first derivative into account and which, like the Treanor distribution for the vibrational levels of a molecule, depends on only two parameters T_e and Θ_1. The electron temperature characterizes both the external conditions and the distribution function over the levels. The temperature Θ_1 characterizes the inner properties of the distribution for fixed T_e. Without dwelling on the questions of its external effect on a relaxing system we note only the obvious simiplicity of considering relaxation over characteristic times for changes in the system that are greater than the times to establish a quasiequilibrium distribution.

Up to this time we have been considering comparatively dense plasmas where collisional processes dominate. In an optically thin plasma with a low electron density the population kinetics of the excited states will be determined by radiative decay of the levels in addition to atom–electron collisions of the first and second type.

Let A(n, m) be the probability per unit time of a spontaneous radiative transition from level n to level m. In the population balance equation for the n-th level we shall restrict ourselves to considering only the flux from the (n + 1)-st level to the n-th level out of the radiative flux downward to n from higher levels and to the flux from n to (n − 1) out of the radiative flux from level n to all lower-lying levels. Then the system of equations (2.3) takes the form

$$\begin{aligned} dN_n/dt = {} & V(n+1, n) N_{n+1}N_e - V(n, n+1) N_n N_e + V(n-1, n) N_{n-1}N_e - \\ & - V(n, n-1) N_n N_e + A(n+1, n) N_{n+1} - A(n, n-1) N_n. \end{aligned} \tag{2.32}$$

The solution of Eqs. (2.32) is completely analogous to the solution of Eq. (2.3). Making the substitution

$$\frac{df_k}{dt} \to \frac{df_k}{dt} + A(k, k-1) f_k - \frac{g_{k+1}}{g_k} \exp\left(-\frac{E_{k+1,k}}{kT_e}\right) A(k+1, k),$$

on the right-hand side of Eq. (2.15) we have

$$f_n = f_1 + \sum_{m=2}^{n} \frac{\exp(E_m/kT_e)}{g_m V(m, m-1)} \sum_{k=1}^{m-1} g_k \exp\left(-\frac{E_k}{kT_e}\right) \frac{df_k}{d\tau} - \sum_{m=2}^{n} \frac{A(m, m-1)}{N_e V(m, m-1)} f_m. \tag{2.33}$$

This equation is a recurrence relation for the unknown distribution function f_n. Expanding it and assuming the derivatives df_k/dt to be given, we obtain

$$f_n = Q_1^n f_1 + \sum_{m=2}^{n} \frac{\exp(E_m/kT_e)}{g_m V(m, m-1)} Q_{m-1}^n \sum_{k=1}^{m-1} g_k \exp\left(-\frac{E_k}{kT_e}\right) \frac{df_k}{d\tau}, \tag{2.34}$$

where

$$Q_m^n = \prod_{k=m}^{n-1} \left(1 + \frac{A(k+1, k)}{N_e V(k+1, k)}\right)^{-1} \quad (Q_1^1 = 1).$$

In the first iteration we set $f_k = f_1$ on the right-hand side of Eq. (2.34). As a result we find the quasistationary distribution function including the radiative decay of the levels:

$$N_n = N_1 g_n \exp\left(-\frac{E_n}{kT_e}\right)(Q_1^n + \beta_n^1 \varkappa); \tag{2.35}$$

$$\beta_n^1 = \sum_{m=2}^{n} \frac{\exp(E_m/kT_e)}{g_m V(m, m-1)} Q_{m-1}^n \sum_{k=1}^{m-1} g_k \exp\left(-\frac{E_k}{kT_e}\right), \qquad \varkappa = \frac{1}{N_e} \frac{d}{dt} \ln N_1. \tag{2.36}$$

Here and in the following computations the zero level for the excitation energy of the atom is taken to be its ground state.

In the limiting case of high electron densities Eq. (2.35) transforms to Eq. (2.16). For small N_e the distribution function (2.35) may deviate strongly from Eq. (2.16). Without analyzing the distribution function in detail or the conditions for realizing an inverted level population, which will be given in the following chapter, we note only that in an optically thin hydrogen plasma at low electron densities there may be a population inversion due to radiative depopulation of the lower laser level.

In conclusion we shall examine the question of the applicability of the single-quantum approximation to ionization and recombination problems. In general, the single-quantum approximation is applicable if the number of two-quantum ($\Delta n = \pm 2$) transitions per unit time is much less than the number of single-quantum ($\Delta n = \pm 1$) transitions. For small electron temperatures, when the distance between neighboring energy levels is less than or on the order of T_e, single-quantum transitions play the basic role and the single-quantum approximation is satisfactory. In relaxation problems with relatively large electron temperatures and more accurate accounting for radiative decay of the levels, it becomes necessary to take multiquantum processes into account.

We now rewrite the system of equations (2.1) which includes all possible transitions in the following form:

$$f_n = \frac{1}{V(n)} \left[\sum_{n'} V(n, n') f_{n'} - \frac{1}{N_e} \frac{df_n}{dt}\right], \qquad V(n) = \sum_{n'} V(n, n'). \tag{2.37}$$

In the quasistationary stage $df_n/dt = df_1/dt$. The multiquantum nature of the process will be taken into account by the iteration method. In the first step of this method it is necessary to take the single-quantum-approximation distribution function (2.16) as the test distribution function on the right-hand side of Eq. (2.37). Substituting Eq. (2.16) in (2.37) yields

$$f_n^{(1)} = f_1 + \frac{1}{V(n)} \left[\sum_{n'} V(n, n') \beta_{n'}^{(0)} - 1\right] \frac{df_1}{d\tau} = f_1 + \beta_n^{(1)} \frac{df_1}{d\tau}.$$

In the i-th step of the successive approximations we find

$$f_n^{(i)} = f_1 + \beta_n^{(i)} \frac{df_1}{d\tau},$$
$$\beta_n^{(i)} = \frac{1}{V(n)} \left[\sum_{n'} V(n, n') \beta_{n'}^{(i-1)} - 1 \right], \tag{2.38}$$

where $f_n^{(i)}$ and $\beta_n^{(i)}$ denote the distribution function and coefficient of the first derivative obtained in the i-th step of the iteration.

In just this manner it is possible to obtain a distribution function taking all possible radiative transitions into account.

Successive iterations make it possible to find the exact distribution function. The convergence of the iteration method depends on the choice of the (initial) test distribution function.

2. Level Populations in Atoms. Plasma Lasers

Compared to hydrogen, the structure of the discrete levels of any other atom is rather complicated. For simplicity we shall assume that the level with principal quantum number n is split into sublevels with various values of the orbital quantum number l.

The balance equations, which take inelastic collisions of atoms with electrons into account, have the form

$$\frac{dN_{nl}}{dt} = N_e \sum_{n'l'} V(n'l', nl) N_{n'l'} - N_e N_{nl} \sum_{n'l'} V(nl, n'l'), \tag{2.39}$$

where N_{nl} is the population of the energy level nl of the atom and $V(nl, n'l')$ is the probability of the transition $nl \to n'l'$ due to electron–atom collisions averaged over the electron distribution.

In a plasma with a Maxwellian velocity distribution, we have

$$V(nl, n'l') g_{nl} \exp\left(-\frac{E_{nl}}{kT_e}\right) = V(n'l', nl) g_{n'l'} \exp\left(-\frac{E_{n'l'}}{kT_e}\right), \tag{2.40}$$

where $g_{nl} = 2l + 1$ and E_{nl} are the statistical weight and energy of the level nl.

For a whole series of elements, for example, the lithium-like atoms and ions, the splitting in the orbital quantum number is unimportant. Thus, the probability of transitions with a change in l is much greater than the probability of transitions with a change in the principal quantum number. For such atoms the system of equations (2.39) may be solved assuming a Boltzmann distribution over l (the spectroscopic notation s, p, d, ... is introduced for the values $l = 0, 1, 2, \ldots$),

$$N_{nl} = N_{ns}(2l + 1) \exp\left[-\frac{\Delta E_l(n)}{kT_e}\right], \quad \Delta E_l(n) = E_{nl} - E_{ns}. \tag{2.41}$$

We shall solve the system (2.39) in the single-quantum approximation taking into account the optically allowed transitions ($\Delta l = \pm 1$). Substituting Eq. (2.41) into Eq. (2.39) and going to the variable τ, we obtain

$$\frac{dN_{ns}}{d\tau} = -V(ns, n+1p) N_{ns} + V(n+1p, ns) 3 \exp\left[-\frac{\Delta E_p(n+1)}{kT_e}\right] N_{n+1s} +$$
$$+ V(n-1p, ns) 3 \exp\left[-\frac{\Delta E_p(n-1)}{kT_e}\right] N_{n-1s} - V(ns, n-1p) N_{ns}. \tag{2.42}$$

These equations are analogous to Eq. (2.3) for the hydrogen atom; however, here, due to branching of the relaxation flux, the factors in front of the populations (regarded as effective probabilities) are no longer related by equations of the type (2.2) for the direct and reverse processes. We shall examine relaxation of a lithium plasma as an example. As in the case of hydrogen, we seek a solution in the form of a series,

$$N_{ns} = \sum_{i=0}^{n-2} \alpha_n^i \frac{d^{(i)} N_{2s}}{d\tau^i}.$$

As a result there exist recurrence relations for α_n^i, from which we find

$$\alpha_m^0 = \exp\left[-\frac{E_{m,2}(s)}{kT_e}\right], \qquad E_{m,n}(s) = E_{ms} - E_{ns}, \tag{2.44}$$

$$\alpha_m^1 = \alpha_m^0 \sum_{n=2}^{m-1} \frac{1}{V(ns,\ n+1p)} \left[1 + \sum_{k=3}^{n} \prod_{q=k}^{n} \frac{V(qs,\ q-1p)}{V(q-1s,\ qp)}\right] = \alpha_m^0 \beta_m^1. \tag{2.45}$$

In this case the parameter that controls the recombination and ionization kinetics will be the temperature Θ_2 between the 2s and 3s levels:

$$\varkappa = \frac{1}{N_e} \frac{d}{dt} \ln N_{2s} = V(2s,\ 3p) \left\{\exp\left[E_{3,2}(s)\left(\frac{1}{T_e} - \frac{1}{\Theta_2}\right)\right] - 1\right\}.$$

Finally, we write the expression for the population of the discrete levels and the temperatures between the (n + 1)s and np levels:

$$N_{nl} = N_{2s} g_{nl} \exp\left(-\frac{E_{nl} - E_{2s}}{kT_e}\right)(1 + \beta_n^1 \varkappa); \tag{2.46}$$

$$\Theta_n = \frac{kT_e}{1 + \dfrac{kT_e}{E_{n+1s} - E_{np}} \ln \dfrac{1 + \beta_n^1 \varkappa}{1 + \beta_{n+1}^1 \varkappa}}. \tag{2.47}$$

Due to the branching of the relaxation flux during recombination of the plasma a population inversion may arise between the (n + 1)s and np levels if

$$\frac{1 + \beta_{n+1}^1 \varkappa}{1 + \beta_n^1 \varkappa} > \exp\left[\frac{E_{n+1s} - E_{np}}{kT_e}\right]. \tag{2.48}$$

This condition is particularly simple for the 3s – 2p transition:

$$\Theta_2 > \frac{E_{3,2}(s)}{\Delta E_p(2)} T_e.$$

Once a Boltzmann distribution over l is specified, the problem of the relaxation of any atom reduces to the same problem for the ns levels of a hydrogen atom with V(ns, n + 1p) being taken instead of the probability of excitation of the (n + 1)s level from the ns level and V(n + 1s, np) in place of the probability of extinction from the (n + 1)s level to ns. For $T_e \gg \Delta E_p(n)$ this assumption does not lead to significant errors. For electron temperatures $T_e \ll \Delta E_p(n)$, however, the probability of exciting the (n + 1)s level from ns is considerably reduced due to the large difference (relative to T_e) in the energies of the (n + 1)s and (n + 1)p levels. Then the error in the calculation may be large. This shortcoming may be corrected by replacing V(ns, n + 1p) with the effective probability of excitation of the (n + 1)s level from the ns level,

$$V^*(ns,\ n+1p) = V(ns,\ n+1p) \exp \frac{\Delta E_p(n+1)}{kT_e}.$$

Calculations show that this substitution improves the results.

We now consider an optically thin plasma. In Eq. (2.39) there appears a new term associated with radiative decay of the levels,

$$\frac{dN_{nl}}{dt}=N_e\sum_{n'l'}V(n'l',\ nl)N_{n'l'}-N_eN_{nl}\sum_{n'l'}V(nl,\ n'l')+$$
$$+\sum_{E_{n'l'}>E_{nl}}A(n'l',\ nl)N_{n'l'}-N_{nl}\sum_{E_{n'l'}<E_{nl}}A(nl,\ n'l'), \tag{2.49}$$

where $A(nl, n'l')$ is the probability of the radiative $nl \to n'l'$ transition. We assume that

$$N_{nl}=g_{nl}\exp\left(-\frac{E_{nl}}{kT_e}\right)f_{nl}. \tag{2.50}$$

As before we shall solve the system of equations (2.49) in the single-quantum approximation for the optically allowed transitions while assuming a Boltzmann distribution over the orbital quantum number except for the s–p transitions, that is, $f_{ns} \neq f_{np} = f_{nd} = \dots$. In the quasi-stationary stage the distribution function may be sought in the form

$$f_{ns}=\alpha_{ns}f+\beta_{ns}\frac{df}{d\tau}, \tag{2.51}$$

$$f_{np}=\alpha_{np}f+\beta_{np}\frac{df}{d\tau}, \tag{2.52}$$

where f is the reduced population of the ground state of the atom (ion).

We shall take account of radiative decay in the following manner:

$$\sum_{E_{n'l'}>E_{np}}A(n'l',\ np)N_{n'l'}=A(n+1s,\ np)N_{n+1s}+A(n+1d,\ np)N_{n+1d},$$

$$A(np)=\sum_{E_{n'l'}<E_{np}}A(np,\ n'l')=A(np,\ ns)+A(np,\ n-1s)+A(np,\ n-1d),$$

$$\sum_{E_{n'l'}>E_{ns}}A(n'l',\ ns)N_{n'l'}=A(np,\ ns)N_{np}+A(n+1p,\ ns)N_{n+1p},$$

$$A(ns)=\sum_{E_{n'l'}<E_{ns}}A(ns,\ n'l')=A(ns,\ n-1p).$$

Substituting Eqs. (2.51) and (2.52) in the level population balance equations (2.49) for the ns and np levels and neglecting the flux between the ns and (n + 1)p levels by comparison with the flux between the ns and np levels, we match the coefficients of f and df/dt in the left and right parts of our equations. Then we obtain the following recurrence relations for the unknown coefficients α and β:

$$1+R(ns)\beta_{ns}=V(ns,\ n-1p)\beta_{n-1p}+R(ns,\ np)\beta_{np}; \tag{2.53}$$

$$1+R(np)\beta_{np}=V(np,\ n-1s)\beta_{n-1s}+V(np,\ ns)\beta_{ns}+R(np,\ n+1s)\beta_{n+1s}+$$
$$+V(np,\ n-1d)\beta_{n-1p}+R(np,\ n+1d)\beta_{n+1p}, \tag{2.54}$$

where we have introduced the notation

$$R(nl)=V(nl)+N_e^{-1}A(nl),\qquad R(nl,\ n'l')=V(nl,\ n'l')+N_e^{-1}A(nl,\ n'l'),$$

$$V(np)=\sum_{n'=n,\ n\pm1}V(np,\ n's)+\sum_{n'=n\pm1}V(np,\ n'd),$$

$$V(ns)=V(ns,\ n-1p)+V(ns,\ np),$$

$$A(nl,\ n'l')=\begin{cases}A(nl,\ n'l'), & E_{n'l'}<E_{nl}\\ \dfrac{g_{n'l'}}{g_{nl}}\exp\left(-\dfrac{E_{n'l'}-E_{nl}}{kT_e}\right)A(n'l',\ nl), & E_{n'l'}>E_{nl}.\end{cases}$$

If we discard the number 1 on the left-hand sides of Eqs. (2.53) and (2.54), then we obtain analogous expressions for the undetermined coefficients α on replacing β by α.

It can be shown that in the case of purely collisional relaxation all $\alpha_{ns} = \alpha_{np} = 1$ for $A = 0$.

Knowledge of the coefficients α and β depends strongly upon which column (s or p) the ground state (state with the least energy) lies in. For example, if the ground state of Li I is the 2s state, then $\alpha_{2s} = 1$ and $\beta_{2s} = 0$.

When these assumptions for studying the various processes are inadequate, the distribution function obtained here may be used as a test function in the method of successive approximations. The relation between the distribution functions in the m-th stage and those in the (m − 1)-st stage is given by the equality

$$f_{nl}^{(i)} = \frac{1}{P(nl)}\left[N_e \sum_{n'l'} V(nl, n'l') f_{n'l'}^{(i-1)} + \sum_{E_{n'l'}>E_{nl}} A(nl, n'l') f_{n'l'}^{(i-1)} - \frac{df}{dt}\right], \qquad (2.55)$$

where

$$P(nl) = N_e \sum_{n'l'} V(nl, n'l') + \sum_{E_{n'l'}<E_{nl}} A(nl, n'l').$$

The results of a calculation of the populations in Li using these formulas are in good agreement with a numerical calculation [55] of relaxation in a dense lithium plasma as well as in qualitative agreement with the results for helium [56] and argon [57] atoms. This approach evidently may be used to compute the discrete level populations of any atom, as well as in obtaining a population inversion in various nonequilibrium regimes.

3. The Level Population Distribution Function of Atoms in Mixtures

In a plasma containing a mixture of various elements the distribution function of the atomic level populations is naturally more complicated than for a plasma of simple chemical composition. Here there are certain analogies with the population distribution functions of the vibrational levels in a mixture of gases. In mixtures there are several new prospects for creating an inverted population. A number of computer calculations analyzing mixtures with the aid of the balance equations and the relevant literature are discussed by Gudzenko, Shelepin, and Yakovlenko in this volume [7]. In the following we shall analyze atomic mixtures (some problems in kinetics involving molecules are examined in [58]).

If we neglect resonance processes during atomic collisions, in the zeroth approximation the motion of the electrons over the levels inside each type of atom may be considered independently. The overall distribution function is created due to resonance effects in atomic collisions (interchange, charge exchange, ionization effects). Without claiming any kind of complete analysis, we shall examine the features of the distributions which arise here using the example of a binary mixture of elements A and B when charge exchange is important.

We shall assume that in the mixture of two gases A and B, consisting primarily of neutral atoms and their singly charged ions, the controlling processes in populating the excited levels of both the atoms and ions are inelastic collisions of the first and second kind between atoms and electrons, as well as charge exchange between ground-state neutrals and ions in excited s-states,

$$A + B^+(n_0 s) \rightleftarrows A^+(n_1 s) + B + \Delta E.$$

At energies close to the thermal energy the efficiency of charge exchange is determined by the (small) energy defect (the change in energy of the electron during the transition) ΔE and the nonadiabaticity of the transition.

With these assumptions the system of balance equations for the populations of the energy levels of the A^+ and B^+ ions has the form

$$\frac{dN_{nl}^{A,B}}{dt}=N_e\sum_{n'l'}V(n'l',\ nl)N_{n'l'}^{A,B}-N_eN_{nl}^{A,B}\sum_{n'l'}V(nl,\ n'l')+Q_{nl}^{A,B}+\left(\frac{dN_{nl}^{A,B}}{dt}\right)_R, \tag{2.56}$$

where the subscript A refers to the level system of the A^+ ion, and B to that of the B^+ ion; N_{nl} is the population of the energy level nl; $N_eV(nl,\ n'l')$ is the frequency of the $nl\rightarrow n'l'$ transition due to inelastic collisions of ions with electrons; Q_{nl} is the rate of repopulation of the nl level due to charge exchange; and (dN_{nl}/dt) is the rate of change in the population of the nl level due to recombination or ionization.

To Eq. (2.56) we must also add the normalization and plasma quasineutrality conditions:

$$N_i^{A,B}=\sum_{nl}N_{nl}^{A,B};\quad N_e=N_i^A+N_i^B;\quad N=N_0^A+N_0^{B'}+N_e. \tag{2.57}$$

Here $N_0^{A,B}$ and $N_i^{A,B}$ are the densities of neutral atoms A and B, and of ions A^+ and B^+.

The system of equations (2.56) will be solved in the single-quantum approximation ($\Delta n = 0,\ \pm1$) taking the optically allowed transitions into account and assuming a Boltzmann distribution over the orbital quantum number l. We shall seek a distribution function in the form

$$N_{nl}^{A,B}=g_{nl}\exp\left(-\frac{E_{nl}^{A,B}}{kT_e}\right)f_{nl}^{A,B}, \tag{2.58}$$

where g_{nl} and E_{nl} are the statistical weight and energy of the nl level (the zero energy reference level for excitation of the ions is taken to be the energy of their ground state) and T_e is the electron temperature.

Substituting Eq. (2.58) in the population balance equations (2.56) for the ns configurations and using the principle of detailed balance for the rates of the forward and reverse processes, we have

$$\frac{df_{ns}^{A,B}}{dt}=N_eV(ns,\ n-1p)(f_{n-1s}^{A,B}-f_{ns}^{A,B})+N_eV(ns,\ n+1p)(f_{n+1s}^{A,B}-f_{ns}^{A,B})+\exp\left(\frac{E_{ns}^{A,B}}{kT_e}\right)\left[Q_{ns}^{A,B}+\left(\frac{dN_{ns}^{A,B}}{dt}\right)_R\right]. \tag{2.59}$$

These equations differ from Eq. (2.42) by the presence on the right-hand side of two terms associated with charge exchange and the rate of change of the ion density due to ionization or recombination. Then the rate of charge exchange is not an independent quantity but is determined by the level population distribution functions of the ions.

We shall now examine charge exchange between a definite pair of near-resonance energy levels such that

$$Q_{ns}^A=Q\delta_{nn_1};\qquad Q_{ns}^B=-Q\delta_{nn_0}.$$

For the reduced populations $f_{A,B}$ of the ground state of the A^+ and B^+ ions in the quasi-stationary stage of the relaxation process we have

$$\frac{df_{ns}^{A,B}}{dt}=\frac{df^{A,B}}{dt}.$$

We assume that the ions are lost or created only in the ground states. Then taking the s states of the ions with principal quantum numbers n_A and n_B to be the ground states, summation of Eq. (2.56) yields

$$\frac{dN_i^A}{dt}=\sum_{nl}\frac{dN_{nl}^A}{dt}=Q+\left(\frac{df^A}{dt}\right)_R=\frac{df^A}{dt},$$
$$\frac{dN_i^B}{dt}=\sum_{nl}\frac{dN_{nl}^B}{dt}=-Q+\left(\frac{df^B}{dt}\right)_R=\frac{df^B}{dt}, \tag{2.60}$$

where for simplicity the statistical sums of the ions have been taken equal to unity.

With these assumptions the balance equations (2.59) can be solved analogously to the equations for a single-component plasma. As a result, the desired distribution function has the form

$$f_{ns}^A=f^A+\beta_{ns}^A Q N_e^{-1}, \qquad n\leqslant n_1,$$
$$\beta_{ns}^A=\sum_{m=n_A}^{n-1} V^{-1}(ms,\ m+1p)\left(1+\sum_{i=n_A}^{m-1}\prod_{k=i}^{m-1}\frac{V(k+1s,\ kp)}{V(ks,\ k+1p)}\right). \tag{2.61}$$

We obtain an equivalent expression for the function f_{ns}^B on replacing the symbols A by B, n_1 by n_0, and Q by −Q.

We now write an expression for the charge exchange rate assuming the populations of the ground states of both the ions and neutrals to be equal to their densities:

$$Q=W[A,\ B^+(n_0s)]\,N_0^A N_{n_0s}^B-W[A^+(n_1s),\ B]\,N_{n_1s}^A N_0^B, \tag{2.62}$$

where $W[A, B^+(n_0s)] = \langle\sigma(A, B^+)v\rangle$ is the averaged (over the velocity distribution of the heavy particles) product of the charge-exchange cross section and the relative velocity of the colliding particles.

In an equilibrium state corresponding to the electron temperature, Q = 0. From this it is possible to find the relationship between the charge-exchange rates for the forward and reverse processes:

$$\frac{W[A,\ B^+(n_0s)]}{W[A^+(n_1s),\ B]}=\left(\frac{N_{n_1s}^A N_0^B}{N_0^A N_{n_0s}^B}\right)_0=\exp\frac{\Delta E}{kT_e}, \tag{2.63}$$

where $\Delta E = I^B - I^A + E_{n_0s}^B - E_{n_1s}^A$ is the energy defect and I^A and I^B are the ionization potentials of A and B atoms.

Keeping Eq. (2.63) in mind, we now substitute the distribution functions (2.61) in Eq. (2.62). As a consequence, on solving the resulting equation for the unknown Q, we find

$$Q=\frac{W\exp\left(-\frac{E_{n_0s}^B}{kT_e}\right)\left[N_0^A N_i^B-N_0^B N_i^A\exp\left(\frac{\Delta I}{kT_e}\right)\right]}{1+W\exp\left(-\frac{E_{n_0s}^B}{kT_e}\right)\left[N_0^A N_e^{-1}\beta_{n_0s}^B+N_0^B N_e^{-1}\beta_{n_1s}^A\exp\left(\frac{\Delta I}{kT_e}\right)\right]}, \tag{2.64}$$
$$\Delta I=I^A-I^B;\qquad W=W[A,\ B^+(n_0s)].$$

The sign of the charge-exchange rate determines the direction of energy transfer from one type of ion to the other. For Q > 0 charge exchange is a positive pumping source for the level system of the A^+ ions. Then for the levels lying below the charge-exchange level the distribution for A^+ is equivalent to a quasistationary distribution for a plasma consisting only of A and A^+ but with more intense recombination. Thus, for these levels, recombination is a distinctive mechanism for enhancing the recombination flux. In this case a population inversion may develop due to branching of the recombination flux from the level to which charge exchange takes place. The ns → (n − 1)p transitions may serve as working transitions in such a scheme. This mechanisms is clearly important in the experimentally observed laser oscillations [59, 60]. A population inversion evidently may arise as well in the case of a negative charge-exchange source which depopulates the lower working level.

In addition to pulsed lasing, cw lasing can also be realized with mixtures. To do this there must be continuous preferential creation of ions of one type and loss of ions of the other type in order to keep the neutral and ion densities constant. As a specific example of using the resulting quasistationary distributions we now examine the physical kinetics of a helium–selenium laser in which lasing takes place on the 5p−5s transitions of the selenium ion and is produced by charge exchange of He^+ ions on neutral Se atoms forming Se^+ ions in an excited 5p state [61] by the reaction

$$\mathrm{He}^+ + \mathrm{Se} \rightleftarrows \mathrm{He} + \mathrm{Se}^+(5p) + \Delta E.$$

We shall regard the plasma as optically thin and singly ionized. In the balance equations for the populations of the energy levels of Se^+ we shall include radiative decay of the levels in addition to inelastic ion−electron collisions and charge exchange, i.e.,

$$\frac{dN_{ns}}{dt} = N_e \sum_{n'l'} V(n'l',\, nl)\, N_{n'l'} - N_e N_{nl} \sum_{n'l'} V(nl,\, n'l') + \sum_{E_{n'l'} > E_{nl}} A(n'l', nl) N_{n'l'} - \\ - \sum_{E_{n'l'} < E_{nl}} A(nl,\, n'l')\, N_{nl} + Q_{nl} + \left(\frac{dN_{nl}}{dt}\right)_R. \qquad (2.65)$$

Here $A(nl,\, n'l')$ is the rate of the radiative $nl \to n'l'$ transition, and the remaining notation is the same as in Eq. (2.56).

Summation of Eqs. (2.65), by analogy with Eq. (2.60), yields the expression

$$\sum_{nl} \frac{dN_{nl}}{dt} = Q + 3\left(\frac{df}{dt}\right)_R = 3\frac{df}{dt}, \qquad (2.66)$$

where f is the reduced population of the ground state of the selenium ion. The statistical weight of Se^+ was taken equal to that of the ground state $g_{4p} = 3$.

As before, we shall first seek a solution to Eqs. (2.65) for the reduced populations f_{nl} taking the selection rules $\Delta n = 0, \pm 1$, and $\Delta l = \pm 1$ into account and assuming a Boltzmann distribution over l except over the s−p transitions, that is, $f_{ns} \neq f_{np} = f_{nd} = \ldots$. In the quasistationary stage we seek a distribution function for $n \le n_0$ (where n_0p is the charge-exchange level) in the form

$$f_{ns} = \alpha_{ns} f + \beta_{ns} Q\,(3N_e)^{-1}; \qquad (2.67)$$

$$f_{np} = \alpha_{np} f + \beta_{np} Q\,(3N_e)^{-1}. \qquad (2.68)$$

After exactly the same calculations as those used to find the distribution functions (2.51) and (2.52), we find that the unknown factors α_{nl} and β_{nl} are given by the recurrence relations (2.53) and (2.54).

We assume that the populations of the ground states of the neutral atoms and ions are equal to their (total) densities and substitute the distribution function (2.68) in the expression for the charge exchange rate,

$$Q = W(\mathrm{He}^+,\, \mathrm{Se})\, N_i^{\mathrm{He}} N_0^{\mathrm{Se}} - W(\mathrm{He},\, \mathrm{Se}^+)\, N_0^{\mathrm{He}} N_{5p}.$$

Then using the relation between the rates of the forward and reverse processes,

$$\frac{W(\text{He, Se}^+)}{W(\text{He}^+,\text{ Se})}=\frac{1}{3}\exp\left(\frac{\Delta E}{kT_e}\right),\quad \Delta E=I^{\text{Se}}-I^{\text{He}}+E_{5p},$$

we obtain

$$Q=\frac{W\left[N_0^{\text{Se}}N_i^{\text{He}}-\frac{1}{3}\exp\left(\frac{\Delta I}{kT_e}\right)\alpha_{5p}N_0^{\text{He}}N_i^{\text{Se}}\right]}{1+\frac{1}{3}W\exp\left(\frac{\Delta I}{kT_e}\right)\beta_{5p}N_e^{\text{He}}N_e^{-1}},$$

$$W=W(\text{He}^+,\text{ Se});\qquad \Delta I=I^{\text{Se}}-I^{\text{He}}. \tag{2.69}$$

An analysis of the level population distribution function of Se^+ has permitted us to conclude that a population inversion is realized between the 5p and 5s levels due to rapid radiative depopulation of the lower laser level.

Continuous-wave lasing is possible in a helium–selenium laser because the selenium ions are mainly created by charge exchange and lost on the walls of the discharge tube. The density of neutral selenium atoms, which also take part in the reaction, is replenished from a heated piece of selenium located near the anode. The helium pressure is so great that its density is practically unchanged by charge exchange. The charge exchange rate is very much less than the rates of ionization of helium atoms by electrons and recombination on the walls of the discharge tube.

The electron temperature is determined by the condition for the discharge to be stationary, i.e., by the equality of the rates of electron impact ionization of helium and of recombination on the walls due to ambipolar diffusion,

$$N_0^{\text{He}}N_e\langle\sigma_i v\rangle=\frac{D}{\Lambda^2}N_e=\frac{k\mu_i}{e\Lambda^2}(T_e+T_i)N_e,$$

where D is the ambipolar diffusion coefficient, $\Lambda = R/2.4$ (where R is the radius of the discharge tube), μ_i is the mobility of the helium ions, T_e and T_i are the electron and ion temperatures, and k and e are Boltzmann's constant and the electronic charge.

If we use Seaton's formula (see, for example, [62]) for the rate of ionization from the l_0^m shell,

$$\langle\sigma_i v\rangle=4.3\cdot10^{-8}m\left(\frac{\text{Ry}}{I}\right)^{3/2}\exp\left(-\frac{I}{kT_e}\right)\left(\frac{kT_e}{I}\right)^{1/2},\quad \text{Ry}=13.6\text{ eV},$$

and the expression

$$\mu_i=\frac{e}{M\nu_{in}},\qquad \nu_{in}=4\pi a^2N_0^{\text{He}}\sqrt{\frac{3kT_i}{M}},$$

for the mobility of the helium ions, where M is the mass of the helium atom and $a = 0.95$ Å is the gas-kinetic radius of helium, then for the parameters of the experimental apparatus in [61] ($N_0^{\text{He}} = 10^{17}$ cm^{-3}, R = 0.27 cm, and T_i = 500°K) the electron temperature is T_e = 4.14 eV, the same as the experimentally observed value.

A calculation shows that as the helium density is increased the electron temperature falls.

Using the optimum discharge parameters for lasing, $N_i^{\text{He}} = 2\cdot10^{12}$ cm^{-3}, $N_0^{\text{Se}} = 2\cdot10^{14}$ cm^{-3}, $N_i^{\text{Se}} = 5\cdot10^{11}$ cm^{-3}, $N_e = 2\cdot10^{12}$ cm^{-3}, and $W \sim \sigma v_{T_i} = 5\cdot10^{-10}$ cm^3/sec, the population inversion was $5.2\cdot10^7$ cm^{-3}, the output power (at λ = 0.5227 μ) was 5.6 mW/cm^3 (the experimentally observed value with 6-8 wavelengths is 7 mW/cm^3), and the gain coefficient of the laser with Doppler line broadening was $5.8 \cdot 10^{-4}$ cm^{-1}. Here we have used the electron impact excitation cross sections for the ions and radiative transition probabilities computed numerically in [62].

Knowing the dependences of the charged and neutral particle densities on the total pressure of the mixture it is possible to analyze the optimum conditions for creation of a population inversion. However, the existence of optimum helium pressures is already obvious from Eq. (2.69). In fact, as the helium pressure is increased the density of its ions rises and, with that, the charge exchange rate and the inverted population. With further increase in the helium pressure the inverse of direct charge exchange will predominate so the upper laser level begins to be emptied. In addition, as the helium pressure rises, there is a reduction in the electron temperature which may produce a drop in the charged particle concentrations, the electron density, and the populations of the excited levels of Se^+. We also note that it is impossible to greatly increase the neutral selenium atom density or the discharge current since this can lead to resonant capture of the radiation from the selenium ions and cessation of lasing.

These calculations for a helium–selenium laser have therefore demonstrated the completely satisfactory agreement between theoretical data obtained on the basis of the quasistationary distributions and experiment.

CHAPTER III

THE KINETICS OF MOVING PLASMA LASERS

A technique is developed for the design of plasma-dynamic lasers and pinch discharge lasers using quasistationary analytic distribution functions of the populations in the discrete levels. The freezing effect and radiative depopulation of the lower working level are examined for an expanding hydrogen plasma. The possible existence of an efficient collisional mechanism which would permit use of large plasma volumes for lasing with lithium, helium, and beryllium plasmas is demonstrated. The basic experimental relationships for pinch lasers are analyzed.

1. Plasma-Dynamic Lasers

A moving plasma with rapidly changing density and temperature is a medium in which substantial disequilibrium is realized. Both during rapid compression and during expansion (escape) of the plasma, favorable conditions are created for lasing. However, analysis of the population inversion in such lasers is hindered because of the complex kinetics of the processes involved and the additional need to take macroscopic motion of the plasma into account. In this chapter, an analytic method is developed for finding the electron distribution function that is a generalization of the approach of the previous chapter. This method is applicable both to compression in pinch discharges and to expansion of a plasma (used in plasma-dynamic lasers). It makes it possible to analyze both the conditions for creating an inverted population and the kinetic processes in moving plasma lasers.

There is presently a rapid growth in research on gas-dynamic lasers in which lasing takes place on vibrational–rotational transitions of molecules (mainly CO_2). Various types of lasers have been developed, including the electrical gas-dynamic laser, in which an auxiliary discharge is used to augment the efficiency. Large cw powers have been obtained.

At the same time, plasma-dynamic lasers, in which a population inversion is created among atomic levels as a plasma expands, are much less developed. The idea of a plasma-dynamic laser was discussed in [63], where the possibility was studied of creating an inverted population due to recombination as a highly ionized plasma (field free or magnetized) expands adiabatically.

A population inversion in argon among the lower-lying levels of the 5p group relative to the levels of the 4p group and in helium between the n = 4 and n = 3 levels (n is the principal quantum number) was obtained experimentally in [14] in plasma as they expanded through a nozzle. Spectroscopic measurements [64] of an arc-heated mixture of helium and methane or argon and methane at high pressures as it expanded through a nozzle revealed a population inversion on the neutral carbon line corresponding to the transition $2p^{2\,1}S_0 \rightarrow 2p3s\,^1P_1^0$ (λ = 2478.6 Å). Besides these materials a population inversion has been observed in neon [65] and in hydrogen at low electron densities [66, 67].

Relaxation in an expanding plasma has been examined theoretically in [68-71]. Numerical calculations in [70, 71], respectively, have demonstrated the possibility of obtaining a population inversion on a neutral nitrogen line (λ = 1745 Å) and on xenon atomic lines (7p − 5d, 4*f* − 5d).

The significance of plasma lasers (compared with ordinary gas-dynamic ones) is primarily that it may be possible to extend the wavelength range into the visible and ultraviolet ranges. However, considerable difficulties arise in their development: objective difficulties associated with the speed of relaxation in atomic systems; and subjective, due to the difficulty and awkwardness of analyzing the creation of an inversion in plasma streams and, even more, in the choice of optimum conditions and materials for lasing.

We shall consider electron−ion recombination caused by cooling of the electrons as a plasma expands. It is assumed that the plasma is initially in a state of thermodynamic equilibrium with an electron temperature $T_e(0)$. Starting at some time the degree of ionization of the plasma begins to deviate from the equilibrium value. As the recombination flux moves over the discrete levels of the atom (or ion), a population inversion may be produced.

In general, to analyze the process of creating an inversion as a plasma expands, it is necessary to solve jointly the equations for the population relaxation, electron temperature, and gas dynamics. We shall assume that the characteristic time for changes in the electron temperature is considerably greater than the time to establish a quasistationary distribution, so there is a quasistationary distribution for the excited states of the atom [70].

In order to present a clearer account of this new technique for designing plasma-dynamic lasers we have avoided detailed discussions of a number of problems in gas dynamics and the recombination heating of electrons; rather, we have taken T_e and N to be given functions of time. Computing T_e and N is similar to computing the gas temperature and density in ordinary gas-dynamic lasers. Most consideration has been given to the quasistationary population distribution function method.

We shall examine the system of population balance equations for the discrete levels of an atom:

$$\frac{dN_{nl}}{dt} + N_{nl}\,\mathrm{div}\,\mathbf{v} = N_e \sum_{n'l'} V(n'l',\, nl)\, N_{n'l'} - N_e N_{nl} \sum_{n'l'} V(nl,\, n'l') + \\ + \sum_{E_{n'l'} > E_{nl}} A(n'l',\, nl)\, N_{n'l'} - \sum_{E_{n'l'} < E_{nl}} A(nl,\, n'l')\, N_{nl}, \quad \frac{dN}{dt} + N\,\mathrm{div}\,\mathbf{v} = 0, \tag{3.1}$$

where the previous notation is used for the populations of the energy levels and the probabilities of various elementary processes, N is the total density of heavy particles in the plasma, and **v** is the hydrodynamic velocity of the gas.

In the following we shall limit ourselves to examining a spatially homogeneous plasma. In view of this, the total derivative $d/dt = \partial/\partial t + (\mathbf{v}\nabla)$ in Eq. (3.1) equals the partial derivative with respect to time $\partial/\partial t$.

Normalizing the populations in Eq. (3.1) to the total density of heavy particles in the plasma,

$$N_{nl} = N g_{nl} \exp\left(-\frac{E_{nl}}{kT_e}\right) f_{nl} \tag{3.2}$$

(the energy of the ground state is taken as the zero energy reference level) and using the relation between the forward and reverse transitions, we obtain

$$\frac{df_{nl}}{dt} = N_e \sum_{n'l'} V(nl,\ n'l')(f_{n'l'} - f_{nl}) + \sum_{E_{n'l'} > E_{nl}} A(n'l',\ nl)\frac{f^0_{n'l'}}{f^0_{nl}} f_{n'l'} - \sum_{E_{n'l'} < E_{nl}} A(nl,\ n'l') f_{nl}; \tag{3.3}$$

$$f^0_{nl} = g_{nl} \exp\left(-\frac{E_{nl}}{kT_e}\right).$$

Equations (3.3) coincide with Eqs. (2.49) of Chapter II, where the distribution function was obtained for constant N and T_e; hence, in seeking a solution to this equation we shall use the results of that chapter.

A number of simple patterns in the variation of the population distribution function as the plasma expands are made clear in the example of an optically thin hydrogen plasma. The distribution function is given by Eqs. (2.35) and (2.36), in which

$$\varkappa = \frac{1}{N_e} \frac{d}{dt} \ln\left(\frac{N_1}{N}\right). \tag{3.4}$$

In this equation N_e and the population of the ground state are determined by conservation of charge and the number of heavy particles:

$$N = N_0 + N_i, \qquad \sum_{n=1}^{n_0} N_n = N_0, \qquad N_e = N_i. \tag{3.5}$$

We shall be considering recombination at small temperatures T_e such that

$$\varkappa \gg Q^n_1/\beta^1_n \qquad (n = 2,\ 3,\ \ldots). \tag{3.6}$$

This inequality is clearly true for $\varkappa \gg V(1, 2)$. Using Eq. (3.6) the expression for the population inversion between the n + 1 and n levels takes the form

$$\frac{N_{n+1}}{g_{n+1}} - \frac{N_n}{g_n} = \varkappa N_1 \left[\beta^1_{n+1} \exp\left(-\frac{E_{n+1}}{kT_e}\right) - \beta^1_n \exp\left(-\frac{E_n}{kT_e}\right)\right]. \tag{3.7}$$

This implies that a population inversion between levels 3 and 2 may be realized if

$$N_e < V^{-1}(3,\ 2)\left(\frac{g_2}{g_3} A(2, 1) - A(3, 2)\right). \tag{3.8}$$

For electron densities that are not very small (i.e., such that $N_e \gg A(m^*, m^* - 1)/V(m^*, m^* - 1)$), the location of the bottleneck $m = m^*$ in the energy spectrum and $\beta^1_{n_0}$ (as it is a function of T_e) are independent of N_e and are completely determined by the electron temperature.

For the populations of the levels lying below the bottleneck we can, as before, limit ourselves to the last term with m = n. Then, since at recombination temperatures $\sum_{k=1}^{m-1} g_k \exp(-E_k/kT_e) = 1$, we obtain an expression for the population inversion between neighboring levels of the hydrogen atom:

$$\frac{N_{n+1}}{g_{n+1}} - \frac{N_n}{g_n} = \varkappa N_1 \left[\frac{Q_n^{n+1}}{g_{n+1}V(n+1,\ n)} - \frac{Q_{n-1}^{n}}{g_nV(n,\ n-1)}\right]. \tag{3.9}$$

From this equation it is clear that population inversions can be realized in a hydrogen plasma only due to radiative depopulation of the lower laser level with

$$N_e < \frac{g_nA(n,\ n-1) - g_{n+1}A(n+1,\ n)}{g_{n+1}V(n+1,\ n) - g_nV(n,\ n-1)}. \tag{3.10}$$

Matching the distribution function with the Saha distribution for the upper levels and assuming that the density of neutral atoms N_0 is equal to the population N_1 of the ground state, we come to [using Eqs. (2.31), (3.4), and (3.5)] an equation for the degree of ionization of the plasma ($x = N_e/N$):

$$\frac{dx}{dt} = -bN^2x^3T_e^{-11/2}\ (\text{eV}), \qquad b = 2.4 \cdot 10^{-28}. \tag{3.11}$$

In the common case of constant N and T_e (free relaxation) integration of Eq. (3.11) yields

$$N_e^2(t) = \frac{N_e^2(0)}{1 + \dfrac{2bN_e^2(0)}{T_e^{11/2}}t}, \tag{3.12}$$

$$\gamma = -\frac{1}{N_e^2}\frac{dN_e}{dt} = bN_eT_e^{-11/2}\ (\text{eV}), \tag{3.13}$$

where $N_e(0)$ is the initial electron density.

Thus, in the stationary-sink regime the electron density remains constant for times such that

$$t \ll \frac{T_e^{11/2}(\text{eV})}{2bN_e^2(0)}.$$

The recombination coefficients calculated using Eq. (3.13) in this temperature (T_e) range for electron densities such that $10^{13} \le N_e \le 10^{16}$ cm^{-3} are, in all, only two or three times less than those given by the numerical calculations of [3]. In addition, Eq. (3.13) is exactly the same as the expression for the recombination rate found by numerical calculations of relaxation in a lithium plasma in [55].

We now assume that during expansion the density N of heavy particles in the plasma and the electron temperature T_e vary according to

$$\begin{aligned} N &= N(0)\,(t_0/t)^{\alpha} \qquad (t > t_0), \\ T_e &= T_e(0)\,(t_0/t)^{\beta} \qquad (\alpha,\ \beta > 0), \end{aligned} \tag{3.14}$$

where N(0) is the initial plasma density, $T_e(0)$ in the initial electron temperature, α and β are some given parameters, and t_0 is the time scale for changes in N and T_e.

Then, the result of integrating Eq. (3.11) is

$$x^2 = \frac{x_0^2}{1 + \frac{\xi}{\varepsilon}\left[(t/t_0)^{\varepsilon} - 1\right]}, \qquad \xi = \frac{2bt_0 N^2(0)\, x_0^2}{T_e^{11/2}(0)}, \tag{3.15}$$

where $\varepsilon = 11/2\beta - 2\alpha + 1 \neq 0$ and x_0 is the initial degree of ionization of the plasma.

It is clear from this that for $\varepsilon > 0$ the degree of ionization approaches zero as $t \to \infty$, while for $\varepsilon < 0$ the plasma expands to infinity still partially ionized (the freezing effect), with the residual ionization determined by the formula

$$x(\infty) = \frac{x_0}{\sqrt{1 + \xi/\varepsilon}}. \tag{3.16}$$

Clearly, the residual ionization increases with the initial electron temperature and decreases with increased initial plasma density or characteristic scale for changes in the density and electron temperature.

For spherical ($N \propto t^{-3}$), cylindrical ($N \propto t^{-2}$), and plane ($N \propto t^{-1}$) adiabatic expansion of a gas with the same electron and gas temperature ($T \propto N^{\gamma-1}$, where γ is the adiabatic index) into a vacuum, "freezing" will be observed for $\gamma < 1\frac{10}{33}$, $\gamma < 1\frac{3}{11}$, and $\gamma < 1\frac{2}{11}$ in the cases of spherical, cylindrical, and plane expansion, respectively.

Knowing the dependence of the degree of ionization on the time, it is possible to analyze the time variation of the population inversion, $\Delta N/g$, given by Eq. (3.9) as well. Let the plasma be in a state of thermodynamic equilibrium at the initial time with an adiabatic index $\gamma = 1.2$ and $\alpha = 2$ (cylindrical expansion); $\Delta N/g$ was computed in accordance with these initial data. Figure 4 shows the population inversion between levels 3 and 2 at time $t = 4.3t_0$ for $N_e = 5 \cdot 10^{14}$ cm^{-3}. The initial growth in the inverted population is due to more effective radiative decay of the lower laser level; later it falls due to the reduced density of the expanding plasma.

A study of the variation of the inverted population with $N(0)$ and $T_e(0)$ showed that it depends very strongly on these parameters of the plasma. If we increase $T_e(0)$ with $N(0)$ constant, then the initial degree of ionization of the plasma rapidly approaches its limiting value $x_0 = 1$, and the inverted population [as is clear from Eq. (3.9)] drops. On reducing $T_e(0)$ the inverted population will also fall since the initial degree of ionization is rapidly reduced. Thus, there is an optimum initial electron temperature at which the inverted population attains its maximum value.

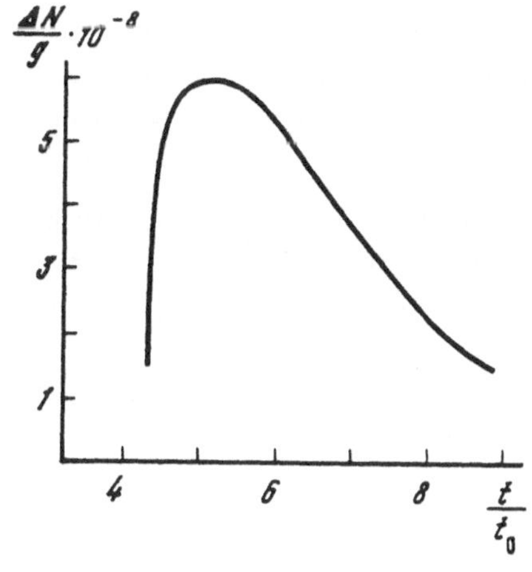

Fig. 4. The inverted population as a function of time during expansion of a hydrogen plasma. $T_e^{(0)} = 8000°K$; $N_0 = 2 \cdot 10^{19}$ cm^{-3}.

We note that use of hydrogen in plasma-dynamic lasers is not promising because of the upper limit on the electron density and the requirement that the optical thickness of the plasma be small.

As opposed to hydrogen, collisional depopulation of the lower working level may be possible in more complex atoms (ions) due to branching of the recombination flux. For definiteness we shall examine the energy level system of lithium. The single-quantum-approximation solution of the population balance equations (3.3) in the case of a dense lithium plasma, when radiative decay may be neglected and a Boltzmann distribution over l is assumed, is given by Eq. (2.46), in which one must set

$$\varkappa = \frac{1}{N_e}\frac{d}{dt}\ln\frac{N_{2s}}{N}, \tag{3.17}$$

$$\beta_n^1 = \sum_{m=2}^{n-1}\frac{1}{V^*(ms,\ m+1p)}\left(1+\sum_{i=3}^{m}\prod_{k=i}^{m}\frac{V(ks,\ k-1p)}{V^*(k-1s,\ kp)}\right), \tag{3.18}$$

where $V^*(ms,\ m+1p)$ are the effective ionization probabilities of the transition $ms \to (m+1)s$.

We shall examine recombination when $N_i = N_e$ for those times beginning with which the population distribution function (2.46) becomes strongly different from the Boltzmann distribution ($\beta^1\varkappa \gg 1$) and the population of the ground state of lithium N_{2s} substantially exceeds the populations of the excited states and is on the order of the neutral density ($N_{2s} = N_0 = N - N_e$); that is, we shall assume that the following inequalities are satisfied:

$$\frac{1}{\beta_n^1} \ll \varkappa \ll \frac{\exp(E_{ns}/kT_e)}{\beta_n^1}. \tag{3.19}$$

Matching the distribution function (2.46) with the Saha distribution for the upper levels ($n \geq n_0$), we obtain an equation for the degree of ionization in the form

$$\frac{dx}{dt} = -\frac{K(T_e)}{\beta_{n_0}^1}N^2x^3, \qquad K(T_e) = Z_1^{-1}\left(\frac{2\pi\hbar^2}{mkT_e}\right)^{3/2}\exp\left(\frac{I}{kT_e}\right), \tag{3.20}$$

where Z_1 is the statistical sum of the lithium ion and I is the ionization potential of lithium.

If the bottleneck in the spectrum of the energy levels, which determines the recombination rate, lies in the neighborhood of the hydrogenlike levels, then Eq. (3.11) for the degree of ionization, as well as Eq. (3.13), may be used in the case of recombination of any atomic plasma. The closer the energy spectrum of the atom is to hydrogenlike, the better Eq. (3.11) represents the recombination process. If the spectrum differs greatly from hydrogenlike, then the dependence of the sum (3.18) on T_e must be analyzed. As T_e is reduced, the bottleneck is shifted toward the region of hydrogenlike levels. Hence, for small T_e the conditions for applicability of Eq. (3.11) are improved.

We shall now study a population inversion in the 4s and 3p levels of a lithium plasma. Assuming that the probability for ionization of the ground state of the lithium atom $V^*(2s, 3p) \ll V(3s, 2p)$, we find from Eqs. (2.46), (3.18), and (3.19) that

$$\begin{gathered}\frac{\Delta N}{g} = \frac{N_{4s}}{g_{4s}} - \frac{N_{3p}}{g_{3p}} = \varkappa N_{2s}G(T_e),\\ G(T_e) = [3V(3p, 2s)]^{-1}\left[\frac{V(3s,\ 2p)}{3V(4p,\ 3s)} - \exp\left(-\frac{E_{3p}-E_{3s}}{kT_e}\right) + \exp\left(-\frac{E_{4s}-E_{3s}}{kT_e}\right)\right].\end{gathered} \tag{3.21}$$

The value of the function $G(T_e)$ depends on the specific energy spectrum of the plasma and the corresponding transition probabilities. Using the cross sections for inelastic collisions of atoms with electrons calculated in [62], it can be shown that at small recombination temperatures G is practically constant and the population inversion as a function of time and the initial electron temperature and density of the plasma is determined by the factor $\varkappa N_{2s}$.

We shall assume that the bottleneck in the energy-level spectrum of the lithium atom lies in the region of the hydrogenlike levels. Then the factor $\varkappa N_{2s}$ can be computed using Eqs. (3.17) and (3.11) to yield

$$\varkappa N_{2s}=\frac{\xi}{2t_0}f\left(\frac{t}{t_0}\right),\qquad f(x)=\frac{x^{\varepsilon-1}}{1+\xi^{\varepsilon-1}(x^{\varepsilon}-1)}. \tag{3.22}$$

An analysis of the behavior of the function $f(x)$ allows us to conclude that the population inversion will increase in time at the beginning of the expansion of the plasma if $\xi < \varepsilon - 1$. The time t^* at which the population inversion reaches its maximum values is evaluated using the formula

$$t^*=t_0[(\varepsilon-1)(\varepsilon-\xi)\xi^{-1}]^{1/\varepsilon}. \tag{3.23}$$

Then the maximum population inversion is

$$\left(\frac{\Delta N}{g}\right)_{\max}=\left(\frac{\varepsilon-1}{2t_0}\right)^{\frac{\varepsilon-1}{\varepsilon}}\left(\frac{\xi}{\varepsilon-\xi}\right)^{1/\varepsilon}G(T_e). \tag{3.24}$$

Thus, the greater the initial density of the plasma, the sooner the inverted population reaches its maximum and the larger that maximum is.

The time dependence of $\Delta N/g$ is shown in Fig. 5. The expansion of the plasma was modeled by a cylindrical (α = 2) expansion law with an adiabatic index γ = 16/11 and with N_0 = $4 \cdot 10^{14}$ cm^{-3}, $T_e(0)$ = 0.4 eV, and $t_0 = 2.5 \cdot 10^{-5}$ sec. The initial growth in $\Delta N/g$ is explained by an increase in the recombination flux to the discrete levels due to cooling of the electrons. The fall in the population inversion is caused both by a reduction in the recombination flux and by a reduction in the density of the plasma as it expands.

Examination of $(\Delta N/g)_{max}$ as a function of $T_e(0)$ leads to the discovery of optimum initial electron temperatures, $T_{e0}(0)$. The existence of $T_{e0}(0)$ may be explained by the fact that as $T_{e0}(0)$ increases, on the one hand, the initial degree of ionization of the plasma and, therefore, the electron density increase, and, on the other hand, the recombination flux which populates

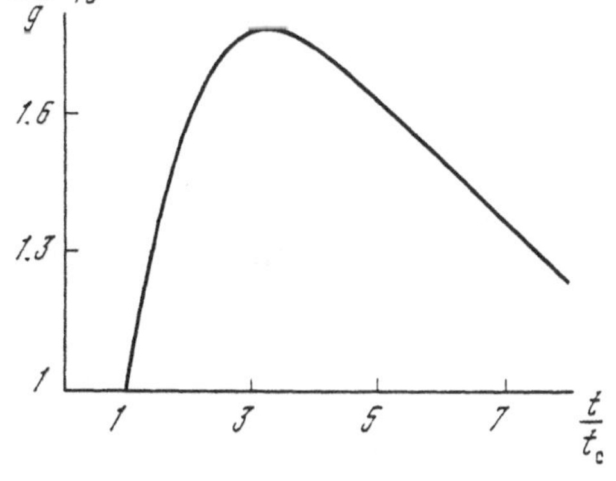

Fig. 5. The time dependence of the population inversion in an expanding lithium plasma.

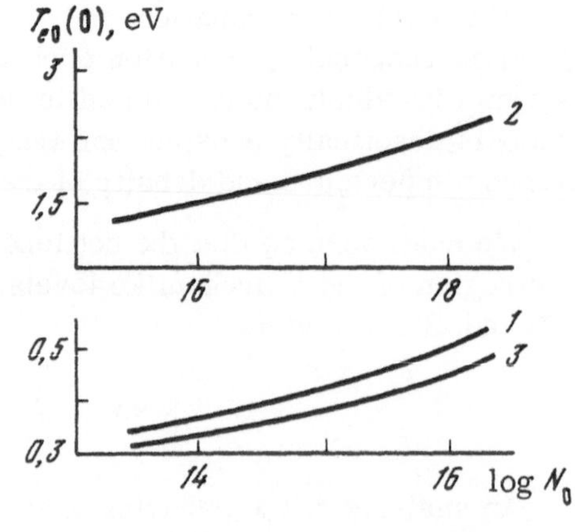

Fig. 6. The dependence of the optimal initial electron temperature on the density of a lithium plasma (curve 1), of a Be II plasma (curve 2), and of an orthohelium plasma (curve 3).

the upper discrete levels decreases. The optimal electron temperatures are found from the condition $\partial\xi/\partial T_e(0) = 0$. Since $11/4T_e(0) = \partial \ln x_0/\partial T_e(0)$, we have

$$\frac{2}{11}\left(\frac{I}{kT_e(0)} - \frac{5}{4}\right) = \frac{y}{1+2y-\sqrt{1+4y}}, \qquad y = K(T_e(0))\,N(0). \tag{3.25}$$

The results of a calculation of the optimal initial electron temperatures as a function of the initial density of a lithium plasma are shown in Fig. 6. It is clear that the $T_{e0}(0)$ lie in a relatively narrow interval of electron temperatures and increase with N_0.

By taking into account the dependence of $T_{e0}(0)$ on N_0 we have also obtained the inverted population $\Delta N/g$ as a function of N_0. This function is shown in Fig. 7 on a logarithmic scale where it is almost a straight line. Thus, as the initial density of the plasma is increased, the inverted population increases according to a power law $\Delta N/g \propto N_0^{\alpha}$, where α is the slope of this line (for lithium $\alpha = 0.8$), up to the limits of applicability of the theory as given by the inequality (3.19).

Analogous calculations may be done in the second ionization region of an atom if we assume that the plasma consists of singly and doubly charged ions.

Estimates for Be II show that for $N_0 = 3.6 \cdot 10^{16}$ cm^{-3}, $T_{e0}(0) = 1.6$ eV, and $t_0 = 10^{-5}$ sec, the population inversion over the 4s − 3p levels has a maximum equal to $4.6 \cdot 10^{11}$ cm^{-3} at $t^* = 2.6 \cdot 10^{-5}$ sec. The optimum electron temperatures as a function of the initial density for a beryllium plasma are shown in Fig. 6.

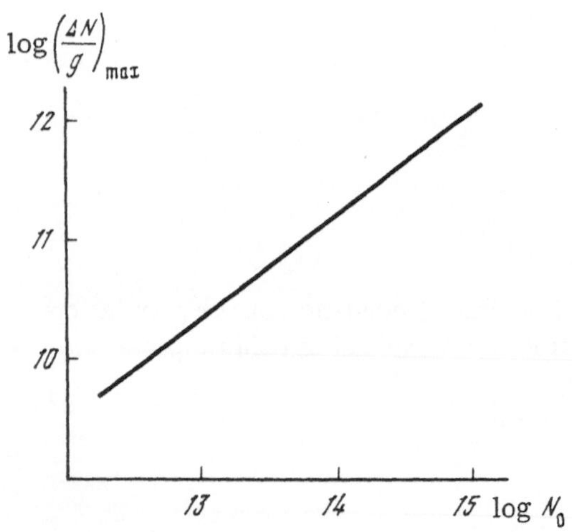

Fig. 7. The maximum population inversion $\Delta N/g$ as a function of the initial density (for lithium).

The optimum electron temperatures for He I (Fig. 6) were calculated assuming the metastable 2^3S state of orthohelium to be the ground state. For $N_0 = 2.2 \cdot 10^{14}$ cm^{-3}, $T_e(0) = 0.35$ eV, and $t_0 = 10^{-5}$ sec, the maximum population inversion between the 4^3S and 3^3P states is $(\Delta N/g)_{max} \simeq 7 \cdot 10^{11}$ cm^{-3} which occurs at the time $t^* = 6 \cdot 10^{-5}$ sec.

An analysis of the population distribution function for parahelium showed that no population inversion is observed over the $3^1S - 2^1P$ and $4^1S - 3^1P$ transitions. This is due to the presence in parahelium of the highly excited metastable 2^1S state. During recombination the 2^1S state becomes overpopulated and thereby increases the populations of the lower laser levels, 2^1P and 3^1P.

If the plasma expands through an aperture (slit, nozzle) of characteristic dimension a, then the time scale t_0 for changes in the density and electron temperature is on the order of a/v, where v is the escape velocity of the plasma from the aperture. Then the distance from the aperture at which the inverted population reaches a maximum is roughly $a(t^*/t_0)$; that is, the greater the escape velocity of the plasma from the aperture, the greater the inverted population and the earlier it reaches its maximum value.

Therefore, from the examples of lithium, beryllium, and helium plasmas it is clear that a collisional mechanism can be realized and used for lasing in large plasma volumes. This is an indication of the prospects for plasma-dynamic lasers. Using the method developed here, it is possible to analyze further the optimum regimes for plasma streams and to design specific systems for various plasma compositions.

2. Pinch Discharge Lasers

Lasing on ions in a pinch discharge has been obtained and studied by a number of authors [13, 72-78]. A detailed review of the experimental results has been written by Sutovskii [13]. Lasers using pinches yield substantial pulsed powers and make it possible to move into the short-wavelength region. To estimate the prospects for and feasibility of such lasers, as well as to improve and optimize them, one must study the kinetics of the processes responsible for lasing. However, up to this time this group of problems has not been worked out; even the very mechanism for producing the population inversion is not completely explained.

A complete analysis of the relaxation phenomena in a pinch discharge is rather involved. However, by making a number of assumptions the problem can be reduced to an analytic method for finding the population distribution function of the discrete levels. Having constructed the appropriate distribution function, which is determined by the plasma parameters, and knowing how these parameters vary, it is possible to derive the experimental characteristics of the laser.

The physical reason for the formation of an inverted population [13] is related to the fact that the probability for radiative decay of the lower laser level is large compared to the analogous quantity for the upper level. This leads (for example, in argon lasers) to overpopulation of the upper laser level of Ar II during electron impact ionization.

We shall examine the population kinetics of the discrete levels of Ar II during a rapid rise in the plasma density while the electron temperature T_e varies slowly compared to the relaxation time. We shall assume the plasma to be optically thin and spatially uniform.

The system of balance equations for the reduced dimensionless populations has the form (3.3). To solve it we shall use the single-quantum approximation including the optically allowed transitions and assuming a Boltzmann distribution between three neighboring levels ($f_{nl-1} = f_{nl} = f_{nl+1}$). Radiative decay will be taken into account as follows:

$$\sum_{E_{n'l'} < E_{nl}} A(nl,\ n'l') = A(nl,\ n-1l-1) + A(nl,\ n-1l+1) = A(nl),$$

$$\left.\begin{aligned}\sum_{E_{n'l'}>E_{nl}} A(n'l',\ nl)\frac{f^0_{n'l'}}{f^0_{nl}} f_{n'l'} = A^*(n+1l) f_{n+1l},\\ A^*(n+1l) = \frac{f^0_{n+1l}}{f^0_{nl}} A(n+1l).\end{aligned}\right\} \tag{3.26}$$

Then Eq. (3.3) breaks down into a system of level population balance equations with specific values of the orbital quantum number l.

Introducing the effective transition probabilities

$$\begin{aligned} R_{n,\,n+1}(l) = V(nl,\ n+1l-1) + V(nl,\ n+1l+1),\\ R_{n+1,\,n}(l) = V(n+1l,\ nl-1) + V(n+1l,\ nl+1)\end{aligned} \tag{3.27}$$

and assuming that the principle of detailed balance holds for $R_{n,n+1}(l)$ and $R_{n+1,n}(l)$ we find that the system of level population balance equations with that value of l coincides with Eq. (2.32). Thus, for Ar II, where the principal quantum number takes the values n = 3, 4, ..., we may write

$$N_{nl} = N_{3p} g_{nl} \exp\left(-\frac{E_{nl}}{kT_e}\right)[Q^n_3(l) + \beta^1_{nl}\varkappa], \qquad l = s,\ p,\ d, \tag{3.28}$$

$$Q^n_m(l) = \prod_{k=m}^{n-1}\left[1 + \frac{A(k+1l)}{N_e R_{k+1,\,k}(l)}\right]^{-1}, \tag{3.29}$$

$$\beta^1_{nl} = \sum_{m=4}^{n} \frac{Q^n_{m-1}(l)}{f^0_{ml} R_{m,\,m-1}(l)} \sum_{k=3}^{m-1} f^0_{kl}, \tag{3.30}$$

$$\varkappa = \frac{1}{N_e}\frac{d}{dt}\ln\frac{N_{3p}}{N}. \tag{3.31}$$

We now consider the expression for the population inversion between the 4p and 4s configurations when $R_{3,4}(s) = R_{3,4}(p)$:

$$\frac{\Delta N}{g} = \frac{N_{4p}}{g_{4p}} - \frac{N_{4s}}{g_{4s}} = N_{3p}\left[1 + \frac{\varkappa}{R_{3,\,4}(p)}\right]\left[\exp\left(-\frac{E_{4p}}{kT_e}\right)Q^4_3(p) - \exp\left(-\frac{E_{4s}}{kT_e}\right)Q^4_3(s)\right]. \tag{3.32}$$

From this equation it is clear that for purely collisional relaxation no inversion is observed between the 4p and 4s levels.

A definitive expression for the distribution function and the population inversion can be obtained when the specific time dependences of the ground state population of Ar II and the electron density of the plasma are known. We shall assume that during ionization of Ar II the populations of the upper levels are small compared to those of the lower levels. Then the parameter $\varkappa$ is determined by equating the population of some discrete level n_0p to zero, i.e.,

$$\varkappa^{-1} = -\sum_{m=4}^{n_0} \frac{1}{f^0_{mp} R_{m,\,m-1}(p)\, Q^{m-1}_3(p)} \sum_{k=3}^{m-1} f^0_{kp}. \tag{3.33}$$

Let a plasma consist of Ar II and Ar III at densities N_1 and N_2, respectively. Then, using the charge-conservation condition and the expression for the total density of heavy particles in the plasma, we have

$$N_e = N_1 + 2N_2, \qquad N = N_1 + N_2 \tag{3.34}$$

and assuming that $N_{3p} = N_1$, Eqs. (3.31) and (3.34) are easily integrated together. As a result we obtain

$$N_{3p} = 2\eta N\left[\eta + \exp\left(2\int_0^t |\varkappa| N(t')dt'\right)\right]^{-1}, \tag{3.35}$$

$$N_e = 2N\left[1 + \eta\exp\left(-2\int_0^t |\varkappa| N(t')\,dt'\right)\right]^{-1}, \tag{3.36}$$

where $\eta = N_1(0)/N_2(0)$; and $N_1(0)$ and $N_2(0)$ are the initial densities of Ar II and Ar III.

The population inversion produced in a pinch discharge depends on the plasma parameters N_e, T_e, and N, and the initial parameters as shown in Eqs. (3.31)-(3.32). In order to determine these parameters and their time variations the dynamics of a pinch discharge must be known. The distinctive features of such a discharge are high values of the discharge current $i \sim 10^3$-10^6 A and high rates of current rise $di/dt \sim 10^{10}$-10^{12} A/sec.

The available experimental data permit us to divide the process of formation of a pinch discharge into three stages. During the electrical breakdown state the electron temperature and conductivity of the plasma rise and the magnetic field is forced out of the central portion of the discharge (a skin layer is formed). During the compression state the plasma column is squeezed toward the axis of the discharge due to the interaction of the discharge current with its intrinsic magnetic field. Then the electron temperature may be regarded as practically unchanged. Finally, the last stage of a pinch discharge is characterized by thermalization of the ion component of the plasma and equilibration of the electron and ion temperatures. In this stage the electron temperature rises rapidly.

In general the compression dynamics of the plasma column are described by a complicated system of magnetohydrodynamic equations. To explain the basic phenomena associated with creation of a population inversion we shall use the simple model of plasma compression proposed in [79-80]. In this model, when a definite skin effect is present, the cylindrical gas volume is compressed by the skin layer. An expression was obtained in [79] for the time to compress the plasma column with a linearly rising current $i = t(di/dt)$:

$$t_1 = a_0 (MN_0)^{1/4} c^{1/2} \frac{di}{dt}^{-1/2}, \tag{3.37}$$

where a_0 is the initial radius of the gaseous discharge chamber, N_0 is the initial density of the gas, M is the mass of a heavy gas particle, c is the speed of light, and di/dt is the rate of rise of the discharge current.

The compression of a plasma column in the absence of a skin effect for the magnetic field was studied in [81], where it was shown that for a power-law current rise $i = i^{(n)} t^n$, the compression time is given by

$$t_1 \approx \left[\frac{ca_0^2 \sqrt{MN_0}}{i^{(n)}}\right]^{1/n+1}. \tag{3.38}$$

We shall study a population inversion between the 4p and 4s levels as the plasma column is compressed, when the electron temperature depends weakly on the time and may be regarded as practically constant. We note that, as a rule, $R_{3,4}(p) \ll R_{4,5}(p)$ and the principal contribution to the sum in Eq. (3.33) is from the first term. Limiting ourselves to the first two terms in Eq. (3.33), we have

$$\begin{gathered}\frac{\Delta N}{g} = N_{3p}\left[1 - \frac{Q_3^4(s)}{Q_3^4(p)}\exp\frac{\Delta E}{kT_e}\right]\frac{R_{3,4}(p)}{R_{4,5}(p)}\left[1 + \exp\left(-\frac{E_{4p}}{kT_e}\right)\right],\\ \Delta E = E_{4p} - E_{4s}.\end{gathered} \tag{3.39}$$

As can be seen from this equation the population inversion is completely determined by the electron temperature at small electron densities. At high electron densities the inversion [82] between the 4p and 4s levels may disappear since relaxation becomes collisional. The critical values of the electron density are found by setting Eq. (3.39) equal to zero to yield

$$N_{e\,\max}=\left[\frac{A\,(4s)}{R_{4,\,3}\,(s)}-\frac{A\,(4p)}{R_{4,\,3}\,(p)}\exp\frac{\Delta E}{kT_e}\right]\left[\exp\left(\frac{\Delta E}{kT_e}\right)-1\right]^{-1}. \tag{3.40}$$

For $T_e = 4.3$ eV $= 5 \cdot 10^{4}$°K and the radiative decay probabilities for these levels given in [57], the value of the limiting electron density was found to be $5 \cdot 10^{16}$ cm^{-3}. Here and in the following calculations we have used the probabilities for inelastic collisions of Ar II with electrons given in [83].

For electron densities $N_e \ll A(4s)/R_{4,3}(s)$, the population inversion is completely determined by the population of the upper laser level.

From this it is clear that in this case the fundamental pattern of producing an inversion has much in common [57, 82-83] with the usual argon ion laser. The excited levels of Ar II are populated by electron impact. The lower laser level is depopulated by radiative decay. The greater the electron temperature and ground state population N_{3p}, the greater the ionization flux to upper levels from the ground state and the greater the populations of the excited levels. For electron densities that are not very large and when the radiative flux which depopulates the upper excited levels is comparable in magnitude to the ionization flux, negative temperature states are possible in a plasma of singly and multiply ionized argon atoms.

In the following we shall present the results of a calculation of the population inversion as a function of time, the electron temperature, and the initial gas densities. In order to qualitatively describe the time variation of the inverted population we shall consider a plasma whose density varies in time as

$$N=N_0\left(\frac{N_m}{N_0}\right)^{t/t_1}, \tag{3.41}$$

where N_m is the plasma density at the time of maximum compression of the plasma column.

Figure 8 shows the time dependence of the inverted population for $T_e = 4.3$ eV, $N_0 = 3 \cdot 10^{14}$ cm^{-3}, $N_m = 4 \cdot 10^{15}$ cm^{-3}, $a_0 = 10$ mm, di/dt $= 10^{10}$ A/sec, and $|\varkappa| = R_{3,4}(p)$ = const. The initial growth in the inverted population is explained by the increase in the plasma density due to compression of the current layer toward the center. Its later fall is due to further ionization of singly ionized argon. The location of the maximum of the curve depends on the ratio of the compression time to the characteristic ionization time of Ar II. The most favorable conditions for lasing are obtained when during the time the Ar II is compressed it does not be-

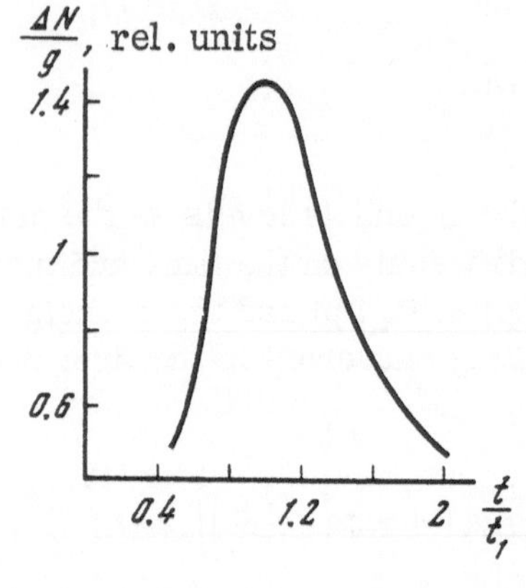

Fig. 8. Time variation of the population inversion in Ar II.

come strongly ionized, that is, for small compression times or large rates of rise of the current. By studying the population of the ground state it is possible to show that the population inversion is a maximum when $t = t_1$ if

$$N^{-1}\frac{dN}{dt} > 2|\varkappa|N. \tag{3.42}$$

In fact, for $t > t_1$ a buildup (cumulation) stage begins during which the electron temperature rises rapidly due to transfer of thermal energy from the ions to the electrons. In the end this leads to accelerated ionization of Ar II and cessation of lasing. In addition, lasing may be cut off during the rapid rise in the ion temperature (which increases the Doppler broadening). Thus, lasing on Ar II cuts off rapidly at the moment of buildup, but may take place on more highly ionized ions.

Figure 9 shows the maximum population inversion at $t = t_1$ as a function of T_e [it is assumed that the condition (3.42) is satisfied] for the device parameters listed previously. At small T_e the probabilities of excitation of the upper laser level $R_{3,4}(p)$ increase more rapidly than the probabilities of its being emptied while, at the same time, they remain small in magnitude (i.e., they cause practically no change in the ground state population). This is the reason for the growth in the population inversion at small T_e. At large T_e the fall in $\Delta N/g$ is explained by ionization of Ar II and a reduction in the population of the ground state.

With increasing initial gas densities there is, on the one hand, an increase in the number of active lasing centers and, on the other, more rapid ionization of Ar II, which reduces the population of the ground state; thus, the dependence of $\Delta N/g$ on N_0 has an optimum.

Analyzing Eq. (3.35) for an extremum at $t = t_1$ while assuming that the plasma density N_m at the moment of maximum compression is a function of the initial gas densities with the form

$$N_m = BN_0^{\alpha}, \quad B = B(\alpha, a_0, T_e, \ldots), \tag{3.43}$$

and also assuming that $\int_0^{t_1} N dt = N_m t_1$, we obtain a formula for the initial gas densities at which the population of the ground state is a maximum. With these assumptions, we also find the population inversion to have a maximum as well at

$$(N_0)_{opt} = \frac{z_0 (di/dt)^{1/2}}{2|\varkappa| B a_0 M^{1/4} c^{1/2}}, \tag{3.44}$$

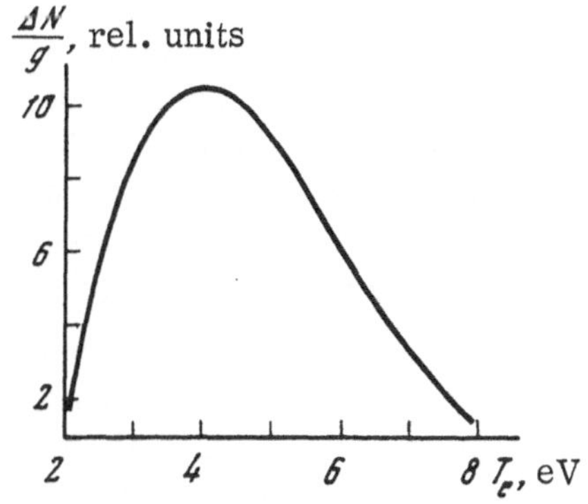

Fig. 9. The population inversion as a function of the electron temperature.

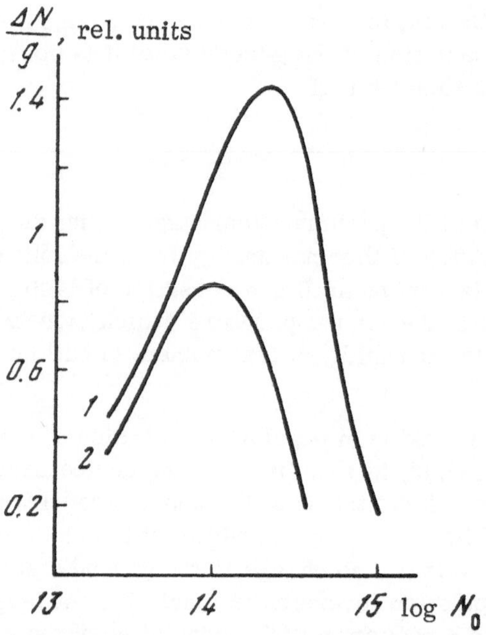

Fig. 10. Dependence of the population inversion on the initial gas density. Curves 1 and 2 refer to gaseous discharge tubes of radius 10 and 20 mm, respectively.

where z_0 is the root of the equation $1 + \exp(-z) = [(4\alpha + 1)/4\alpha]z$. Thus for $\alpha = 3/4$ and the parameters of the apparatus in [13], $(N_0)_{opt} = 2 \cdot 10^{14}$ cm^{-3} ($P_0 \simeq 10^{-2}$ torr). The value of the constant B was found from the condition that for $N_0 = 3 \cdot 10^{14}$ cm^{-3}, $N_m = 4 \cdot 10^{15}$ cm^{-3}.

In Fig. 10, $\Delta N/g$ is plotted as a function of N_0 for two values of the radius of the discharge tube and $\alpha = 3/4$ when the constant B is independent of a_0. As the radius of the tube is increased the optimum initial density decreases while the maximum population inversion also decreases, in agreement with experimental results. We note that these curves are valid for small initial densities obeying Eq. (3.42). If this condition is not fulfilled, then it is difficult to study the dependence of the maximum inversion on a_0 because of the need to determine the time at which $\Delta N/g$ has a maximum, and because the constant B does not depend on a_0.

When the rate of rise of the current is increased the population inversion goes up but the duration of lasing decreases. However, di/dt cannot be increased without limit since as the rate of rise of the current is increased the plasma density N_m becomes so large that relaxation becomes purely collisional.

The mechanism for lasing in the case of ions of higher charge state during the compression stage of pinch discharges will evidently be the same as in the case of Ar II. On going to more highly charged ions, the probabilities for inelastic collisions of ions with electrons become smaller and the optimum initial densities become larger. The growth in the radiative probabilities increases the negative absorption coefficient, and the limiting value of the electron density at which relaxation is no longer purely collisional is also increased.

The basic approaches to increasing the efficiency of pinch discharge lasers are use of higher ionization states for lasing (preionization is then necessary), increasing the repetition rate (limited due to overheating of the system), and, chiefly, studying possible collisional mechanisms for emptying the lower working level.

CHAPTER IV

SOME PROBLEMS IN THE KINETICS OF PLASMAS WITH MULTIPLY CHARGED IONS

Processes in a "hot" plasma are considered. The equations of magnetohydrodynamics including the radiant flux are employed as the basis of a study of the use of a high-current self-constricted discharge with a return current in optically opaque and semitransparent plasmas as a powerful light source. The theory of self-constricted high-current ac discharges is presented. A quasistationary distribution function for the multiplicity of ionization in a plasma with multiply charged ions is proposed.

1. The Physical Processes in a "Hot" Plasma and Their Uses

Recently there has been increasing interest in research on high-temperature plasmas, in particular, multiply charged ion plasmas. To produce them requires the release of large concentrations of energy over short periods of time. There are three basic ways of injecting large amounts of energy into materials: lasers, electron beams, and high-current electrical discharges. At present, solid state lasers have powers of up to 10^{12} W and can deliver radiant energy fluxes of up to about 10^{16} W/cm^2, producing electric fields of up to 10^7 V/cm and temperatures of about 10^{6}°K. Developments in pulsed high-voltage technology make it possible to deliver powers of up to 10^{14} W with an energy flux of about 10^{13} W/cm^2 using an electron beam.

Kinetic theory is quite inadequately developed for dealing with the multiply charged ion media which are produced when these amounts of energy are deposited in matter. The relaxation pattern in a multiply charged ion plasma is very different from that in a low-temperature plasma (cf. [7]). The roles of both radiative transitions in ions and electron bremsstrahlung are increased (in electron beams up to 10% of the energy may go into x-ray bremsstrahlung). Charge exchange and the Auger effect accompanying expansion of the plasma are very important. At present we do not have a composite picture of the relaxation processes in a multiply charged ion plasma; even the foundations of the kinetics are clearly not established. Thus, the discussion in this chapter will have a fragmented character.

We note that the nonequilibrium multiply charged ion medium created by the introduction of large amounts of energy is of interest from the standpoint of three problems: thermonuclear fusion, x-ray lasers, and powerful radiation sources. The results of research on the first problem are summarized in [84, 85], and on the second, in [7, 86]. Here our attention will be focused on the problem of creating high-intensity radiation sources with the aid of high-current electrical discharges.

We shall examine high-current self-constricting discharges in a dense plasma ($N \sim 10^{18}$-10^{20} cm^{-3}) at relatively low temperatures ($T \sim 3$-10 eV) when radiation plays an important role in the development of the discharge. Such discharges are studied in connection with the problem of making powerful light sources for "pumping" lasers. Then the choice of discharge parameters is dictated by the parameters of the required output radiation and duration of illumination. The basic requirement of such discharges is that the radiation be close to that from a "black" body in a given wavelength region and that the discharge be transparent to shorter wavelength radiation. This circumstance indicates the appropriateness of studying the equilibrium and stability of high-current pinch discharges in both optically opaque and transparent plasmas.

Theoretical [87-97] and experimental [98-104] studies of simple cylindrical discharges (z-pinches) and of discharges with return current carriers (such a system consists of a coaxial cylindrical current-carrying layer and a massive metal conductor on the axis along which a current flows in the opposite direction) have shown that such discharges have high emissivities. It has been shown that discharges in optically opaque and semitransparent plasmas are subject to pinch and kink type instabilities [104], and that discharges in optically transparent plasmas are also subject to an overheating (thermal) instability [101]. The physical reason for the development of a thermal instability in a transparent discharge is the inability of the volume emission to compensate for the growth in temperature fluctuations produced by ohmic heating of the plasma. The development of instabilities causes a reduction in the duration of the discharge, makes its structure highly nonuniform, and impairs its quality as a radiation source.

The equilibrium and stability of a high-current pinch discharge when radiation plays a controlling role in the energy balance was analyzed in the framework of a single-fluid magnetohydrodynamic model including radiative transport in [89-94]. With these parameters the plasma could be regarded as a completely ionized ideal gas with an effective charge z and isotropic transport coefficients. The system of MHD equations, in which only the flux of radiant energy was included, while the energy of the radiation was neglected in comparison to the thermal energy of the particles and the thermal conductivity of the plasma and the viscous terms were neglected in comparison to radiative energy transfer, has the form

$$\left.\begin{array}{c}
\operatorname{div}\mathbf{B}=0, \quad \operatorname{rot}\mathbf{B}=\frac{4\pi}{c}\mathbf{j}=\frac{4\pi}{c}\sigma\left\{\mathbf{E}+\frac{1}{c}[\mathbf{vB}]\right\},\\
-c\operatorname{rot}\mathbf{E}=\frac{\partial\mathbf{B}}{\partial t}=\operatorname{rot}[\mathbf{vB}]-\frac{c^2}{4\pi}\operatorname{rot}\left(\frac{1}{\sigma}\operatorname{rot}\mathbf{B}\right),\\
\rho\left[\frac{\partial\mathbf{v}}{\partial t}+(\mathbf{v}\nabla)\mathbf{v}\right]=-\nabla P+\frac{1}{4\pi}[\operatorname{rot}\mathbf{B}\cdot\mathbf{B}],\\
\frac{\partial\rho}{\partial t}+\operatorname{div}\rho\mathbf{v}=0,\\
\frac{\partial}{\partial t}\left(\frac{\rho v^2}{2}+\rho\varepsilon\right)+\operatorname{div}\left\{\rho\mathbf{v}\left(\frac{v^2}{2}+\varepsilon+\frac{P}{\rho}\right)\right\}=\mathbf{jE}-\operatorname{div}\mathbf{S},\\
\varepsilon=\frac{3}{2}\frac{P}{\rho}=\frac{3}{2}v_s^2, \quad P=\frac{(1+z)kT}{M}\rho=\rho v_s^2, \quad \rho=MN,\\
\sigma=\frac{\alpha}{z}T^{3/2}, \quad \alpha=4\cdot 10^7,
\end{array}\right\} \tag{4.1}$$

where **E** and **B** are the electric and magnetic (induction) field strengths, **v** is the hydrodynamic velocity, σ is the electrical conductivity for a completely ionized plasma, P is the pressure, ε is the specific energy, v_s is the speed of (isothermal) sound, ρ is the density of the plasma, M is the mass of an ion, T is the temperature of the plasma (electrons and ions), and S is the radiant energy flux.

Very often the energy equation for this system is written in the entropy form,

$$\rho T\left[\frac{\partial s}{\partial t}+(\mathbf{v}\nabla)s\right]=\frac{j^2}{\sigma}-\operatorname{div}\mathbf{S}, \tag{4.2}$$

$$s=\frac{(1+z)k}{M}\ln\frac{T^{3/2}}{\rho}, \tag{4.3}$$

where s is the entropy of an ideal gas.

The system of equations (4.1) is supplemented by the radiative transport equation,

$$\mathbf{\Omega}\nabla I_\nu=\varkappa_\nu'(I_{\nu\mathrm{eq}}-I_\nu), \tag{4.4}$$

which determines the flux,

$$\mathbf{S}=\int_0^\infty \mathbf{S}_\nu d\nu=\int_0^\infty d\nu \int d\mathbf{\Omega\Omega} I_\nu. \tag{4.5}$$

Here Ω is the unit vector in the direction of propagation of the photons, I_ν is the spectral emission intensity, and $I_{\nu eq}$ and $\varkappa'_\nu$ are the equilibrium spectral intensity and the absorption coefficient, both including the effect of "re-emission", i.e.,

$$I_{\nu eq}=\frac{2h\nu^3}{c^2}\frac{1}{\exp(h\nu/kT)-1}; \tag{4.6}$$

$$\varkappa'_\nu=\varkappa_\nu[1-\exp(-h\nu/kT)]. \tag{4.7}$$

The absorption coefficient is determined by the specific mechanism for absorption of light in the discharge. In the case of the (inverse) bremsstrahlung mechanism [89],

$$\varkappa_\nu=4.1\cdot 10^{-23}\frac{z^3N^2}{T^{7/2}}\left(\frac{kT}{h\nu}\right)^3. \tag{4.8}$$

For multiply ionized atoms [89],

$$\varkappa_\nu=10^{-7}T^{-2}\sum_m (m+1)^2 N_m \exp\left(-\frac{I_m}{kT}\right)F_m\left(\frac{h\nu}{kT}\right), \tag{4.9}$$

where

$$F_m(x)=\begin{cases} x^{-3}\exp(x), & x<I_m/kT,\\ \frac{2I_m}{kT}x^{-3}\exp\left(\frac{I_m}{kT}\right), & x>I_m/kT,\end{cases}$$

and where N_m is the distribution function over the charge of the ions and I_m is the ionization potential of an m-fold ion.

In an optically opaque plasma the intensity of the emission is close to equilibrium, that is,

$$I_\nu=I_{\nu eq}+\Delta I_\nu. \tag{4.10}$$

Substituting Eq. (4.10) in Eq. (4.4), we obtain the radiative heat transfer approximation,

$$I_\nu=I_{\nu eq}-\frac{\Omega}{\varkappa'_\nu}\nabla\left(I_{\nu eq}-\frac{\Omega}{\varkappa'_\nu}\nabla I_{\nu eq}\right); \tag{4.11}$$

$$\mathbf{S}=-\frac{16}{3}\sigma^* T^3\nabla T, \tag{4.12}$$

where σ^* is the Stefan–Boltzmann constant and $l = l(\rho, T)$ is the Rosseland mean free path of the photons.

In an optically transparent plasma the mean free path of the photons is greater than the size of the discharge and $I_\nu \ll I_{\nu eq}$. In this case

$$\operatorname{div}\mathbf{S}=\int_0^\infty d\nu\int d\Omega\varkappa'_\nu I_{\nu eq}=q_s(\rho, T). \tag{4.13}$$

The quantities l and q_s depend on the specific absorption mechanism for light in the plasma; the corresponding values are given in [92].

We now proceed to a specific analysis using the physical picture of the basic processes in a pinch discharge developed above.

2. An Analysis of High-Current Discharges with a Return Current

We shall discuss a return current discharge in a dense optically opaque and semitransparent plasma on the basis of the system of equations (4.1) and establish the relationship between the required discharge parameters (temperature, characteristic plasma dimensions) and the external parameters (total number of particles in the discharge, discharge current). We shall also establish the range over which the discharge exists. The study of such discharges is important since they have considerably more stability than simple cylindrical discharges and can serve as efficient radiation sources.

We are interested mainly in the stationary equilibrium characteristics of the discharge since information on the stability of these discharges can be obtained by a trivial generalization of the results of an analysis of an opaque discharge [89, 94]. Thus, we rewrite Eqs. (4.1) in a cylindrical coordinate system as follows:

$$\left.\begin{aligned} &\frac{1}{r}\frac{\partial}{\partial r}(rB)=\frac{4\pi}{c}j=\frac{4\pi}{c}\sigma E; \quad \frac{\partial P}{\partial r}+\frac{1}{4\pi r}B\frac{\partial}{\partial r}(rB)=0,\\ &P=\frac{(1+z)\,k}{M}\rho T=(1+z)\,kNT, \quad \sigma=\frac{\alpha}{z}T^{3/2}, \quad \alpha=4\cdot 10^{7},\\ &\sigma E^{2}=\frac{j^{2}}{\sigma}=q_{s}(N,\,T), \end{aligned}\right\} \tag{4.14}$$

where the corresponding quantities refer to the stationary equilibrium ($\mathbf{v}$ = 0) state of the discharge. Here τ is the conductivity of the plasma, ρ is the density, M is the ion mass, and z is the average ionic charge. The emissivity q_s of the plasma, as in [94], is defined in the following way:

$$q_{s}=\operatorname{div}\mathbf{S}_{1}+\operatorname{div}\mathbf{S}_{2}, \tag{4.15}$$

where $\mathbf{S}_1$ is the radiant flux of long-wavelength photons with frequencies $\nu < \nu_0$ for which the medium is completely opaque and $\mathbf{S}_2$ is the radiant flux of short-wavelength photons for which the medium is transparent. The boundary frequency ν_0 is chosen from the condition that the optical thickness of the discharge τ is equal to unity, i.e.,

$$\tau=\varkappa_{\nu}'(\nu_{0})\,x_{\text{eq}}=1, \tag{4.16}$$

where x_{eq} is the equilibrium thickness of the plasma layer.

For concreteness we shall assume bremsstrahlung absorption in the plasma. For it we can write an exact expression for $\varkappa_\nu'$ [89]:

$$\varkappa_{\nu}'=\frac{4.1\cdot 10^{-23}z^{3}N^{2}}{T^{7/2}}\cdot\frac{1-\exp(-x)}{x^{3}} \qquad (x=h\nu/kT). \tag{4.17}$$

Then for $\mathbf{S}_1$ and $\mathbf{S}_2$ we obtain [94]

$$\mathbf{S}_{1}=\int_{0}^{\nu_{0}}\mathbf{S}_{\nu}d\nu=-\frac{4\pi}{3}\frac{kT}{h}\int_{0}^{x_{0}}\frac{\nabla I_{\nu\text{eq}}}{\varkappa_{\nu}'}\,dx=-\beta_{0}\frac{T^{13/2}f(x_{0})}{N^{2}z^{3}}\nabla T, \tag{4.18}$$

$$\operatorname{div} \mathbf{S}_2 = \frac{1}{r}\frac{\partial}{\partial r}(rS_2) = \int_{\nu_0}^{\infty} d\nu \int d\Omega \varkappa_\nu' I_{\nu_{eq}} = \gamma_0 N^2 T^{1/2} z^3 \exp(-x_0);$$

where

$$I_{\nu_{eq}} = \frac{2(kT)^3}{c^2h^3} \cdot \frac{x^3}{\exp(x) - 1}, \qquad f(x_0) = \int_0^{x_0} \frac{\exp(2x)\, x^7 dx}{[\exp(x) - 1]^3}, \tag{4.19}$$

$$\beta_0 = 2.8 \cdot 10^{17}, \qquad \gamma_0 = 1.4 \cdot 10^{-27}.$$

First we shall consider some characteristic features of a pinch with a return current conductor in an optically opaque plasma. The limit of an opaque discharge corresponds to $\nu_0 \to \infty$ and $S_2 \to 0$. Such a discharge was examined in [89] where general expressions were obtained for the distributions of the basic quantities over the discharge. It is easy to show that in the case of a quasiplane discharge, i.e., one with

$$r = R_0 + x, \quad |x| \ll R_0$$

(R_0 is the mean radius of the plasma shell), all the quantities that characterize the discharge can be expressed in terms of the discharge current i, the total number of particles per unit length of the discharge N_0, and the radius R_0. Specifically,

$$B(x) = \frac{4\pi}{c}\sigma E = \sqrt{8\pi P(0)}\frac{x}{x_{eq}},$$
$$P(x) = P(0)\left(1 - \frac{x^2}{x_{eq}^2}\right), \qquad \rho(x) = \rho(0)\left(1 - \frac{x^2}{x_{eq}^2}\right), \tag{4.20}$$

and the temperature, which is practically uniform over the cross section due to the high radiative heat conductivity, is given by

$$T(x) \simeq T(0) = \frac{i^2}{3(1+z)kc^2N_0}\frac{x_{eq}}{R_0}. \tag{4.21}$$

Here P(0), $\rho(0)$, and N(0) are the values of the corresponding quantities at the center of the plasma layer. In addition,

$$T(0) \simeq 2.07 \cdot 10^{-2}\left(\frac{z}{1+z}\right)^{2/13} \frac{i^{8/13}}{N_0^{3/13}R_0^{6/13}}; \tag{4.22}$$

$$x_{eq} \simeq 7.9 \cdot 10^3 \frac{z^{2/13}}{(1+z)^{11/13}} \frac{R_0^{7/13}}{N_0^{11/13} i^{18/13}}; \tag{4.23}$$

$$N(0) = 1.5 \cdot 10^{-5} \frac{i^{18/13} N_0^{2/13}}{z^{2/13}(1+z)^{11/13} R_0^{20/13}}. \tag{4.24}$$

In deriving these expression it was assumed that the radiation leaves only through the other surface of the plasma layer and not from both surfaces as was the case in [89], since in an experiment either the radiation is not absorbed at all in the plasma gap or the discharge itself moves within a cylindrical reflector. The transition to the case where radiation leaves both furfaces is achieved simply by replacing the constant σ^* by $2\sigma^*$ in the equations.

The region of applicability of Eqs. (4.20)-(4.24) is determined, on the one hand, by the requirement that the Rosseland mean free path be small compared to the thickness of the plasma layer, i.e., $l \ll x_{eq}$, and, on the other, by the requirement that the characteristic size scale for changes in the plasma density x_{eq} be small compared to the characteristic size scale for temperature inhomogeneities, i.e., $x^2_{eq} \ll x^2_T$. These requirements place the following restriction on the discharge current:

$$1.1 \cdot 10^{-31}\Phi > i > 6 \cdot 10^{-33}\Phi, \tag{4.25}$$

where

$$\Phi(z, N_0, R_0) = \frac{z^3 N_0^{22/10}}{(1+z)^{4/10} R_0^{12/10}}. \tag{4.26}$$

When the left-hand inequality is violated, the plasma in the discharge becomes optically transparent, and when the right-hand inequality is violated, the temperature in the discharge will be as nonuniform as the density. Finally, for the discharge to be quasiplane, i.e., in order that the inequality $x_{eq} \ll R_0$ hold, the total discharge current must be sufficiently large, i.e.,

$$i > 1.5 \cdot 10^2 \frac{z^{2/18}(1+z)^{11/18} N_0^{11/18}}{R_0^{6/18}}. \tag{4.27}$$

These equations show that a return-current discharge differs greatly from the usual linear pinch. This difference lies in the weaker dependence of the temperature and characteristic thickness of the discharge on the external parameters (the discharge current and the total number of particles in the discharge). It follows from Eqs. (4.21) and (4.23) that in a return current discharge $T \propto i^{18/13} N_0^{-2/13}$ and $x_{eq} \propto N_0^{11/13} i^{-18/13}$, while in a linear pinch $T \propto i^2 N_0^{-1}$ and $r_{eq} \propto i^{-3} N_0^{11/6}$. Another interesting feature of return-current discharges is that there is a transition from an opaque to a transparent discharge when the total current is increased and a transition to a discharge with a strongly nonuniform temperature when the total current is decreased. In a linear pinch, the situation is just the opposite. This behavior is explained by the dependence of the optical thickness of the discharge on the discharge current. In fact, in a return-current discharge $\tau \propto x_{eq}/l \propto i^{-10/13}$; i.e., it falls with increasing i, while in a linear pinch it increases strongly with the discharge current ($\tau \propto i^2$).

We now turn to the more general case of a semitransparent plasma. A return current discharge in a semitransparent plasma is analyzed in a way similar to that used in [94] for a linear pinch. Assuming that the temperature in the discharge is uniform, the first equations of the system (4.14) yield pressure, density, and magnetic field distributions which coincide exactly with Eq. (4.20).

The temperature distribution along the axis of the discharge is obtained by substituting Eqs. (4.20), (4.18), and (4.19) in the energy balance equation of the system of equations (4.14). As a result we obtain

$$T(x) = T(0)\left[1 + \frac{B}{2C}\frac{x^2}{x^2_{eq\,0}}\left(1 - \frac{4}{3}\frac{x^2}{x^2_{eq\,0}} + \frac{38}{45}\frac{x^4}{x^4_{eq\,0}} - \frac{1}{6}\frac{x^6}{x^6_{eq\,0}} + \frac{1}{25}\frac{x^8}{x^8_{eq\,0}}\right) - \frac{A}{2C}\frac{x^2}{x^2_{eq\,0}}\left(1 - \frac{x^2}{x^2_{eq\,0}} + \frac{1}{3}\frac{x^4}{x^4_{eq\,0}}\right)\right], \tag{4.28}$$

where

$$A = \sigma E^2 = \frac{P_0(0)\,c^2}{2\pi\sigma x^2_{eq\,0}},$$

$$B = \gamma_0\left[\frac{3}{8\pi}\frac{N_0}{R_0 x_{eq\,0}}\right]^2 \sqrt{T_0(0)}\, z^3 \exp(-x_0), \tag{4.29}$$

$$C = \frac{\beta_0 T_0^{15/2}(0) f(x_0) R_0^2}{z^3 N_0^2}\left(\frac{8\pi}{3}\right)^2,$$

and $T_0(0)$, $x_{eq\,0}$, and $N_0(0)$ are given by Eqs. (4.22)-(4.24).

From this there follows a criterion for the applicability of the uniform temperature approximation used here:

$$A \ll 2C, \quad B \ll 2C. \tag{4.30}$$

The thickness of the plasma layer is determined, as usual, from the energy balance condition at the surface of the discharge,

$$S(x_{eq}) = \frac{j^2}{\sigma} 2x_{eq}. \tag{4.31}$$

The flux $S(x_{eq})$ from the outer surface of the plasma layer was found by solving the radiative transport equation with mirror reflection at the inner surface of the discharge. A calculation similar to that in [94, 95] yields

$$S(x_{eq}) = \sigma^* T_0^4(0)\,\xi(x_0), \quad \xi(x_0) = \frac{32}{\pi^4}\frac{x_0^3}{\exp(x_0)-1} + \frac{15}{\pi^4}\int_0^{x_0}\frac{x^3 dx}{\exp(x)-1}. \tag{4.32}$$

Using Eqs. (4.31) and (4.32), as well as Eq. (4.21), we find that

$$x_{eq} = \frac{x_{eq\,0}}{[\xi(x_0)]^{2/13}}, \quad T(0) = \frac{T_0(0)}{[\xi(x_0)]^{2/13}}, \quad N(0) = N_0(0)\,[\xi(x_0)]^{2/13}. \tag{4.33}$$

Finally, to completely determine all these quantities, it is necessary to find x_0 from Eq. (4.16) which may be rewritten in the form

$$1.25\,(\xi\delta)^{13/10}\left[\frac{1-\exp(-x_0)}{x_0^3}\right]^{13/10}\Phi(z, N_0, R_0) = i, \tag{4.34}$$

where $\delta = 4.1 \cdot 10^{-23}$ and $\Phi(z, N_0, R_0)$ is defined in Eq. (4.26).

We now analyze the region of applicability of the expressions we have obtained. As in the case of the linear pinch, for a semitransparent plasma [94] this region is determined by the inequalities (4.30) which ensure temperature uniformity in the discharge. Then the first inequality may be rewritten in the following form

$$\xi(x_0) \ll \frac{43_0\delta}{\sigma^*}\frac{1-\exp(x_0)}{x_0^3} f(x_0) = F(x_0) \simeq 0.81\frac{1-\exp(x_0)}{x_0^3} f(x_0). \tag{4.35}$$

Graphs of the functions $\xi(x_0)$ and $F(x_0)$ are given in Fig. 11. It is clear that this inequality is satisfied for values of x_0 which obey the inequality

$$2.5 \simeq x_{0\,min} < x_0 < x_{0\,max} \simeq 16. \tag{4.36}$$

The second inequality (4.30) takes the form

$$1.2\psi(x_0) \ll F(x_0). \tag{4.37}$$

and, as can be seen from Fig. 11, places no new limitations on the range of possible values of x_0. Substituting the resulting values of x_{0min} and x_{0max} in Eq. (4.34) we find the range of variation of the discharge current in a semitransparent discharge to be

$$1.9 \cdot 10^{-31}\,\Phi > i > 6 \cdot 10^{-34}\,\Phi. \tag{4.38}$$

Comparing this inequality with (4.25) we see that in the semitransparent discharge the upper

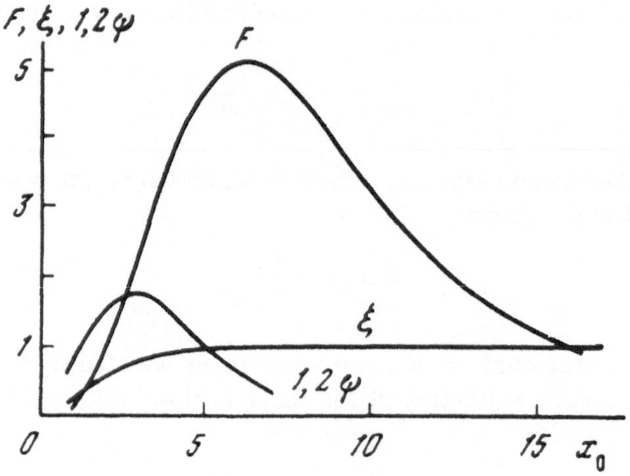

Fig. 11. The graphical solution of inequalities (4.35) and (4.37)

bound on the discharge current has increased by about a factor of 5. However, the conversion factor for radiation near the transparency limit increases relative to the opaque value by $[\xi(x_{0\,min})]^{-1} \simeq 1.1$ in all.

It follows from this analysis of a return-current discharge that the way its characteristics depend on the external parameters is substantially different from a linear z-pinch. This is, first of all, the case for the limiting currents whose magnitudes depend strongly on the number of particles in the discharge and the radius of the return pinch, as opposed to the case of linear discharges, in which the limiting currents are constants depending only on the character of the radiation. Here, as opposed to the linear discharge case, an opaque or gray reverse pinch is realized for currents below the maximum limiting current. In addition, a weaker dependence of its parameters on the discharge current i and the total number of particles is characteristic of a discharge with a return current. These features impose rather high requirements on experimental devices for creating opaque or optically semitransparent reverse-current discharges.

We now make some estimates. We shall proceed from the need to make a reverse pinch with $R_0 = 10$ cm and a radiating surface temperature of $T = 2 \cdot 10^4$°K. In order that the temperature in the discharge be uniform and it be possible to assume $T \simeq T(0)$, the current must be sufficiently large. We shall specify the value $i = 10^6$ A, which is attainable with modern pulsed current sources, in particular, with the experimental apparatus described in [102]. Now for the specified i, R_0, and T we can determine the required number of particles in the discharge using Eq. (4.22). Assuming $z \simeq 1$, we find $N_0 \simeq 10^{20}$ cm^{-1}. We note that the value $N_0 = 10^{20}$ cm^{-1} can be obtained by evaporation of, for example, a silver foil of thickness $\sim 10^{-4}$ cm. The technique of exploding such a thin foil in a vacuum is rather difficult, so it is more convenient to produce the required coaxial layer of working material either by exploding several wires or by simply injecting vapors or powders of the working material into the vacuum, as was done in [104]. To increase N_0 further would be impractical since to achieve the previous temperature one would have to increase the current. In addition, the discharge regime would be shifted toward the lower current limit, that is, toward increased temperature inhomogeneity in the discharge.

In conclusion we shall show that the analysis of the stability of a semitransparent discharge with a reverse current is similar to [94]. The semitransparent discharge is subject only to force instabilities, of which the most harmful long-wavelength pinch type mode has a growth rate of $\gamma^{-1} \sim R_0/v_s$, where v_s is the speed of sound. For the above parameters the lifetime of the reverse discharge is about 50 μsec.

3. The Theory of a High-Current ac Pinch Discharge

We now turn to an analysis of a high-current ac pinch discharge with the simple cylindrical discharge as an example and show that the theoretical results for a dc discharge are transferable to the ac case simply by replacing the physical quantities by their corresponding effective values.

The equilibrium and stability theory for radiating dc discharges in dense plasmas is applicable if the characteristic time for changes in the current is much greater than the time required to establish hydrodynamic equilibrium (a/v_s), that is,

$$(d \ln i/dt)^{-1} \gg a/v_s. \tag{4.39}$$

For a number of reasons this condition is not always fulfilled in some reports [102] dealing with the experimental verification of this theory. The principal reason is that this condition can only be satisfied in devices with a sufficiently long period. Then, first of all, high discharge currents are difficult to achieve, and, second, various kinds of instability may develop during the main phase of the discharge. Thus, in [102], experiment and theory are compared only over a comparatively short time interval near the first maximum in the discharge current during which the current channel of the discharge is compressed by its own magnetic field and is fairly stable (the entire process lasts only a few periods).

These circumstances, as well as the proposal in [103] that a rapidly oscillating discharge may be more stable due to a process of dynamic stabilization, make the problem of constructing a theory of ac pinch discharges of immediate interest.

In the following by an ac discharge we shall mean one in which

$$v_s/a \ll \omega_0 \ll c^2/4\pi\sigma a^2, \tag{4.40}$$

where ω_0 is the frequency of current oscillations in the external circuit, a is the characteristic size of the discharge, σ is the conductivity of the plasma, which is assumed fully ionized, and c is the speed of light.

These inequalities express the fact that the discharge current may be regarded as having a high frequency compared to the hydrodynamic processes and as practically constant compared to the electrodynamic processes (diffusion of the magnetic field in the plasma, absence of a current skin effect).

Below we examine the case of an optically opaque plasma when the Rosseland mean free path of the light is much shorter than the characteristic dimension of the problem and the radiative heat conduction approximation (4.12) is valid for the flux S, and the opposite case of an optically transparent plasma with an emissivity [91] given by

$$q_s = \operatorname{div} S = \gamma_0 T^{1/2} N^2 z^3, \tag{4.41}$$

where γ_0 depends on the light absorption mechanism in the plasma.

We shall consider the quasiequilibrium state of the discharge and its stability. We now define the concept of a quasiequilibrium state after making the following comments. In view of the largeness of the hydrodynamic frequency scale compared to that of the current in the external circuit, the magnetic field is able to follow the changes in the discharge current; that is, if $i = i_0 \sin\omega_0 t$, then

$$j_e = j_{e0} \sin \omega_0 t, \quad B_e = B_{e0} \sin \omega_0 t. \tag{4.42}$$

The energy balance in the discharge is determined, on one hand, by ohmic heating, and, on the other, by removal of heat from the discharge by radiation, i.e.,

$$j_e^2/\sigma_e = q_s(\rho_e, T_e). \tag{4.43}$$

Now the discharge temperature determined by Eq. (4.43) oscillates strongly. The hydrodynamic pressure of the plasma, the magnetic pressure, and the velocity are also oscillatory. But, since $\omega_0 \gg v_s/a$, the plasma density remains practically constant and the velocity oscillations are so small that they can be neglected. We will refer to such a state as quasiequilibrium, and the parameters characterizing it will be labeled e.

This assumption about the way the hydrodynamic variables change may be confirmed by a more rigorous calculation as well. In fact, let us consider the solution of the equations of motion and continuity ignoring the specific time dependences of j_e, B_e, and P and assuming they oscillate to the same extent as the current. We shall seek solutions of these equations in the form of a sum of a term that varies smoothly in time and a small term that oscillates rapidly whose average over the period $2\pi/\omega_0$ is zero:

$$\begin{aligned} &\rho = \rho_e + \rho_e', \quad \rho_e' \ll \rho_e, \quad \int_0^{2\pi/\omega_0} \rho_e' dt = 0, \\ &v = v_e + v_e', \quad v_e' \ll v_e, \quad \int_0^{2\pi/\omega_0} v_e' dt = 0. \end{aligned} \tag{4.44}$$

Substituting Eq. (4.44) in the equations of motion and continuity in cylindrical coordinates and retaining terms of first order in ρ_e' and v_e', we obtain a system of four equations, two for the smoothly varying and two for the oscillating terms:

$$\begin{aligned} &\frac{\partial \rho_e}{\partial t} + \frac{1}{r}\frac{\partial}{\partial r}(r\rho_e v_e) = 0, \quad \rho_e\left(\frac{\partial v_e}{\partial t} + v_e \frac{\partial v_e}{\partial r}\right) = 0, \\ &\frac{\partial \rho_e'}{\partial t} + \frac{1}{r}\frac{\partial}{\partial r}(r\rho_e v_e') + \frac{1}{r}\frac{\partial}{\partial r}(r\rho_e' v_e) = 0, \\ &\rho_e \frac{\partial v_e'}{\partial t} + \rho_e' \frac{\partial v_e}{\partial t} + \rho_e v_e \frac{\partial v_e'}{\partial r} + \rho_e v_e' \frac{\partial v_e}{\partial r} + \rho_e' v_e \frac{\partial v_e}{\partial r} = -v_s^2(t)\frac{\partial \rho_e}{\partial r} - v_s^2(t)\frac{\partial \rho_e'}{\partial r} - \frac{1}{c} j_e B_e. \end{aligned} \tag{4.45}$$

It automatically follows from the second equation of this system that $v_e = 0$, and from the first, that the equilibrium density ρ_e is constant in time. The radial density distribution is found from the last equation of this system upon averaging over the time, and from it we find that in the quasiequilibrium state the average hydrodynamic and magnetic pressures are equal.

Therefore, the quasiequilibrium state of the discharge must obey the following system of equations:

$$\begin{aligned} &\frac{1}{r}\frac{\partial}{\partial r} rB_e = \frac{4\pi}{c} j_e = \frac{4\pi}{c}\sigma_e E_e, \\ &\frac{\partial \langle P_e \rangle}{\partial r} + \frac{1}{c}\cdot\langle j_e B_e \rangle = 0, \\ &\frac{j_e^2}{\sigma_e} = q_s(\rho_e, T_e) = \frac{1}{r}\frac{\partial}{\partial r} rS(\rho_e, T_e). \end{aligned} \tag{4.46}$$

This is suitable for studies of direct and inverse discharges both in optically opaque and optically transparent plasmas.

We shall consider the quasiequilibrium state of a discharge in an optically opaque plasma assuming the temperature to be almost uniform over the discharge diameter because of the high radiant thermal conductivity and replacing the radiant flux from the surface of the plasma by the black body radiant flux. Then

$$\frac{j_e^2}{\sigma_e}\pi r_e^2 = \sigma^* T_e^4 2\pi r_e, \tag{4.47}$$

where r_e is the discharge radius.

From this equation it is easy to obtain the time dependence of the temperature,

$$T_e = T_{e0} \sin|^{4/11}(\omega_0 t). \tag{4.48}$$

Hence, in an optically opaque discharge the temperature oscillates in time with the frequency of the external field; however, near the current maximum it varies much more slowly than harmonically.

By making the substitutions

$$\begin{gathered} i_{ef} = \frac{i_{e0}}{\sqrt{2}}, \quad E_{ef} = \frac{E_{e0}}{\sqrt{2}}, \quad B_{ef} = \frac{B_{e0}}{\sqrt{2}}, \quad P_{ef} = \langle P_e \rangle = \delta_1 P_{e0}, \\ \delta_1 = \frac{1}{2\pi}\int\limits_{-\pi}^{\pi} \sin^{4/11}(x)\,dx = \frac{\Gamma(15/22)}{\sqrt{\pi}\,\Gamma(13/11)}, \end{gathered} \tag{4.49}$$

where $\Gamma(x)$ is the gamma function, it is possible to reduce the first two equations of the system (4.46) to the form

$$\frac{1}{r}\frac{\partial}{\partial r} r B_{ef} = \frac{4\pi}{c} j_{ef} = \frac{4\pi}{c}\sigma_{e0} E_{ef}, \quad \frac{\partial P_{ef}}{\partial r} + \frac{1}{c} j_{ef} B_{ef} = 0. \tag{4.50}$$

The expressions for j_{ef}, E_{ef} and B_{ef} have the simple physical significance of being effective values of the alternating current density and of the alternating electromagnetic field.

The system of equations (4.50) for a cylindrical discharge must be supplemented by the boundary equation

$$B_{ef}|_{r=r_e} = \frac{2i_{ef}}{cr_e}. \tag{4.51}$$

Equations (4.47) and (4.50) coincide with the system of equations for a constant current [89]; thus, the quasiequilibrium state of an ac discharge is described by the same equations as the equilibrium state of a dc discharge, but with replacement of the physical quantities by their corresponding effective values. Finally, for the quasiequilibrium state with a uniform temperature, we obtain

$$\begin{gathered} B_{ef} = \sqrt{4\pi P_{ef}(0)}\,\frac{r}{r_e}, \quad P_{ef} = P_{ef}(0)\left(1 - \frac{r^2}{r_e^2}\right), \\ \rho_e = \rho_e(0)\left(1 - \frac{r^2}{r_e^2}\right), \quad r_e^2 = \frac{P_{ef}(0)\,c^2}{\pi j_{ef}^2}. \end{gathered} \tag{4.52}$$

All the other relations derived for an opaque discharge may be used, in particular, the expressions for its parameters in terms of the total current and the total number of particles in the discharge (N_0), e.g.,

$$T_{ef} = \delta_1 T_{e0} = \frac{i_{ef}^2}{2(1+z)c^2 k N_0}. \tag{4.53}$$

We now discuss quasiequilibrium of a discharge in an optically transparent plasma. In this case, from the energy balance equation including the plasma emissivity [Eq. (4.41)], we find the time dependence of the temperature to be

$$T_e = T_{e0}\,|\sin\omega_0 t|. \tag{4.54}$$

In a transparent discharge the temperature oscillates in exact conformity with the current oscillations.

The quasiequilibrium state for a transparent discharge obeys the same system of equations (4.50) retaining the definition of the effective values of the electromagnetic equatities but setting

$$P_{ef} = \langle P_e \rangle = \frac{P_{e0}}{2\pi}\int_{-\pi}^{\pi} |\sin x|\,dx = \frac{2}{\pi}P_{e0}. \tag{4.55}$$

Writing the boundary conditions for the transparent discharge case in the form

$$rB_{ef}|_{r\to\infty} = \frac{2i_{ef}}{c}, \quad P_{ef}|_{r\to\infty} = 0 \tag{4.56}$$

and requiring that the solutions be finite at zero, we conclude that the solutions for the effective quantities in an ac discharge must be given by the same expressions as in the case of a dc discharge in a transparent plasma, but with the physical quantities replaced by their effective values. For the case of emission by bremsstrahlung and multiply ionized atoms, we find, following [91], that

$$\begin{gathered} P_{ef} = \frac{P_{ef}(0)}{(1+r^2/r_e^2)^2}, \quad B_{ef} = \sqrt{8\pi P_{ef}(0)}\,\frac{r/r_e}{1+r^2/r_e^2}, \\ r_e = \sqrt{\frac{2}{\pi P_{ef}(0)}\,\frac{\beta_{ef} cz}{aE_{ef}}}, \quad \beta_{ef} = \frac{2}{\pi}\sqrt{\frac{2aE_{ef}^2 k^2(1+z)^2}{\gamma_0 z^4}}, \\ P_{ef} = \beta_{ef} T_{e0}^{3/2}. \end{gathered} \tag{4.57}$$

The corresponding formulas for opaque and transparent return-current discharges are derived in an analogous fashion.

The stability of the quasiequilibrium state was analyzed using relatively small linearized perturbations,

$$\rho \to \rho_e + \rho, \quad T \to T_e + T, \quad P \to P_e + P, \quad B \to B_e + B, \text{ etc.},$$

of the system of equations (4.1) and the method of normal oscillations. Because of the time dependence of the quasiequilibrium quantities this system of equations is a system of differential equations with periodic coefficients (period $2\pi/\omega_0$).

We shall consider an optically opaque discharge. In this case it is possible to neglect the effect of temperature fluctuations on the stability of the discharge if the radiant heat flux exceeds the hydrodynamic heat flux, a condition which is automatically satisfied in a low-conductivity plasma [89]. Primarily this means the absence of a thermal (overheating) instability in this kind of discharge. Since the times for force instabilities to develop cannot be less than the hydrodynamic time, r_e/v_s, we must limit ourselves to consideration of processes for which

$$\omega \lesssim v_s/r_e \ll \omega_0. \tag{4.58}$$

From the nature of the periodic force acting on the system we may conclude that the small perturbations in f near the quasiequilibrium values will be almost periodic functions of period $2\pi/\omega_0$; that is, they may be expanded in a Fourier series with slowly varying amplitudes,

$$f=\sum_{n=-\infty}^{\infty} f_n \exp(in\omega_0 t), \qquad \frac{1}{\omega_0}\left|\frac{\partial \ln f_n}{\partial t}\right| \ll 1. \tag{4.59}$$

However, not all the slowly varying amplitudes f_n are important in the development of instabilities. In fact, only those f_n which enter into the equations of motion and continuity averaged over the period of the current oscillations need be taken as nonzero.

Taking the symmetry of the problem into account we write the time and space dependence of f_n in the form

$$f_n(t, \mathbf{r}) = f_n(r)\exp(-i\omega t + im\varphi + ik_z z). \tag{4.60}$$

Perturbations with $m = 0$ and $k_z \neq 0$, as usual, correspond to constrictions (pinches) and those with $m \neq 0$ and $k_z \neq 0$ correspond to kinks.

From an analysis of the stability of an optically opaque dc discharge it follows that the most harmful instabilities to which the discharge is subject are the fundamental constriction mode and helical-kink instabilities. The growth rates of these instabilities are independent of the conductivity. Thus, the stability problem for an optically opaque discharge reduces to a study of the limiting case $\sigma_e \rightarrow 0$. After the appropriate calculations we find that the force instabilities of an ac discharge are described by the same (to within the differences in notation) system of equations as a dc discharge. It is possible to write immediately the growth rates for instabilities in an ac discharge. For the fundamental constriction mode we have

$$\gamma = \left[2\sqrt{3}\,|k_z|\frac{v_{sef}^2}{r_e}\right]^{1/2}, \qquad k_z r_e < 1. \tag{4.61}$$

For long-wavelength helical instabilities the growth rate is $(k_z r_e)^{-1/2}$ times smaller. Here

$$v_{sef}^2 = \delta_1 v_{s0}^2. \tag{4.62}$$

The external (surface) long-wavelength constrictions have smaller growth rates and, in addition, as the conductivity decreases, their growth rate decreases.

We now consider the stability of an optically transparent discharge. A study of the stability of an optically transparent dc discharge shows that along with the force instabilities a thermal instability develops in a low-conductivity plasma and has a growth rate of

$$\gamma \sim c^2/\sigma a^2. \tag{4.63}$$

Compared to such a rapid process the fact that the current varies in time is unimportant – it may be regarded as constant. Thus, the results on the thermal instability in a dc discharge remain valid in the case of an alternating current of frequency $\omega_0 \ll c^2/\sigma a^2$.

The thermal instability may stabilize with rising temperature since the condition for existence of overheating ($v_s/a \ll c^2/\sigma a^2$) breaks down. Then the stability of the discharge is determined by the times for force instabilities with $\omega \lessapprox v_s/a \ll \omega_0$ to develop. Such processes may be regarded as slow compared to the period of the current in the external circuit; therefore, the method of slowly varying amplitudes developed for studying the force instabilities of an opaque discharge may be applied to them.

The absence of a sharp boundary in a transparent discharge leads to disappearance of the fundamental constriction mode which is independent of the conductivity. However, besides the fundamental mode there are volume force oscillations with growth rates $\gamma \propto \sigma_e \to 0$ in a low-conductivity plasma. Calculations show that to obtain the growth rates for such oscillations it is sufficient to make the substitution

$$v_s^2 \to v_{sef}^2 = \frac{\sqrt{\pi}}{12}\frac{\Gamma(1/4)}{\Gamma(3/4)} v_{s0}^2 \approx v_{s0}^2 \tag{4.64}$$

in the results of [91].

Thus, the growth rates for development of the most harmful long-wavelength volume oscillations in a simple z-pinch are given by

$$\gamma = \frac{k_z^2 r_e^2}{12\left(n+\frac{1}{2}\right)^3} \cdot \frac{\pi\sigma_{s0} v_{sef}^2}{c^2}. \tag{4.65}$$

We emphasize once again that an ac discharge is subject to force instabilities such as pinching and kinks, and an optically transparent plasma is further subject to thermal instabilities. The growth rates obtained here for optically opaque and transparent discharges differ from the old value (for a dc discharge) by a numerical factor of order unity; thus, an ac discharge has about the same degree of instability as a dc discharge.

4. The Ionization State Distribution Function in a Multiply Charged Ion Plasma. Laser-Produced Plasmas

The action of laser radiation on matter is now often used to produce a multiply charged plasma. A detailed review of theoretical and experimental research on laser-produced plasmas is given in [84] which deals with the gas-dynamic motion of a laser plasma as well as its ionization state. A laser plasma is characterized by strong disequilibrium. Ionization multiplicities of over 30 have been observed experimentally. It has been shown that under these conditions the distribution of populations over the levels of multiply charged ions obeys a corona model, in which radiative relaxation predominates. While the kinetics of the ion level population distribution have been discussed repeatedly, the process of formation of the multiply charged ions has not yet been studied adequately.

We shall examine the kinetics of formation of multiply charged ions assuming that the most probable processes during the creation and destruction of ions are electron impact ionization and three-body and radiative recombination in which the charge of an ion is reduced by unity. The velocity distribution of the electrons is assumed to be Maxwellian. The ion density balance equations have the form

$$dN_n/dt = V(n-1, n) N_e N_{n-1} - W(n, n-1) N_e^2 N_n - V(n, n+1) N_e N_n + \\ + W(n+1, n) N_e^2 N_{n+1} + Q(n+1, n) N_{n+1} - Q(n, n-1) N_n, \tag{4.66}$$

where N_n is the density of ions with charge n, V(n, n + 1) is the electron impact ionization rate for an ion with charge n to one with charge n + 1, W(n + 1, n) is the rate of three-body recombination of an (n + 1)-fold ion to an n-fold ion, and N_e is the electron density.

Equations (4.66) must be supplemented by the normalization equation and the condition for quasineutrality of the plasma,

$$\sum_{n=0}^{n_0} N_n = N; \quad \sum_{n=1}^{n_0} nN_n = N_e. \tag{4.67}$$

Here N is the density of heavy particles in the plasma and n_0 is the number of electrons in an atom.

We now turn to the case of high electron densities and neglect radiative recombination in Eq. (4.66). Let the degree of ionization of the plasma be different from that corresponding to the temperature T_e and density N_e of the free electrons at the initial time.

We shall examine the establishment of a thermodynamic equilibrium Saha distribution in a plasma with slowly changing free-electron temperature and density (compared to the characteristic relaxation time).

We shall write the equilibrium ion density distribution function in the form

$$N_n = G_n \exp\left(-\frac{E_n}{kT_e}\right) N_0,$$
$$G_n = \frac{Z_n}{Z_0}\left[\frac{2}{N_e}\left(\frac{mkT_e}{2\pi\hbar^2}\right)^{3/2}\right]^n; \qquad E_n = \sum_{i=0}^{n-1} I_i, \qquad E_0 = 0, \tag{4.68}$$

where Z_n and I_n are the statistical weight and ionization potential of an n-fold ion, k is Boltzmann's constant, m is the electron mass, and $\hbar$ is Planck's constant.

We seek a nonequilibrium distribution function of the form

$$N_n = G_n \exp\left(-\frac{E_n}{kT_e}\right) f_n. \tag{4.69}$$

We now substitute this in Eq. (4.66) and neglect terms proportional to the time derivatives of the electron temperature and density. Then, since for a Maxwellian electron distribution function the rates of ionization and three-body recombination are related to one another by [62]

$$V(n, n+1)\, G_n \exp\left(-\frac{E_n}{kT_e}\right) = N_e W(n+1, n)\, G_{n+1} \exp\left(-\frac{E_{n+1}}{kT_e}\right), \tag{4.70}$$

we have

$$\frac{df_n}{d\tau} \equiv \frac{1}{N_e}\frac{df_n}{dt} = V(n, n+1)(f_{n+1} - f_n) + N_e W(n, n-1)(f_{n-1} - f_n). \tag{4.71}$$

We note that Eqs. (4.66) and (4.71) have much in common with the population balance equations for the discrete levels of an atom [Eq. (2.3)]. This is also true of the methods for solving the kinetic equations (4.66), in particular, the iteration method. Following that method, we rewrite Eq. (4.71) as follows:

$$f_n = f_0 + \sum_{m=0}^{n-1} \frac{\exp(E_m/kT_e)}{G_m V(m, m+1)} \sum_{k=0}^{m} G_k \exp\left(-\frac{E_k}{kT_e}\right)\frac{df_k}{d\tau}. \tag{4.72}$$

To find the distribution in the first step we substitute a test equilibrium distribution function $f_k = f_0$ on the right-hand side of Eq. (4.72). Then we have

$$f_n = f_0 + A_n \frac{df_0}{d\tau},$$
$$A_n = \sum_{m=0}^{n-1} \frac{\exp(E_m/kT_e)}{G_m V(m, m+1)} \sum_{k=0}^{m} G_k \exp\left(-\frac{E_k}{kT_e}\right), \qquad A_0 = 0. \tag{4.73}$$

In the second step of the iteration we again substitute the new distribution function (4.73) on the right-hand side of Eq. (4.72), and so on. Thus we find a distribution function in the form of a series of successive time derivatives of the neutral atom density. In the quasistationary stage of recombination of a multiply charged ion plasma, the neutral density will be a slowly varying function of time; thus, in the series of time derivatives it is sufficient to limit ourselves to the first derivative term alone.

As opposed to the recombination problem, in the ionization problem the slowly varying function of time will be the density of bare nuclei. In accordance with this, since

$$\sum_{k=0}^{m} G_k \exp\left(-\frac{E_k}{kT_e}\right)\frac{df_k}{d\tau} = -\sum_{k=m+1}^{n_0} G_k \exp\left(-\frac{E_k}{kT_e}\right)\frac{df_k}{d\tau},$$

we rewrite Eq. (4.72) in the following way:

$$f_n = f_{n_0} + \sum_{m=n}^{n_0-1} \frac{\exp(E_m/kT_e)}{G_m V(m, m+1)} \sum_{k=m+1}^{n_0} G_k \exp\left(-\frac{E_k}{kT_e}\right)\frac{df_k}{d\tau}. \tag{4.74}$$

From this we find the quasistationary distribution function to be

$$f_n = f_{n_0} + B_n \frac{df_{n_0}}{d\tau},$$
$$B_n = \sum_{m=n}^{n_0-1} \frac{\exp(E_m/kT_e)}{G_m V(m, m+1)} \sum_{k=m+1}^{n_0} G_k \exp\left(-\frac{E_k}{kT_e}\right), \qquad B_{n_0} = 0. \tag{4.75}$$

We introduce the temperatures Θ_n while characterize the nonequilibrium degree of ionization and recombination of n-fold ions and are defined by the equation

$$\frac{N_e N_{n+1}}{N_n} = \frac{2Z_{n+1}}{Z_n}\left(\frac{mkT_e}{2\pi\hbar^2}\right)^{3/2} \exp\left(-\frac{I_n}{k\Theta_n}\right). \tag{4.76}$$

Keeping Eq. (4.69) in mind, we obtain

$$\Theta_n = \frac{T_e}{1 + \frac{kT_e}{I_n}\ln\frac{f_n}{f_{n+1}}}. \tag{4.77}$$

The quasiequilibrium distribution function for the multiplicity of the ions specified by Eqs. (4.76)-(4.77) depends on three parameters N_e, T_e, and $\varkappa$. Unlike the equilibrium Saha distribution this function describes the nonstationary processes of successive stripping of electrons (ionization) and reduction in the charge of ions (recombination).

During recombination of a multiply charged plasma all $\Theta_n > T_e$ since

$$\frac{f_n}{f_{n+1}} = \frac{1 + A_n\varkappa}{1 + A_{n+1}\varkappa} < 1, \qquad \varkappa = \frac{d}{d\tau}\ln f_0 > 0. \tag{4.78}$$

In the case of ionization all $\Theta_n < T_e$ since

$$\frac{f_n}{f_{n+1}} = \frac{1 + B_n\varkappa}{1 + B_{n+1}\varkappa} > 1, \qquad \varkappa = \frac{d}{d\tau}\ln f_{n_0} > 0. \tag{4.79}$$

The parameter $\varkappa$ may be related to one of the temperatures Θ_n. For example, in the recombination problem this parameter is easily linked to the temperature Θ_0 by

$$\varkappa A_1 = \exp\left(\frac{I_0}{kT_e} - \frac{I_0}{k\Theta_0}\right) - 1, \tag{4.80}$$

and in the ionization problem, with the temperature Θ_{n_0-1} by

$$\varkappa B_{n_0-1} = \exp\left(\frac{I_{n_0-1}}{k\Theta_{n_0-1}} - \frac{I_{n_0-1}}{kT_e}\right) - 1. \tag{4.81}$$

From Eqs. (4.78) and (4.79) it is clear that during recombination of a plasma when $\varkappa A_n \gg 1$, which is certainly satisfied if

$$\exp\left(\frac{I_0}{k\Theta_0}\right) \ll \exp\left(\frac{I_0}{kT_e}\right), \tag{4.82}$$

and during ionization when $B_n\varkappa \gg 1$ or

$$\exp\left(\frac{I_{n_0-1}}{k\Theta_{n_0-1}}\right) \gg \exp\left(\frac{I_{n_0-1}}{kT_e}\right), \tag{4.83}$$

the temperatures Θ_n are independent of this parameter and are completely determined by the temperatures and density of the free electrons alone. An analogous situation occurs during motion of a recombination flux over the discrete levels of an atom in the stationary-sink approximation [47]. At this stage of the relaxation, the system of equations (4.66) is equivalent to a differential equation for the neutral density (in the recombination problem) or for the density of bare nuclei (in the ionization problem) together with a set of algebraic equations for the densities of the remaining ions.

It should be noted that the temperatures Θ_n or the ratios of the ion densities in two successive ionization states are independent of the absolute magnitude of the ionization rates $V(n, n+1)$ and are determined solely by the form of their dependences on the ionization potential and electron temperature.

As an example of the application of these formulas we have examined an aluminum plasma. A multiply charged aluminum plasma has been obtained by a number of authors, for example, in [105-107]. The study of this medium is also of importance from the standpoint of building a short-wavelength laser. Thus, based on a comparison of the hard UV absorption and emission spectra of an aluminum plasma produced by a powerful laser beam, Jaegle et al. [108] have concluded that stimulated emission might be obtained at a wavelength of 117.41 Å (using the $2p^6S_0 - 2p^54d^1P_1^0$ transition).

Here we have calculated the temperature Θ_n both in the case of a recombining multiply charged plasma and in the case of ionization. The ionization rate as a function of I_n and T_e was assumed to have the same form as in Seaton's formula [62],

$$V(n, n+1) \sim \left(\frac{Ry}{I_n}\right)^{3/2} \exp\left(-\frac{I_n}{kT_e}\right)\sqrt{\frac{kT_e}{I_n}}, \qquad \text{where} \quad Ry = 13.6 \text{ eV}. \tag{4.84}$$

The statistical sums of the ions were assumed equal to unity.

The results of a calculation of the temperature Θ_n for a recombining plasma with constant $N_e = 10^{20}$ cm^{-3} and $T_e = 5$ eV are shown as plots of Θ_n as a function of the ionic charge n for several values of the parameter in Fig. 12. Curves 1, 2, and 3 give an idea of the sequence of quasistationary distribution functions through which a multiply charged plasma relaxes to an equilibrium state. The temperatures Θ_n decrease one after another until they reach

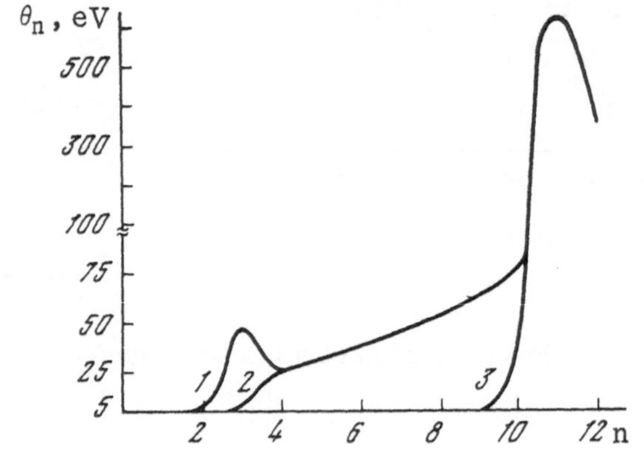

Fig. 12. Recombination of a multiply charged aluminum plasma (T_e = 5 eV, $N_e = 10^{20}$ cm^{-3}). Curves 1, 2, and 3 correspond to temperatures $\theta_0 = 5.01$, $\theta_3 = 6.7$, and $\theta_{10} = 40$ eV.

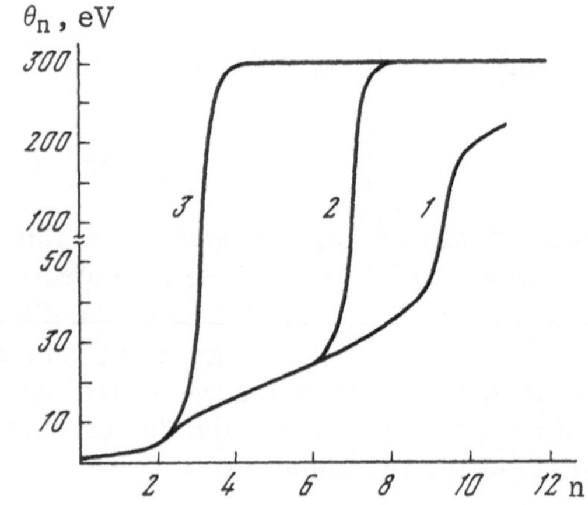

Fig. 13. Ionization of aluminum ($N_e = 10^{21}$ cm^{-3}, T_e = 300 eV). Curves 1, 2, and 3 correspond to temperatures $\theta_{12} \leqslant 230$ eV, θ_8 = 299 eV, and θ_4 = 296 eV.

the equilibrium value T_e = 5 eV. The temperatures Θ_0 and Θ_1 reach equilibrium fastest of all. Then Θ_2 decreases, with temperatures $\Theta_3, \Theta_4, \ldots, \Theta_{12}$ remaining fixed. As soon as the temperature Θ_2 has reached the equilibrium value, Θ_3 begins to fall, with $\Theta_4, \Theta_5, \ldots, \Theta_{12}$ unchanged, and so on. Finally the temperature Θ_{12} relaxes. Thus, during recombination an equilibrium Saha distribution is achieved most rapidly for the ions with lower charges.

The quasistationary distributions of the temperatures Θ_n in an ionizing aluminum plasma with $N_e = 10^{21}$ cm^{-3} and T_e = 300 eV are shown in Fig. 13. For $\Theta_{12} \leqslant 230$ eV the temperatures $\Theta_{11}, \Theta_{12}, \ldots$ are constant and depend only on N_e and T_e (curve 1). At this stage of ionization of the plasma it is possible to judge the plasma parameters (temperature and density of the electrons) by measuring the temperatures Θ_n experimentally and using these quasistationary distributions. As opposed to recombination, here the temperatures Θ_n successively rise to the equilibrium value T = 300 eV, beginning with Θ_{12}. The last one to relax is Θ_0. In this case the highly charged ions reach ionization equilibrium sooner than ions with lower charges.

With the normalization conditions (4.67) and Eqs. (4.76) and (4.77) it is easy to go from the temperature distributions to the absolute magnitudes of the ion densities. The ion density distribution functions corresponding to the first curves in Figs. 12 and 13 are shown in Figs. 14 and 15. The characteristic breaks in these distribution functions correspond to the jumps in the ionization potentials when an electron is removed from a filled shell.

It should be noted that the quasistationary distribution function for ions up to charge 12, inclusively, in Fig. 15 remains constant for $N_{13} \leqslant 8 \cdot 10^{18}$ cm^{-3}. From the figure it is clear that the nonequilibrium plasma consists primarily of heliumlike and hydrogenlike ions, while

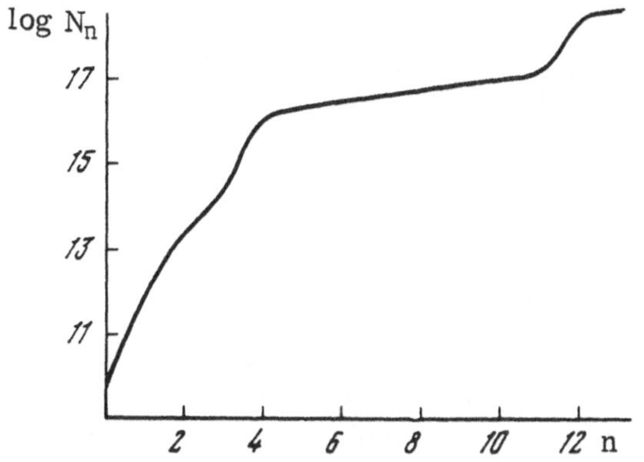

Fig. 14. Absolute values of the ion densities in a recombining plasma. This distribution function corresponds to curve 1 of Fig. 12.

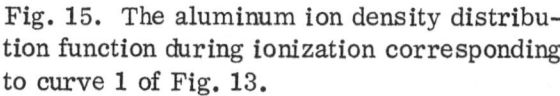

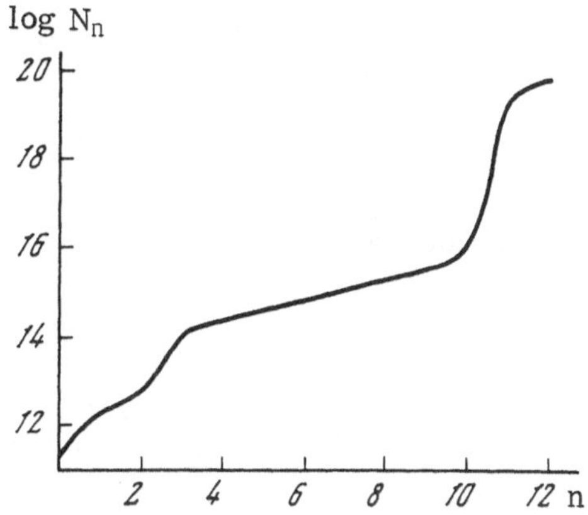

Fig. 15. The aluminum ion density distribution function during ionization corresponding to curve 1 of Fig. 13.

in an equilibrium state at T_e = 300 eV the plasma consists almost entirely of fully stripped nuclei.

In order to know the time variation of the parameter $\varkappa$ and its relation to the initial conditions we now insert the distribution functions (4.73) and (4.75) in the normalization condition (4.67). As a consequence an ordinary differential equation is generated whose solution is

$$f = f^0 + \Delta f \exp(-t/t_0), \tag{4.85}$$

$$f^0 = \frac{N}{\sum_{n=0}^{n_0} G_n \exp(-E_n/kT_e)}, \qquad \Delta f = (f)_{t=0} - f^0, \qquad t_0 = \frac{\sum_{n=0}^{n_0} G_n \exp(-E_n/kT_e)\,\chi_n}{\sum_{n=0}^{n_0} G_n \exp(-E_n/kT_e)},$$

where $f = f^0$ and $\chi_n = A_n$ in the case of recombination, and $f = f_{n_0}$ and $\chi_n = B_n$ in the case of ionization.

In Eq. (4.85) t_0 should be taken to mean the characteristic relaxation time of a multiply charged plasma. Some calculations with a proportionality constant of 10^{-7} in Eq. (4.84) yield the following values of t_0: for a recombination plasma with $N_e = 10^{20}$ cm^{-3} and T_e = 5 eV, $t_0 \sim 6 \cdot 10^{-5}$ sec; for ionization with N = 10^{21} cm^{-3} and T = 300 eV, $t_0 \sim 10^{-7}$ sec; and, with $N_e = 10^{21}$ cm^{-3} and T_e = 600 eV, $t_0 \sim 3 \cdot 10^{-9}$ sec.

Of special interest from the standpoint of diagnostics of multiply charged plasmas is the possibility of using quasistationary distributions in problems on the interaction of intense laser light with matter. When laser light is sharply focused onto a material target with an energy flux q ~ 10^{12-14} W/cm^2, a flare with a dense, hot core with $N_e \sim 10^{20-21}$ cm^{-3} and $T_e \sim 10^2$-10^3 eV is formed. Quasistationary distributions of the ion densities over the multiplicities of ionization may be used if the lifetime of the plasma in the dense, hot core region (the lifetime is on the order of magnitude of the ratio of the radius of the focal spot to the isothermal sound speed in the plasma) is less than or on the order of the relaxation time t_0.

In the case of optically thin plasmas with relatively low electron densities it is necessary to include radiative recombination in Eq. (4.66). Upon doing some calculations fully analogous to those for taking radiative decay of the energy levels of an atom into account, we obtain the following quasistationary distribution functions:

$$f_n = f_0[\varepsilon_0^{(n)} + A_n \varkappa], \qquad \varkappa = \frac{d}{d\tau}\ln f_0,$$
$$A_n = \sum_{m=0}^{n-1} \frac{\exp(E_m/kT_e)}{G_m V(m,\, m+1)} \varepsilon_m^{(n)} \sum_{k=0}^{m} G_k \exp\left(-\frac{E_k}{kT_e}\right) \tag{4.86}$$

for recombination and

$$f_n = f_{n_0}\left[\frac{1}{\varepsilon_n^{(n_0)}} + B_n \varkappa\right], \qquad \varkappa = \frac{d}{d\tau}\ln f_{n_0},$$
$$B_n = \sum_{m=n}^{n_0-1} \frac{\exp(E_m/kT_e)}{G_m V(m,\, m+1)} \frac{1}{\varepsilon_n^{(m)}} \sum_{k=m+1}^{n_0} G_k \exp\left(-\frac{E_k}{kT_e}\right) \tag{4.87}$$

for ionization, where

$$\varepsilon_m^{(n)} = \prod_{k=m}^{n-1}\left[1 + \frac{Q(k+1,\, k)}{N_e^2 W(k+1,\, k)}\right]^{-1}, \qquad \varepsilon_n^{(n)} = 1.$$

We now consider the relaxation kinetics of a multiply charged plasma when there is a photoionization source. The ion density balance equations then take the form

$$dN_n/dt = V(n-1,\, n) N_e N_{n-1} - W(n,\, n-1) N_e^2 N_n - V(n,\, n+1) N_e N_n +$$
$$+ W(n+1,\, n) N_e^2 N_{n+1} + P(n-1,\, n) N_{n-1} - P(n,\, n+1) N_n. \tag{4.88}$$

Including photoionization processes is fully analogous to including radiative recombination; thus, we immediately write the desired quasitationary distribution functions:

$$f_n = \mu_0^{(n)} f_0 + A_n \frac{df_0}{d\tau},$$
$$A_n = \sum_{m=0}^{n-1} \frac{\exp(E_m/kT_e)}{G_m V(m,\, m+1)} \mu_{m+1}^{(n)} \sum_{k=0}^{m} G_k \exp\left(-\frac{E_k}{kT_e}\right) \tag{4.89}$$

for recombination and

$$f_n = \frac{f_{n_0}}{\mu_n^{(n_0)}} + B_n \frac{df_{n_0}}{d\tau},$$
$$B_n = \sum_{m=n}^{n_0-1} \frac{\exp(E_m/kT_e)}{G_m V(m,\, m+1)} \frac{1}{\mu_n^{(m+1)}} \sum_{k=m+1}^{n_0} G_k \exp\left(-\frac{E_k}{kT_e}\right) \tag{4.90}$$

for ionization, where

$$\mu_m^{(n)} = \prod_{k=m}^{n-1} \left[1 + \frac{P(k,\ k+1)}{N_e V(k,\ k+1)}\right], \quad \mu_n^{(n)} = 1. \tag{4.91}$$

Therefore, these results provide a basis for analyzing the ionization kinetics of multi-electron atoms and the recombination kinetics of multiply charged ions including both collisional and radiative transitions. The ion distribution functions over the charge states show great similarity with the population distribution functions for the energy levels of atoms. These functions can be used both in analyses of the effect of laser radiation on matter and in discussions of high-current discharges and accelerator technology.

CONCLUSION

The analysis done here has demonstrated the generality of the quasiequilibrium distribution method. The distributions obtained for the rotational and electronic levels, as well as those previously known for the vibrational levels, are closely interrelated. It might be expected that a single approach to this problem in kinetics would be developed, with a single method for analyzing arbitrary multilevel systems.

On the other hand, our results show the effectiveness of the quasiequilibrium distribution technique as it allows us to study in detail the physical kinetics of lasers and various non-equilibrium media.

This study has pointed out a number of unutilized prospects for lasers using rotational and electronic transitions and the special promise of plasma lasers.

In conclusion I wish to express my deep gratitude to my research adviser, L. A. Shelepin, for his extensive advice, as well as to B. F. Gordiets, A. A. Rukhadze, and A. F. Aleksandrov for help with Sections 1 and 2 of Chapter 1 and Sections 2 and 3 of Chapter IV.

LITERATURE CITED

1. L. I. Gudzenko and L. A. Shelepin, Dokl. Akad. Nauk SSSR, 160:1296 (1965).
2. L. I. Gudzenko and L. A. Shelepin, Zh. Éksp. Teor. Fiz., 46:1445 (1963).
3. D. R. Bates, A. E. Kingston, and R. W. P. McWhirter, Proc. Roy. Soc., A267:297 (1962).
4. L. P. Pitaevskii, Zh. Éksp. Teor. Fiz., 42:1326 (1962).
5. S. T. Belyaev and G. N. Budker, in: Plasma Physics and the Problems of Controlled Thermonuclear Reactions [in Russian], Vol. 3, Izd. Akad. Nauk SSSR, Moscow (1958).
6. L. M. Biberman, V. S. Vorob'ev, and I. T. Yakubov, Usp. Fiz. Nauk, 107:353 (1972).
7. L. I. Gudzenko, L. A. Shelepin, and S. I. Yakovlenko, Tr. FIAN, 83:100 (1975) (previous article in this volume).
8. B. F. Gordiets, A. I. Osipov, E. V. Stupochenko, and L. A. Shelepin, Usp. Fiz. Nauk, 108:655 (1972).
9. B. F. Gordiets, N. N. Sobolev, V. V. Sokovikov, and L. A. Shelepin, Phys. Lett., A25:173 (1967).
10. C. E. Treanor, J. W. Rich, and R. G. Rehm, J. Chem. Phys., 48:1798 (1968).
11. B. F. Gordiets, Sh. S. O. Mamedov, A. I. Osipov, and L. A. Shelepin, Teor. Éksp. Khim., 9:460 (1973).
12. M. Goldman, Spin Temperature and Nuclear Magnetic Resonance in Solids, Oxford University Press (1970).
13. V. M. Sutovskii, Tr. FIAN, 56:66 (1971).
14. V. M. Gol'dfarb, E. V. Il'ina, I. E. Kostygova, and G. A. Luk'yanov, Opt. Spektrosk., 27:204 (1969).
15. L. A. Shelepin, Tr. FIAN, 83:3 (1975) (first article in this volume).

16. S. A. Reshetnyak and L. A. Shelepin, Zh. Prikl. Mekh. Tekh. Fiz., No. 4, p. 18 (1972).
17. S. A. Reshetnyak and L. A. Shelepin, Zh. Prikl. Spektrosk., 21:45 (1974).
18. S. A. Reshetnyak and L. A. Shelepin, Kratk. Soobschch. Fiz., No. 5, p. 26 (1972).
19. S. A. Reshetnyak and L. A. Shelepin, Zh. Tekh. Fiz., 44:1724 (1974).
20. A. F. Aleksandrov and S. A. Reshetnyak, Zh. Prikl. Mekh. Tekh. Fiz., No. 2, 21 (1973).
21. S. A. Reshetnyak and L. A. Shelepin, Kvant. Élektron., No. 1, p. 752 (1974).
22. B. F. Gordiets, S. A. Reshetnyak, and L. A. Shelepin, Kvant. Élektron., No. 1, p. 591 (1974); Preprint FIAN No. 134 (1973).
23. S. A. Reshetnyak and L. A. Shelepin, Zh. Prikl. Mekh. Tekh. Fiz., No. 4, p. 14 (1974).
24. W. S. Benedict, M. A. Pollack, and W. J. Tomlinson, IEEE J. Quant. Electron., QE-5:108 (1969).
25. A. F. Krupnov, Izv. Vuz., Radiofizika, 13:961 (1970).
26. T. J. Chang. Appl. Phys. Lett., 17:249 (1970).
27. C. A. Brau and R. M. Jonkman, J. Chem. Phys., 52:477 (1970).
28. B. A. Ditkin and A. P. Prudnikov, Integral Transforms and Operational Calculus [in Russian], Fizmatgiz, Moscow (1961).
29. M. N. Safaryan and E. V. Stupochenko, Zh. Prikl. Mekh. Tekh. Fiz., 4:29 (1964).
30. A. I. Osipov and E. V. Stupochenko, Usp. Fiz. Nauk, 79:81 (1963).
31. A. I. Osipov, Doctoral Dissertation, Moscow State University (1965).
32. B. F. Gordiets, A. I. Osipov, and L. A. Shelepin, Zh. Éksp. Teor. Fiz., 59:615 (1970).
33. R. L. Wilkins, J. Chem. Phys., 57:912 (1972).
34. N. Skribanowitz, I. P. Herman, R. M. Osgood, Jr., M. S. Feld, and A. Javan, Appl. Phys. Lett., 20:428 (1972).
35. T. F. Deutsch, Appl. Phys. Lett., 11:18 (1967).
36. A. S. Biryukov, B. F. Gordiets, and L. A. Shelepin, Zh. Éksp. Teor. Fiz., 55:1456 (1968).
37. R. A. McFarlane and L. H. Fretz, Appl. Phys. Lett., 14:385 (1969).
38. W. M. Houghton and C. J. Jachimowski, Appl. Opt., 9:329 (1970).
39. W. J. Sarjeant, Z. Kwerovsky, and E. Brannen, Appl. Opt., 11:735 (1972).
40. N. V. Karlov, Yu. B. Konev, and G. P. Kuz'min, Kratk. Soobshch. Fiz., 8:17 (1971).
41. A. S. Biryukov, B. F. Gordiets, and L. A. Shelepin, Zh. Éksp. Teor, Fiz., 57:585 (1969).
42. B. M. Smirnov, Atomic Collisions and Elementary Processes in Plasmas [in Russian], Atomizdat, Moscow (1968).
43. E. V. Stupochenko, S. A. Losev, and A. I. Osipov, Relaxation Processes in Shock Waves [in Russian], Nauka, Moscow (1965).
44. E. E. Nikitin, The Theory of Elementary Atomic and Molecular Processes in Gases [in Russian], Khimiya, Moscow (1970).
45. B. F. Gordiets, A. I. Osipov, and L. A. Shelepin, Zh. Éksp. Teor. Fiz., 60:102 (1971).
46. N. G. Basov, V. I. Igoshin, E. P. Markin, and A. N. Oraevskii, Kvant. Élektron., No. 2, p. 3 (1971).
47. B. F. Gordiets, L. I. Gudzenko, and L. A. Shelepin, Zh. Prikl. Mekh. Tekh. Fiz., No. 6, p. 115 (1968).
48. D. R. Bates and A. E. Kingston, Planet. Space Sci. 11:1 (1963).
49. R. W. P. McWhirter and A. T. Hearn, Proc. Phys. Soc., 82:641 (1963).
50. T. Holstein, Phys. Rev., 72:1212 (1947).
51. L. M. Biberman, Dokl. Akad. Nauk SSSR, 59:659 (1948).
52. L. M. Biberman, V. S. Vorob'ev, and I. T. Yakubov, Zh. Éksp. Teor. Fiz., 56:1992 (1969); Teplofiz. Vys. Temp., 5:201 (1967).
53. V. S. Vorob'ev and I. T. Yakubov, Pis'ma Zh. Éksp. Teor. Fiz., 4:43 (1965).
54. V. S. Vorob'ev, Zh. Éksp. Teor. Fiz., 51:327 (1966).
55. B. F. Gordiets, L. I. Gudzenko, and L. A. Shelepin, Zh. Éksp. Teor. Fiz., 55:942 (1968).
56. L. C. Jonson, Phys. Rev., Ser. II, 155:64 (1967).
57. B. F. Gordiets, I. L. Dymova, and L. A. Shelepin, Zh. Prikl. Spektrosk., 15:205 (1971).

58. E. M. Cherkasov, Candidate's Dissertation, Physics Institute, Academy of Sciences of the USSR (1969).
59. E. L. Latush and M. F. Sêm, Zh. Éksp. Teor. Fiz., 64:2017 (1973).
60. E. L. Latush and M. F. Sêm, Pis'ma Zh. Éksp. Teor. Fiz., 15:637 (1972).
61. V. F. Keidan, V. M. Mikhalevskii, M. F. Sêm, and A. P. Shelepo, Kvant. Élektron., No. 1, p. 75 (1973).
62. L. A. Vainshtein, I. I. Sobel'man, and E. A. Yukov, Electron Impact Excitation Cross Sections of Atoms and Ions [in Russian], Nauka, Moscow (1973).
63. L. I. Gudzenko, S. S. Filippov, and L. A. Shelepin, Zh. Éksp. Teor. Fiz., 51:1115 (1966).
64. S. W. Bowen and C. Park, AIAA J., 10:522 (1972).
65. A. Czernichowski, J. Chapelle, and F. Cabannes, C. R. Acad. Sci. Paris, Ser. B, 270:54 (1970).
66. V. M. Gol'dfarb, E. V. Il'ina, I. E. Kostygova, G. A. Luk'yanov, and V. A. Silant'ev, Opt. Spektrosk., 20:1085 (1966).
67. W. L. Bohn and P. Hoffman, Z. Naturforsch., 27A:878 (1972).
68. N. M. Kuznetsov and Yu. P. Raizer, Zh. Prikl. Mekh. Tekh. Fiz., 4:10 (1965).
69. L. I. Gudzenko, S. S. Filippov, and S. I. Yakovlenko, Zh. Prikl. Spektrosk., 13:357 (1970).
70. S. W. Bowen and C. Park, AIAA J., 9:493 (1971).
71. E. L. Stupitskii and G. I. Kozlov, Institute of Applied Mechanics Preprint No. 13 (1972).
72. J. S. Hitt and W. J. Haswell, IEEE J. Quant. Electron., 2:4 (1966).
73. V. M. Likhachev, M. S. Ravinovich, and V. M. Sutovskii, Pis'ma Zh. Éksp. Teor. Fiz., 5:55 (1967).
74. A. N. Vasil'eva, V. M. Likhachev, and V. M. Sutovskii, Zh. Tekh. Fiz., 39:341 (1969).
75. R. Illingworth, J. Phys. D (Appl. Phys.), 3:924 (1970).
76. R. Illingworth, J. Phys. D (Appl. Phys.), 5:686 (1972).
77. A. Papayoanou, and J. Gumeiner, J. Appl. Phys., 42:1914 (1971).
78. A. Papayoanou, R. G. Buser, and J. M. Gumeiner, IEEE J. Quant. Electron., QE-9:580 (1973).
79. M. A. Leontovich and S. M. Osovets, At. Énerg., 3:8 (1956).
80. M. Rosenbluth, in: Magnetohydrodynamics [Russian translation], Atomizdat, Moscow (1958).
81. A. A. Rukhadze and S. A. Triger, Preprint FIAN No. 168 (1968).
82. V. F. Kitaeva, A. I. Odintsov, and N. N. Sobolev, Usp. Fiz. Nauk, 99:381 (1969).
83. I. L. Beigman, L. A. Vainshtein, P. L. Rubin, and N. N. Soholev, Pis'ma Zh. Éksp. Teor. Fiz., 6:919 (1967).
84. V. A. Boiko, O. N. Krokhin, and G. V. Sklizkov, Preprint FIAN No. 121 (1972).
85. Yu. V. Afanas'ev, É. M. Belenov, O. N. Krokhin, and I. A. Poluéktov, Zh. Éksp. Teor. Fiz., 65:121 (1972).
86. A. G. Molchanov, Usp. Fiz. Nauk, 106:165 (1972).
87. B. B. Kadomtsev, in: Reviews of Plasma Physics, Vol. 2, Consultants Bureau (1966).
88. V. B. Rozanov, Dokl. Akad. Nauk SSSR, 182:2 (1968).
89. A. A. Rukhadze and S. A. Triger, Zh. Prikl. Mekh. Tekh. Fiz., 3:11 (1968); Preprint FIAN No. 168 (1968).
90. A. A. Rukhadze and S. A. Triger, Zh. Éksp. Teor. Fiz., 56:1029 (1969).
91. V. I. Rozanov, A. A. Rukhadze, and S. A. Triger, Zh. Prikl. Mekh. Tekh. Fiz., 5:18 (1968).
92. A. A. Rukhadze and S. A. Triger, Preprint FIAN No. 26 (1969).
93. A. A. Rukhadze, V. S. Savodchenko, and S. A. Triger, Zh. Prikl. Mekh. Tekh. Fiz., 6:58 (1965).
94. A. F. Aleksandrov, E. P. Kaminskaya, and A. A. Rukhadze, Zh. Prikl. Mekh. Tekh. Fiz., 1:33 (1971).
95. V. V. Pustovalov and V. B. Rozanov, Preprint FIAN No. 52 (1969).

96. V. B. Rozanov and A. A. Rukhadze, Preprint FIAN No. 132 (1969).
97. A. F. Aleksandrov and A. A. Rukhadze, Usp. Fiz. Nauk., 112:193 (1974).
98. A. D. Klimentov, G. V. Mikhailov, F. A. Nikolaev, and V. B. Rozanov, Abstracts of Reports at the Second All-union Conference on the Physics of Low-Temperature Plasmas, Institute of Physics, Academy of Sciencies of the Belorussian SSR, Minsk (1968), p. 163.
99. E. A. Smurs, Ark. Fys. 29:27 (1965).
100. A. D. Klementov, G. V. Mikhailov, F. A. Nikolaev, V. B. Rozanov, and Yu. P. Sviridenko, Teplofiz. Vys. Temp., 8:736 (1970).
101. F. A. Nikolaev, V. B. Rozanov, and Yu. P. Sviridenko, Kratk. Soobshch. Fiz., No. 4, p. 59 (1971).
102. A. F. Aleksandrov, V. V. Zosimov, A. A. Rukhadze, V. I. Savoskin, and I. B. Timofeev, in: Brief Contents of Talks at Third All-Union Conf. on the Physics of Low-Temperature Plasmas, Moscow (1971), p. 176.
103. A. F. Aleksandrov, V. V. Zosimov, A. A. Rukhadze, and I. B. Timofeev, Kratk. Soobshch. Fiz., No. 8, p. 72 (1970).
104. A. F. Aleksandrov, V. V. Zosimov, and I. B. Timofeev, Kratk. Soobshch. Fiz., No. 2, p. 25 (1972).
105. E. V. Aglitskii, V. A. Boiko, A. V. Vinogradov, and E. A. Yukov, Preprint FIAN No. 145 (1973).
106. E. V. Aglitskii, O. N. Krokhin, and G. V. Sklizkov, Preprint FIAN No. 57 (1972).
107. W. Seka, C. Breton, J. L. Schwob, and C. Minier, Plasma Phys., 12:73 (1970).
108. P. Jaegle, A. Carillon, P. Dhez, C. Jamelot, A. Sureau, and M. Cukier, Phys. Lett., A36:167 (1971).